THE PRACTITIONER HANDBOOK OF PROJECT CONTROLS

Although projects always carry risk, too many projects run late or exceed their original budgets by eye-watering amounts. This book is a comprehensive guide to the procedures needed to ensure that projects will be delivered on time, to specification and within budget.

Eight expert contributors have combined their considerable talents to explain all aspects of project control from project conception to completion in an informative text, liberally supported where necessary by clear illustrations.

This handbook will benefit all project practitioners, including project managers and those working in project management offices. It will also provide an invaluable guide for students studying for higher degrees in project management and its associated disciplines.

Dennis Lock began his career as an electronics engineer in a research laboratory, but has since served many years in project and administration management in the heavy machine tool and non-ferrous mining industries. Dennis has also carried out successful consultancy assignments in Europe and the United States, and was for eight years an external lecturer in project management to master's degree students at two British universities. He is a fellow of the Association for Project Management and a member of the Chartered Management Institute. As a best-selling author, he has written or edited well over 50 books, many published in multiple languages.

Shane Forth is Fellow of the Association for Project Management and the Association of Cost Engineers with over 40 years' experience in the oil, gas, nuclear power and other industries. As 'GO FORTH' he provides consultancy services to help organizations develop the skills of their project management people. Shane sits on working groups and lectures at universities and events. For his MSc he won the Stephen Wearne Award for best overall performance. He also won APM's Geoffrey Trimble Award for best master's post-graduate dissertation. Shane has been honoured twice by the Engineering Construction Industry Training Board (including a national award for individual leadership and significant contribution to training and development).

THE PRACTITIONER HANDBOOK OF PROJECT CONTROLS

Edited by Dennis Lock

Consulting Editor Shane Forth

LONDON AND NEW YORK

First published 2021
by Routledge
2 Park Square, Milton Park, Abingdon, Oxon OX14 4RN

and by Routledge
52 Vanderbilt Avenue, New York, NY 10017

Routledge is an imprint of the Taylor & Francis Group, an informa business

British Library Cataloguing-in-Publication Data
A catalogue record for this book is available from the British Library

Library of Congress Cataloging-in-Publication Data
A catalog record has been requested for this book

ISBN: 978-0-367-25309-7 (hbk)
ISBN: 978-0-429-28712-1 (ebk)

Typeset in Bembo
by Newgen Publishing UK

To Cleo

CONTENTS

FIGURES

NOTES ON CONTRIBUTORS

Shane Forth is a Fellow of the Association for Project Management (APM) and the Association of Cost Engineers (ACostE). He has over 40 years' experience in the oil and gas, nuclear, power, process and infrastructure industries, having served with global project management, engineering and construction businesses. He is well known and highly regarded for his expertise in project management and controls. As 'GO FORTH', he shares his experience with multiple organizations by providing consultancy services to help them develop the skills of their project management and controls people. Shane also sits on professional, technical and academic working groups, speaks at UK and international events and lectures at universities. Previously, for 27 years he was responsible for strategic and functional leadership and continuous improvement in project management and controls capability. For his MSc in Project Management (Manchester), he won the Stephen Wearne Award for best overall performance. National acclaim followed with the APM Geoffrey Trimble Award for best Master's post-graduate dissertation. Shane has been a driving force in the implementation of national project controls apprenticeship and graduate programmes and has been honoured twice by the Engineering Construction Industry Training Board (ECITB) (including winning a national award for individual leadership and significant contribution to training and development).

Nigel Hibberd, MSc (Cranfield), CEng, FACostE, AIMechE, MAPM) has a manufacturing background in the automotive components industry. His 30 years' experience in project control included head of function for BNFL's Engineering Subsidiary and he was project manager for the development and introduction of SAP (Systems Application Protocol Enterprise system) project control/management functionality into the wider business. Subsequently he became chairman of the Key Users Forum. Nigel was also chairman of ProVoc, the national steering group for project control NVQs (and their subsequent iterations) for 10 years. Nigel serves on the ECITB Project Control Working Group, which has developed further iterations for NVQs (now RQFs) and the Trailblazer Apprenticeships in Project Control. He also shares the representation of ACostE at the BCECA (British Chemicals Contractors Association) PCM meetings, which is an important avenue for engaging with the petrochemical industry and its requirements. Nigel now works part time as an assessor and lectures on project control at two universities. He is a member and past chairman of the ACostE's Accreditation Board and delivers tiered accreditation to Incorporated and Certified Professional level. He is also on the engineering committee, which enables ACostE members to obtain IEng and CEng status via ACostE. Nigel is active on the ACostE NW region committee in delivering continuing professional development events throughout the year, and sometimes giving talks. He serves on the ACostE council, of which he is an ex vice president.

Alison Lawman is a Senior Lecturer in Project Management and Course Leader at York Business School, York St John University. She is an experienced academic in the field of project management, with specialisms in change management and risk analysis. With a background in the financial services sector she is an experienced project practitioner, and has been involved in or responsible for the delivery of a range of projects. She is the academic lead on two major industry-education partnerships with 2020 Project Management at York Business School, and is responsible for the development of the project management education portfolio across undergraduate and postgraduate levels. Her research interests are in the areas of gender, risk management, and artificial intelligence. She is a Fellow of the Higher Education Academy and is a PMI GAC International Onsite Team Reviewer to accredit and validate Project Management Education and Curriculum. She holds a range of project management qualifications including an MSc. Her doctoral research focuses on the professionalization of project management.

Dennis Lock is a freelance writer. He began his career as an electronics engineer in a research laboratory, but subsequently served many years in project and senior administration management. His industrial experience is wide, ranging from guided weapons electronics to giant special machine tool systems and mining engineering. Dennis has carried out consultancy assignments in Europe and the US, and was for several years an external lecturer in project management to Masters degree students at two British universities. He is a best-selling author and has written or edited well over 50 books, many published in multiple languages. Dennis is a Fellow of the APM and a Member of the Chartered Management Institute.

Tony Marks is the CEO of 20/20 Project Management, an international project management training and consultancy company. Tony has worked in project management for over 35 years in industries including oil and gas, engineering and construction, nuclear, utilities, aerospace and defence, transport and logistics, and construction. He is the author of *20:20 Project Management* published by Kogan Page. Tony developed the NHS Scotland *Project Management Guide for Capital Projects* and a new MSc in project management. He has provided consultancy in project management for clients worldwide. He holds an MBA, is a Fellow of the Association for Project Management, Fellow of the Institute of Leadership and Management and a

Fellow of the Chartered Management Institute. Tony holds a range of professional qualifications including APM PMQ (previously APMP), Prince 2 Practitioner (P2P), PMI Professional (PMP) and has lectured on project management for the University of Aberdeen. Tony has led 20/20 Project Management since 2003, which has become the leading specialist provider of project management and controls training in the UK. Its Certificate in Project Controls programme is accredited by the ECITB.

Alan McDougald has led an exceptionally interesting and varied life and it's unfortunate that space is limited here in which to tell more of his story. Alan was raised on a family farm in Texas and now claims to be the kind of rancher who keeps his cows and gives them names. He first worked in a steel foundry for oil pumps, progressing through various jobs in that industry until his career was interrupted by seven years serving in the US armed forces. His early project work included preparing proposals for large petrochemical projects. He has degrees at both Bachelor and Masters levels. He became expert in planning and resource scheduling for projects employing over 500 engineers, beginning those activities before computers became available for that purpose. With subsequent extraordinarily wide experience, Alan has since worked very successfully on all kinds of project developments in countries all over the world, achieving senior management status. He told us that he has earned millions of dollars during his career but 'spent everything on women and whisky'.

Aydin Nassehi is the Head of Department and a Reader in manufacturing systems in the Department of Mechanical Engineering, University of Bristol (UK). He is an Associate Member of the International Academy of Production Engineers (CIRP), the secretary for CIRP UK and the Technical Secretary for the Subject Technical Committee focusing on Production Systems and Organisations in the academy. He is also a member of the Institution of Engineering and Technology (IET). He has a BSc in Industrial Engineering from Sharif University of Technology, an MSc with distinction in Manufacturing Management from Loughborough University; he received his PhD in Innovative Manufacturing Technology in 2007 from the University of Bath followed by an MSc with distinction in software engineering from the University of Oxford. His research interests are in informatics and artificial intelligence in manufacturing. He has a collaborative research portfolio of more than £10m and has more than 100 scientific publications on the topics related to computer integrated and smart manufacturing. He is interested in science fiction, retro computing and cooking.

Lindsay Scott is a director of Arras People, a highly-regarded project management recruitment organisation. She is also founder of PMO Flashmob (a learning and networking community), co-founder of PMO Learning (a training organisation) and creator of the annual PMO Conference. Lindsay is the career columnist for PMI's *PM Network* magazine and she co-edited the *Gower Handbook of People in Project Management* with Dennis Lock.

PART I

Getting started

1
PROJECT FUNDAMENTALS

Dennis Lock

This chapter describes the principal characteristics of projects.

Different types of project

A project is generally different from the 'business-as-usual' work undertaken by organizations in several respects. Projects are usually not repetitive, but each is a unique undertaking requiring new designs, organization and planning, incurring some risk and needing a defined management structure. Projects generally (but not quite always) have recognizable start and finish dates, which encompass a fairly predictable lifecycle pattern (see Figure 1.1). Further definition depends on the kind of project being considered, and I usually categorize projects as belonging to one of four recognizable groups, which are described below (not in any particular order of importance).

Civil engineering, construction, mining and quarrying projects

These projects are usually the first that come to mind when we think of the history of project management. This category includes construction and civil engineering projects to build roads, houses, public buildings, bridges and so on. I include mining, land reclamation, quarrying and tunnelling here. Although designed in offices, the physical work for these projects is carried out at sites that are often exposed to the public gaze but remote from the contractor's headquarters. Projects range in size from single buildings to entire townships.

These projects carry all kinds of risks that could delay completion, increase costs over budget, or result in physical harm or even death to workers. Progress can be delayed by factors outside project management control (particularly the weather). Profit margins are typically low compared with other types of project, and risk of the contractor suffering financial loss or running into serious cash flow problems can be great. Not long before this chapter was written the giant construction organization

Possible lifecycle for a small, very simple manufacturing project

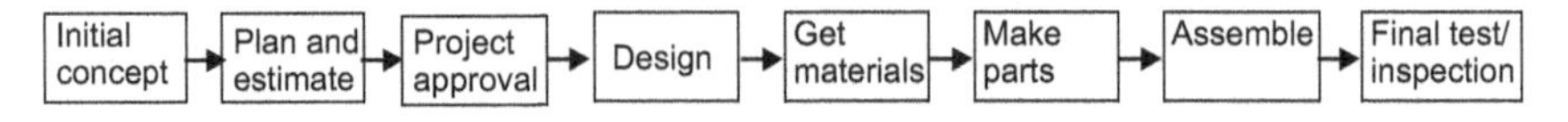

Part of a possible lifecycle for a pure scientific research project with stage gate funding

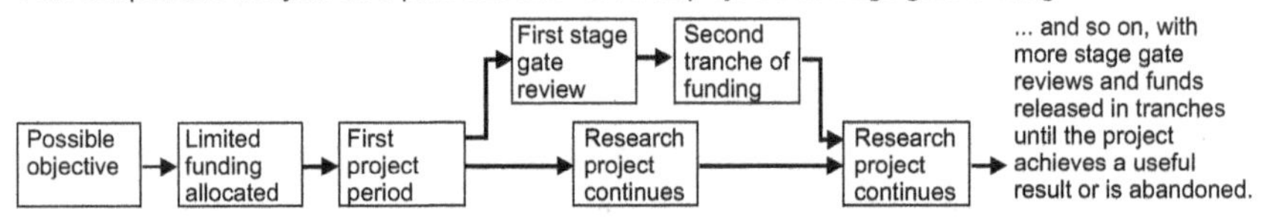

Greatly simplified whole lifecycle of a larger capital project

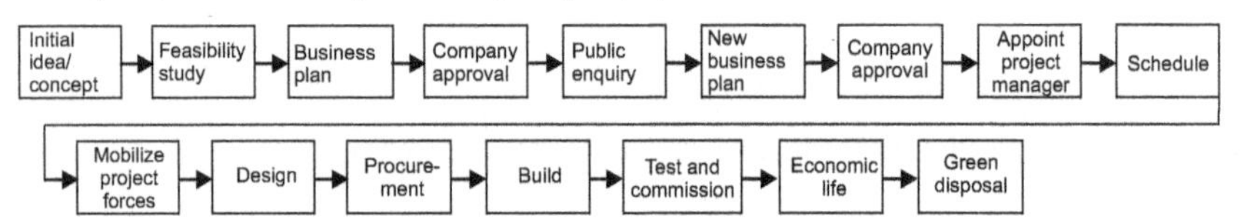

The boundaries between some of these phases can be indistinct with considerable overlapping.

Figure 1.1 Three examples of project lifecycles.

Carillion had collapsed, not through lack of contracts and work but because it ran out of liquid funds and could not pay its bills, so causing many subcontractors serious cash difficulties.

Project organizations tend to be complex, involving not only the main contractor, but many subcontractors and suppliers, providers of funds and possibly a financial guarantor. In the UK very large infrastructure projects can even be subject to parliamentary approval.

Manufacturing projects

Manufacturing projects include everything from micro-technology and the computing sciences to heavy engineering and (arguably) shipbuilding. What they all have in common is that the core activities of these projects can be conducted in enclosed premises, where the contractor is able to exercise greater control over the work, access, working conditions, security and confidentiality. With the exception of shipbuilding, extreme weather conditions are less likely to have a damaging effect. Each of these projects results in a tangible result such as a ship, aircraft, vehicle, machine, electrical or electronic item that will be of practical use to the customer or end user.

The owners of these projects are exposed to commercial and technical risks, particularly from things such as competition from others selling in the same market, inaccurate market forecasts, poor performance or even the collapse of suppliers and subcontractors, political actions (such as increased import tariffs imposed by foreign governments) and so on.

Management or change projects

These are projects that seek to improve business performance by reorganizing companies, updating IT systems, dealing with company mergers and acquisitions and so forth. Their success (or failure) is measured in terms of the resulting benefits, usually forecast in a business plan and quantified by key performance indicators (KPIs). At one time managing such projects was regarded as a separate profession from other project categories but now that distinction is less apparent. A big risk in these projects is that they can affect the status and careers of people working in the changed organizations, sometimes leading to redundancies and often causing personal anxiety or distress.

I include some projects in the entertainment industry in this category because a public performance of any kind is a project that needs some project management skills. However, unlike other projects, these are ephemeral, leaving no permanent structure or physical object after the project ends.

Pure scientific research projects

Setting up a building or other capability for a research project can be a project in itself. However once a research project is authorized to proceed, usual project management methods for planning, measuring and controlling progress are not relevant because the end result cannot be planned or predicted. So spending and work takes place in a continuous stream, often without a positive prediction or expectation of a useful result.

Some companies are willing to accept this apparent lack of control, and even encourage their staff to embark on independent research ventures with the hope (but not positive expectation) of a profitable result. Not many organizations will tolerate expenditure without any checks or controls at all, and here is a good case for stage gate controls. With this process, an individual researcher or a research team is allowed (or encouraged) to pursue work using an allocated budget for a set period (which might be six months or a year). Regular reviews indicate how the research is progressing and can perhaps predict a result. Each periodic review can be regarded as a kind of gate that can be shut if it becomes clear to the provider of funds that the research has no hope of achieving any useful result. Figure 1.1 includes an illustration of this 'stage gate' form of project control.

Stakeholders

A stakeholder is any person, organization or community that will be affected profitably or otherwise during the course of a project or after its completion. Investors in the project are primary stakeholders.

I like to use the case of a project to build a shopping mall on a greenfield site to illustrate the various kinds of project stakeholders. Investors in the project clearly have a direct interest in the outcome. People who have reserved shops and other commercial premises in the new mall are also obvious stakeholders. The local authority can expect to benefit from the increased number of business rate payers, but will also have to provide services such as waste disposal. We can identify all these people and organizations as primary stakeholders. Most primary stakeholders will expect regular progress reports as the project proceeds.

Then there are the secondary stakeholders; people and organizations whose lives and interests will be affected in some way by the project. For example, nearby residents will benefit from the new local shopping and entertainment facilities but might be concerned at the prospect of increased noise and traffic. Environmental pressure groups with objections to the project might organize publicity, demonstrations and other activities that could delay the project. The project manager needs to be aware of all these secondary stakeholders and take care that, as far as possible, they are treated with respect and fairly.

It can be useful to list all the stakeholders well before the project begins and display their interests in a matrix, such as that shown in Figure 1.2.

Multiple projects and programmes

Some companies, especially when they are large organizations, will have work at any given time that mixes everyday business-as-usual activities, internal change projects, and projects for other companies or clients.

The management of simultaneous but similar multiple projects is known as programme management (Lock and Wagner, 2016). When an organization has a mix of business-as-usual plus standalone projects and programmes, the organization becomes involved in project portfolio management (Lock and Wagner, 2018).

Stakeholders	Stakeholders' concerns			
	Time	Cost	Quality	Environment
Project owner				
Project manager				
Bank Guarantor				
Statutory bodies				
Main contractor				
Subcontractors				
Suppliers				
Project workers				
The local authority				
Local residents				
The wider public				
Environmental groups				
... and so on				

Figure 1.2 A matrix for displaying the interests of stakeholders in a public project.

Professional organizations

The following paragraphs list (in alphabetical order) organizations that specialize in promoting and advancing knowledge in the closely associated fields of project management and project controls.

AACE International

This organization was established in 1956 as the American Association of Cost Engineers. In 1992 its legal name changed to AACE International. AACE is an abbreviation for the Association for the Advancement of Cost Engineering. AACE's mission is for its members to drive projects to completion on time, on cost and to meet investment and operational goals. Personal (regular) and other types of membership are available including corporate membership. AACE International is committed to the constructive exchange of ideas between members, development of technical guidance and quality education, and the recognition of subject matter experts. The Association offers certification at technical, professional and expert levels (which are explained further in Chapter 50).

Publications include the journal *Cost Engineering, Recommended Practices (Rps)* which is a valuable reference source. The contents of this journal are subjected to rigorous review and are recommended by the Technical Board and the TCM Framework® (TCM is an abbreviation for total cost management). Professional and technical resources include online communities such as special interest groups and technical subcommittees. Contact details for AACE International are as follows:

AACE International
1265 Suncrest Towne Centre Drive
Morgantown
WV 26505-1876
USA
Telephone: +1 304 296 8444
Email: info@aacei.org

The Association of Controls Management

TAOCM was founded in 2016. It was conceived to recognize the value of good controls management in all its component parts, when these are applied properly and pulled together. TAOCM provides professional recognition and accreditation for project control managers, planning engineers and document controllers and champions best industry standards and practices. Personal membership begins at student level and rises to Fellow. The Association produces an online monthly magazine: *The Association of Controls Management* and (through partnerships with other professional organizations and established training companies) offers a range of project controls training opportunities for its members. Contact details are as follows:

The Association of Controls Management
5 Station Road
Grangemouth
FK3 8DG
Website: www.taocm.co.uk

The Association of Cost Engineers

The ACostE is a member of the International Cost Engineering Council, with its head office at:

Lea House
5 Middlewhich Road
Sandbach
Cheshire, CW11 1XL.
Telephone: 01270 764798
Email: enquries@acoste.org.uk
Website: www.acoste.org.uk

This association was formed as the British Group of the American Association of Cost Engineers (described below). It became an independent body in 1962. The Association represents the professional interests of those with responsibility at all levels for the prediction, planning and control of resources and costs for engineering, manufacturing and construction activities. Its objectives continue to be similar to those of its American parent, which is now known as the Association for the Advancement of Cost Engineering International (AACEI). Personal or corporate membership of the ACostE helps project controls practitioners to broaden their networks, widen their roles and develop their careers. Personal membership begins at Companion level and rises through various grades to Member (MACost E) and Fellow (FACost E).

A tiered accreditation process is provided for project control practitioners, who must be full members of the Association. This is underpinned by National Occupational Standards in project controls, estimating, planning and scheduling and cost engineering. It has been benchmarked against the Engineering Council's professional accreditation programme to allow project controls practitioners to gain professional recognition of their skills and competencies through the various levels up to full certification. There is more about the National Occupational Standards and Tiered Accreditation in Chapter 50.

Further members' services are provided through special interest group events, which are arranged through a network of local branches. Publications include the online journal *Project Controls Professional* and books (see for example Reid, 2009).

Association for Project Management

The APM is the UK corporate member of the International Project Management Association, with its head office at:

Ibis House
Regent Park
Summerleys Road
Princess Risborough
Buckinghamshire, HP27 9LE
Telephone: 0845 458 1944
Email: info@apm.org.uk
Website: www.apm.org.uk

This chartered association arranges seminars and meetings through its network of local branches. Its numerous publications include the periodical journal *Project* and the *APM Body of Knowledge*. Personal or corporate membership of the Association enables project managers and others involved in project management to meet and to maintain current awareness of all aspects of project management. Membership begins at student level and rises through various grades to full member (MAPM) and fellow (FAPM).

The Association has a well-established certification procedure for project managers, who must already be full members. To quote from the Association's own literature, 'the certificated project manager is at the pinnacle of the profession, possessing extensive knowledge and having carried responsibility for the delivery of at least one significant project'. As evidence of competence, certification has obvious advantages for the project manager, and will increasingly be demanded as mandatory by some project purchasers. Certification provides employers with a useful measure when recruiting or assessing staff and the company that can claim to employ certificated project managers will benefit from an enhanced professional image. Certification has relevance also for project clients. It helps them to assess a project manager's competence by providing clear proof that the individual concerned has gained peer recognition of his or her ability to manage projects.

In common with some other associations, the APM has branches throughout the United Kingdom and also has a number of special interest groups.

The Engineering Construction Industry Training Board

This well-regarded training board sponsors the Level 3 Diploma in Project Controls in Practice. For more information visit www.ecitb.org.uk/wp-content/uploads/2019/01/QS-Project-Control-Techniques-RV-1-0-1.pdf

The ECTIB address is as follows:

Engineering Construction Industry Training Board
Blue Court
Kings Langley
Hertfordshire, WD4 8JP.
Telephone: 01923 260000.

Planning Planet

Planning Planet is the online community of project control people. It includes cost engineers, forensic planners, planners, schedulers and quantity surveyors from the largest project delivery organizations across all industries from oil and gas to transport, civil engineering, construction, pharmaceuticals and many more. Founded over 15 years ago, this site has representation from over 150 countries. Individual membership is free with the objective that each individual will become better informed and connected by joining and using Planning Planet. There is an accreditation process (known as the Guild of Project Controls), which covers all levels of competence. For more information see Chapter 50. The website is at www.planningplanet.com.

Project Controls Online

Launched in 2009, Project Controls Online (PCO) is a large repository of project controls information with a presence in over 190 countries for the exchange of information on a global scale through its on-line forums, blogs, practitioners' guides and other literature. Since 2011 PCO has also held its Project Controls Expo in the UK, which is an annual event with international speakers that is attended by over 1000 project controls practitioners from a very wide range of industries. This expo features master-class sessions, case studies, technical presentations, subject matter expert (SME) panels, megaproject sessions, along with a partner showcase, project controls job fair and an exclusive awards night. A similar event has been held in Australia since 2018. The website is at www.projectcontrolsonline.com.

Project Management Institute

Founded in the US in 1969, the Project Management Institute (PMI) is the world's leading not-for-profit organization for individuals around the globe who work in, or are interested in, project management. PMI develops recognized standards, not least of which is the widely respected project management body of knowledge guide, commonly known by its abbreviated title the *PMBOK Guide*.

PMI publications include the monthly professional magazine *PM Network*, the monthly newsletter *PMI Today* and the quarterly *Project Management Journal* as well as many project management books. In addition to its research and education activities, PMI is dedicated to developing and maintaining a rigorous, examination-based professional certification programme to advance the project management profession and recognize the achievements of individual professionals. PMI's Project Management Professional (PMP) certification is the world's most recognized professional credential for project management practitioners and others in the profession. For more information, contact PMI at:

PMI Headquarters
Four Campus Boulevard
Newtown Square

PA 19073-3299
USA
Telephone: +610-356-4600
Email: pmihq@pmi.org
Website: www.pmi.org

References and further reading

APM (2012), *APM Body of Knowledge*, 6th edn, Princes Risborough: Association for Project Management.

Joint Development Board (1997), *Industrial Engineering Projects*, West Hanney: Spon.

Lock, D. (2013), *Project Management*, 10th edn, Farnham: Gower.

Lock, D. (2014), *The Essentials of Project Management*, 4th edn, Farnham: Gower.

Lock, D. and Scott. L. (eds.) (2013), *Gower Handbook of People in Project Management*, Farnham: Gower.

Lock, D. and Wagner, R. (eds.) (2016), *Gower Handbook of Programme Management*, Abingdon: Routledge.

Lock, D. and Wagner, R. (eds.) (2018), *The Handbook of Project Portfolio Management*, Abingdon: Routledge.

Meredith, J.R., Mantel, S.J., Jr. and Shafer, S.M. (2017) *Project Management: A Strategic Managerial Approach*, 10th edn, Hoboken, NJ: Wiley.

Morris, P.W.G. (1994), *The Management of Projects*, London: Thomas Telford.

PMI (2013), *A Guide to the Project Management Body of Knowledge (PMBOK Guide)* 5th edn, Newtown Square, PA: Project Management Institute.

Reid, C.R.A. (2009), *Project Controls: A Practitioner's Gide to the Key Topics*, Sandbach: TASC.

Turner, J.R. (ed.) (2014), *Gower Handbook of Project Management*, Farnham: Gower.

2
INTRODUCING PROJECT CONTROLS

Dennis Lock

This chapter introduces, with brief descriptions, factors that have to be controlled to ensure successful project delivery. Most of these subjects are dealt with at greater length later in this handbook.

The project controls environment

Responsibility for project success, although usually centred on the project manager, requires dedication and enthusiasm from everyone concerned. Project managers cannot act in a vacuum, but need encouragement and support from top management at least as much as they depend on the efforts and skills of all those who carry out the project tasks.

The structure of the project organization will influence the degree of control enjoyed by the project manager, being greatest in a dedicated project task force or team and least in a balanced or weak matrix. Turn to Chapter 6 for more about organization structures.

If the project is very tiny, the project manager should be capable of carrying out planning and following progress on a day-to-day basis, perhaps needing the assistance of a project coordinator. However, most projects benefit from the services of a project management office (PMO), staffed with people who are proficient in the arts of planning, scheduling, progressing, reporting and other duties. The role of a PMO is described in Chapter 8.

Questions of priority

Priority decisions arise when allocating scarce resources to tasks (which is part of the process of resource scheduling described in Chapters 18 and 19). More fundamental questions of priority occur when regarding a project as a whole. Dr Martin Barnes recognized this dilemma when he introduced his 'triangle of objectives' many years ago in a BBC programme. His triangle was intended to concentrate the minds of senior managers when asking the question 'Which project aspect should be given highest priority for control: time, cost or quality?'

I once telephoned Dr Barnes to ask him about his triangle and suggested that 'quality' was perhaps not the most appropriate parameter to choose. If we accept the premise that quality is defined as 'fitness for purpose' then quality is not negotiable: the outcome of every project *must* be 'fit for purpose'. Dr Barnes did not disagree and I was allowed to use an adaptation of the triangle in my books and I now substitute 'specification' for 'quality'. Since those days I have seen many variations on the triangle of objectives theme. Figure 2.1 illustrates Dr Barnes' original triangle with two of the many subsequent variants. Different names are now given to these triangles, with 'magic triangle' being very popular.

The three objectives are always interdependent: each objective is influenced by the other two. Time is particularly important because it is the one resource that can never be replenished: once gone it has gone for eternity. It is usually safe to expect that any project that overruns its scheduled time will also overrun its authorized budget. If a project is running late and over budget there is always a danger that quality might suffer owing to work being done in too much of a rush. However, any version of the triangle of objectives can focus the minds of managers and cause us to reflect and make decisions whenever there is conflict between time, cost and specification or scope.

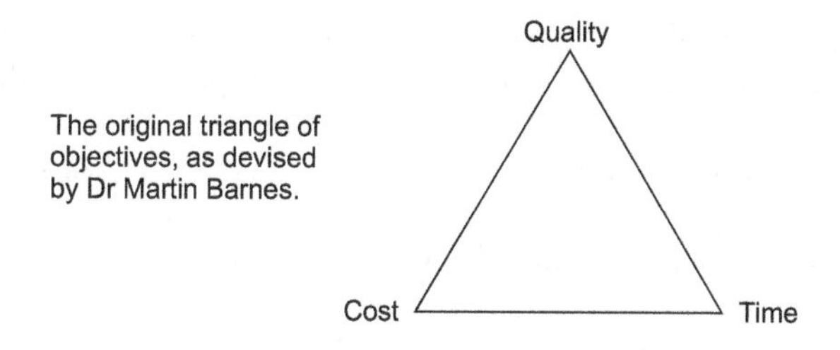

Two of the many later variants devised by other writers

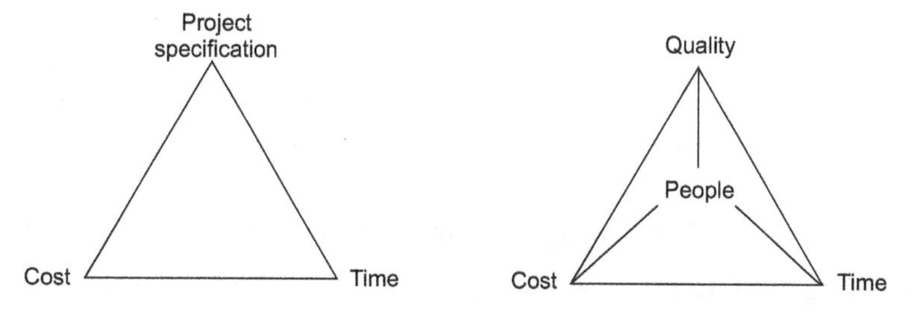

Figure 2.1 Triangles of project objectives. From Lock, 2013.

Project specification and scope

One of the most common reasons for project failure is when sight of the original objectives is lost and the work expands beyond the original intentions. This can happen because of authorized changes but it can also creep up on us gradually unless control is exercised. People ask questions such as 'Wouldn't it be nice or better if we added this', 'Why are we only making two of these: our customer is bound to need more so let's make a couple of spares' or 'These painted panels would look better if we used stainless steel rather than mild steel'. These are small examples of a process known as 'scope creep'. If changes are allowed to happen unchecked, delivery milestones will be missed, with costs exceeding budgets and risk of project failure. Chapter 32 gives essential guidance on the control of changes.

Contract administration

Contract administration includes keeping firm control over the total project price and customers' payments over the duration of the contract, including the effects of authorized changes. This becomes especially important in large projects with long timescales.

I once took over responsibility for a small project that had been running for over a year in which a respected and capable contractor was carrying out maintenance, repairs and small alterations to our office complex of about 40,000 square feet in London. From time to time the contractor's foreman on our site had been agreeing changes or small additions requested by my predecessor. All these minor changes were authorized by my predecessor by signing daywork sheets (slips of paper torn from the contractor's duplicate pads).

The time eventually came when all work was finished and final payment had to be made to the contractor. Total project duration had been about 18 months. Assessing the final price meant going through stacks of daywork sheets, estimating the costs of the additional work and reconciling the money involved with the contractor. This process involved my organization in considerable work and expense. We even had to hire an independent quantity surveyor. The final sum agreed with the contractor was based on estimates. Fortunately everything was resolved amicably but this was only a tiny project.

Cases where poor contract administration has led to disputes over payment between client and contractor are not uncommon. Contract administration is discussed elsewhere in this handbook, particularly in Chapter 28.

Purchasing and materials control

Well over half the total cost of many projects comes from obtaining goods and services from sources outside the organization. Once, when carrying out an assignment for the Chartered Institute of Procurement and Supply, I was astonished to learn that this proportion can be as high as 80 per cent. Those involved in purchasing will know about the 'rights' of procurement which require getting the *right* goods at the *right* price and the *right* quality to the *right* place at the *right* time. Clearly the control of purchasing is vital and one would expect procurement and materials management to feature prominently in every book about project control and management. Astonishingly, most project management writers ignore it completely. However, turn to Chapter 27 in this handbook and you will not be disappointed.

Cash flow

Many designers and engineers are unaware of the dangers that face an organization when cash flows are not controlled. The common misconception is that if a project is forecast to finish at a cost that is less than the total contract price, it must make a profit and succeed.

Consider this. Suppose that you are a small building contractor who meets a wealthy Arab gentleman. He wants to establish a purpose-built luxury residence in the UK. You learn that this Arab gentleman is willing to pay £10 million for this project that you know you could certainly build for less than half that price. The prospect of having an assured £5 million banked within 24 months is very appealing. So here is a project where the revenues will greatly exceed expenditure, with minimum risk and apparently no fear of financial loss. What could possibly go wrong?

Now here's the snag. You will have to find and pay for labour, materials, subcontractors and professional services until the project is finished and handed over to your client. At all times during this project your bank account will show a considerable and growing overdraft – until your bank cries 'No more!' and suspends your account. Your budding little company now has an enormous debt and will probably go bust. You cannot pay your advisers, contractors and labourers. That's an example of a cash flow problem. Such difficulties frequently cause the collapse of huge, famous organizations. **Cash is the life blood of a company**. You have to ensure that you do not go into uncontrollable debt.

Figure 2.2 is one way of illustrating the way in which cash flows can be forecast in a typical medium to large project, where the contract allows for stage payments against the achievement of milestones, or monthly billings against certified measurements of progress.

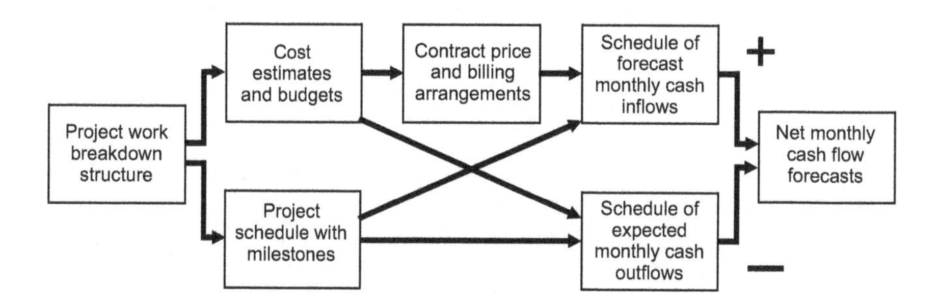

Figure 2.2 Essential elements for forecasting project cash flows. From Lock, 2013.

Progress against schedule

A project that finishes late will anger the client or customer and delay the start of any following projects that are waiting for resources to become available. Late projects damage reputations. They usually cause budgets to be overspent, if only because late projects continue to soak up overhead costs for as long as they occupy people, space and time. Some contracts include penalties for late delivery, where the contractor has to make financial recompense to the client in proportion to the delayed completion time. Any project that is allowed to run late can run into cash flow problems because if work is not done on time milestone targets will be missed, delaying the issue of invoices to the project customer or client, thus resulting in late receipt of revenues and possibly even cash penalties.

The first step in making sure that a project does not run late is to ensure that it begins on time, with adequate resources in place. Progress control depends on checking the status of tasks at all levels against the project schedule. However, simply checking tasks is not enough: any hint of a task running late, especially if it lies on a critical path, demands management action. That might mean using special measures, either to speed the current task or to accelerate following tasks to compensate and regain the schedule. Control requires not only making checks but also taking decisive management action.

Controlling risk

Every project carries risk simply because it is a new venture. Project management writers used to ignore this difficult subject, but now people have learned about how to predict and rank the importance of some risks and how to have mitigating measures ready. When we talk about 'controlling' risk, it has to be acknowledged that there are many kinds of risk events, which simply cannot be predicted and might cause severe damage to a project. Chapters 23, 24, 25 and 26 are devoted to this important subject.

When considering how to mitigate the cost of risk occurrences, never forget that insurance brokers can offer protection to compensate for some risks, but there are always some risks that they will not cover. Figure 2.3, taken from Lock (2013) is how I choose to introduce the subject of risk insurance, which, along with purchasing, tends to be ignored by other writers. It seems that project management writers fight shy of commercial subjects.

Project people

The wellbeing, motivation and satisfaction of people working in or affected by projects are paramount factors in achieving project success. People tend to be ignored when we discuss project controls. However, the way in which a project is organized and managed can clearly have a great influence on motivation, so Chapter 6 is important in that respect. One book that deals in depth with this important aspect of project management is Lock and Scott (2013). There is also some useful advice on

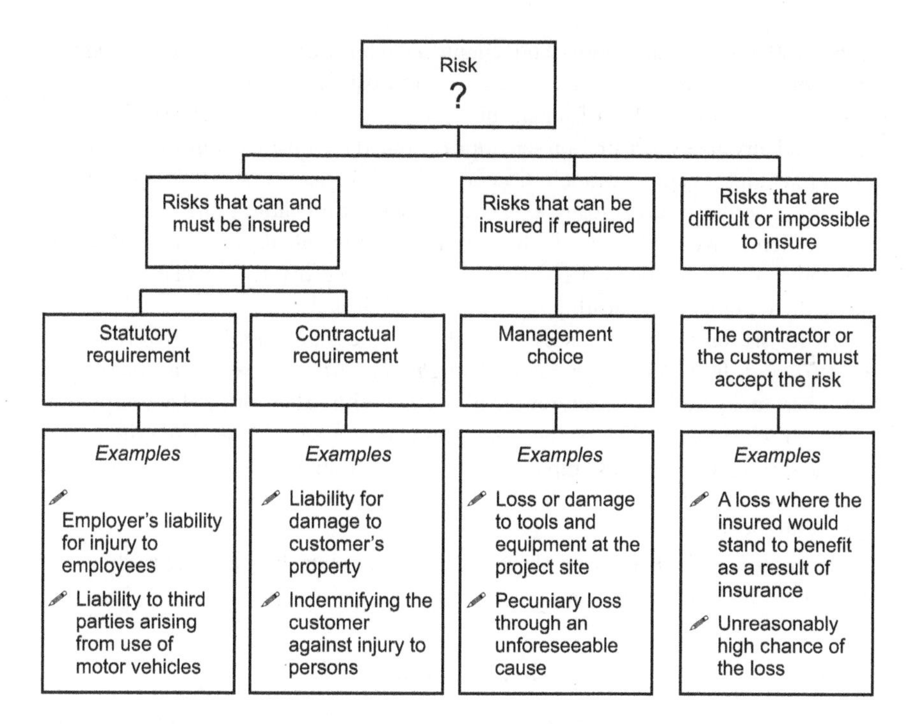

Figure 2.3 Risk and insurance in project management
From Lock, 2013.

motivating people affected by change in IT and management change projects in Lock and Fowler (2006).

Health and safety

In the times of ancient projects, the health and safety of human workers was, to put it mildly, not always top of the project controls agenda. Now there is so much legislation, either home grown in the UK or in the form of directives originally generated by the European Union, that some health and safety rules are seen as 'over the top', leading to accusations of a 'nanny state' being created. However, all those who work on projects have to understand that observing health and safety rules is not only a legal requirement, but is also common sense and a moral obligation. Increased risk to life and limb is always present in some occupations but clearly everything should always be done to minimize those risks and prevent injuries or other health hazards.

Two health and safety events remain prominent in my memory, both concerning accidents below ground level. The first experience dates from the time when I joined the London engineering offices of a mining company, which managed copper mines overseas. During my first month in that job, along with all the other London staff, I attended a moving memorial service at St Bartholomews Church to remember almost 100 miners, supervisors and engineers who were unfortunate enough to be trapped underground when a surface tailings dam failed and flooded the mine workings with slurry. Much later in life, I was a member of a small panel hearing MBA students give short oral presentations about their final examination essays for their project management module. One student, whose day job was a senior manager in a large construction company, explained how the tunnelling project described in his (excellent) essay had a very high safety record *with only three people killed!* When questioned about that being a good result, he explained that this was far better than the predicted rate of one death per mile of tunnel drilled.

UK legislation relevant to health and safety includes the *Health and Safety at Work Act*, 1974 and the *Offices, Shops and Railway Premises Act*, 1963, which is described as 'An Act to make fresh provision for securing the health, safety and welfare of persons employed to work in office or shop premises and provision for securing the health, safety and welfare of persons employed to work in certain railway premises; to amend certain provisions of the Factories Act 1961; and for purposes connected with the matters aforesaid'.

Never ignore health and safety laws and regulations but always adhere to them and ensure that those reporting to you do the same.

Quality and reliability

Quality and reliability of delivered projects is generally outside the direct control of administrative project management and is not included within the scope of this handbook. However, the general management principles governing achievement of satisfactory quality, reliability and performance do need to be known and appreciated by all who work in and manage projects.

At one time it was thought that quality and reliability of any product or part of a project could be achieved simply by inspecting work during the various production processes and ensuring that the design parameters and tolerances were satisfied. Then quality management underwent a dramatic change, largely driven by the Japanese, and we experienced the phenomenon of total quality control (TQM). This change required that all those working in an organization took collective control and responsibility for quality at all stages of design and production. Instead of relying on quality being 'inspected in' we came to learn that quality should be 'designed and built in'. A few quality gurus were prominent in driving these changes. Notable among these were Crosby (1979), Deming (2000) and Juran.

Ask a lay person to describe what is understood by quality and you might be answered that this refers to a product that is made to an extravagant design, using expensive materials and (coincidentally) commanding a high price. However, the essential qualities for quality demand that the product should be fit for purpose, which includes the following properties:

1 Safe.
2 Reliable.

3 Does what the specification says in terms of performance.
4 Environmentally acceptable (both in operation and at end-of-life disposal).

References and further reading

Crosby, P.B. (1979), *Quality is Free: The Art of Making Quality Certain*, New York: McGraw-Hill.

Deming, J. (2000), *Out of the Crisis*, Cambridge, MA: MIT Press.

Fowler, A. and Lock, D. (2006), *Accelerating Business and IT Change: Transforming Project Delivery*, Aldershot: Gower.

Health and Safety at Work Act (1974), Norwich: The Stationery Office.

Juran, J.M. (2016), *Quality Control Handbook: The Complete Guide to Performance Excellence*, 7th edn, New York: McGraw-Hill.

Lock, D. (2013), *Project Management*, 10th edn, Farnham: Gower.

Lock, D. and Scott, L. (eds) (2013), *Gower Handbook of People in Project Management*, Farnham: Gower.

Offices, Shops and Railway Premises Act (1963), Norwich: The Stationery Office.

Turner, J.R. (ed.) (2008), *Gower Handbook of Project Management*, 4th edn, Aldershot: Gower.

3
PROJECT AUTHORIZATION

Shane Forth and Dennis Lock

This chapter explains steps that organizations take when considering new project opportunities. To a very large extent the steps to be followed will depend on the type and size of the proposed project. This is an extremely important subject. A wrong management decision at this stage could commit the organization to a project where the risks of failure are too high or, alternatively, a potentially valuable project opportunity could be missed.

Internal projects

This section is concerned with projects carried out by an organization for its own purposes and at its own expense. These ventures can range from tiny projects for replacing ageing equipment to capital development ventures in the mining, minerals and petrochemical industries.

Low-cost internal projects

For a simple short-term low-cost project lasting for (say) no more than two or three years, a time-scaled graph that simply compares project costs with the expected financial benefits is usually a good indicator of whether or not to authorize expenditure and proceed with the project. Such graphs use this 'payback' method and there is an elementary example in Figure 3.1. Here a small internal project costing an estimated total of £245, 000 is shown to begin repaying its investment after two years and it recoups the investment costs (breaks even) completely after four years and five months. Such projects often arise when plant or equipment needs replacement. Many management change and IT projects can also come into this category. Of course, if the project had to be carried out for reasons of safety, security or a statutory order, then it would have to be authorized in any case.

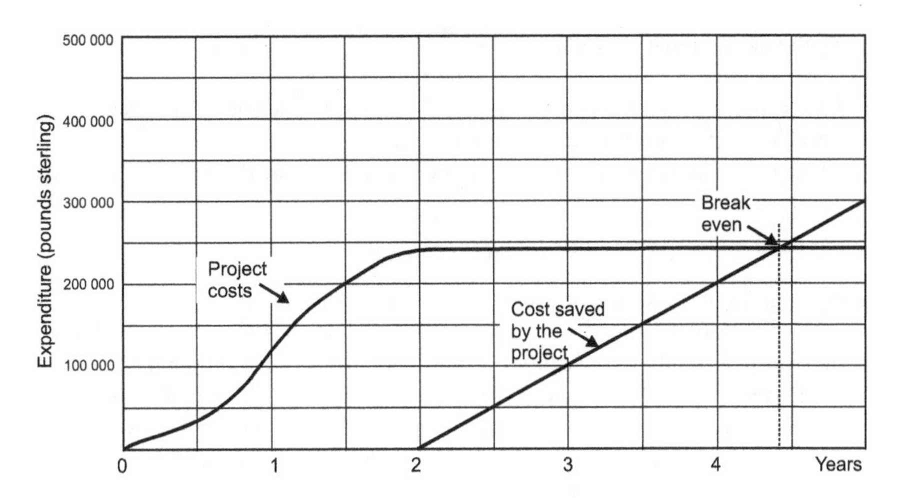

Figure 3.1 A project cost break-even graph.

Making the business case for large capital investment projects

The business case is an important document that is presented to the executive directors of a company when they are asked to commit expenditure to an internal project that would affect the organization, people, operations, finances and procedures of the company in some significant way. A business case might, for example, be concerned with a proposal for company relocation, the addition of new premises or the acquisition of particularly expensive processing or manufacturing plant.

Any proposal to commit capital expenditure to an internal project should be accompanied by a detailed business case document that sets out the estimated capital cost, the expected advantages (return on investment) and the corresponding risks and disadvantages. In other words, the business case should give the C-suite all the information that it is possible to gather to help with making a go or no-go decision on any project or programme that would involve considerable capital expenditure, organizational change and opportunity or risk.

High-cost investment projects

This section is principally concerned with significant capital investment projects. The range of possible projects in this category is very wide indeed. Among many other things, examples include building a new manufacturing plant, reorganizing logistics and distribution (including warehousing and transport), developing a petrochemical plant, establishing a mining complex, or carrying out a significant internal management change project (involving much reorganization and new IT). Some factors common to these projects include:

- High risk from many possible causes.
- Large amount of investment required up front.
- Partners or subcontractors can run out of cash and fail during the project.
- Usually no short-term return on investment.
- Market and other external circumstances can change before the project is delivered.

Feasibility studies

Many large capital investment projects begin with a feasibility study. For very big projects the feasibility study itself can take years and cost millions of pounds. Projects to develop new production facilities in the transport, minerals, mining, oil and gas industries require extensive site exploration, appraisal and predictions. Different project options (cases) will usually have to be investigated and evaluated. The final investment decision could lie somewhere between deciding not to proceed at all, choosing a recommended option, investing in an option that has been modified in some way, or seeking a joint venture partner.

A feasibility study will probably itself require authorization and registration as a project, using the methods described in this chapter. Where several different options are included in the study, each different option must be given a case name or number and it is very important to ensure that the facts, predictions and recommendations in the eventual feasibility study report are clearly tied to their relevant project cases.

Financial forecasts and analyses

A prudent project investor will be interested in many aspects of the proposed project including:

- Duration of the project development period (during which no saving, revenues or other benefits can be expected).
- Predicted project capital and operating costs.
- Forecasts for all cash flows and the expected rate of return on investment.
- Economic, political and market predictions that could affect costs and revenues.
- The timings of key project milestones.
- Reputation of any external partners and contractors who might have to be involved.
- The predicted KPIs.
- Risk.

One important outcome that sums up many of these factors is the expected rate of return on investment (IRR). For an organization's internal management change projects the critical decision factors depend largely on the predicted project outcomes as measured by KPIs. These are improved numbers and other factors in the company's financial and technical performance that will become apparent in future company reports and published accounts.

Discounted cash flows

When considering the balance between project costs and the expected financial savings or gains for projects that are expected to last for more than two or three years before they begin to deliver benefits it is necessary to understand the relationship between costs, revenues and time.

If a sum of (say) £100,000 were to be received and invested immediately in securities at 5 per cent interest per annum, it should be worth £105,000 when the first year is up. But if the receipt of that £100,000 were to be delayed for a year, then it would effectively be worth about £5,000 less because of the loss of interest. So £95,000 is said to be the net present value (npv) of the future £100,000. This effect is unimportant for inexpensive projects of short duration. However it can become highly significant for long duration projects when interest rates are high and large expenditure sums are involved. Then it is necessary to set out all the project cash flows (inflows, outflows and net flows) and discount them.

The npv of a future sum is given by $(100/100+r)^n$ where r is the annual percentage discount rate and n is the number of years. Tables of net present values for a range of future times and percentage discount rates are available, but a computer application such as Microsoft Excel can take the strain and do all the calculations for you. The principles of the method can be demonstrated using a very simple example.

A landowner has decided to cut a road through his forest to provide a much needed short cut for motorists. The road will take one year to construct, including bridging a small river. A toll gate at one end of the road will collect £1 for every vehicle that passes. Construction will take two years for a capital cost of £1 million. Operating and maintenance costs have been estimated at £100,000 per annum. Annual revenues are estimated to be £200,000 gross which, for simplicity, can be regarded as constant throughout the period under review. All local and national taxes have been ignored for this demonstration. Figure 3.2 shows a notional simple payback graph for this project, which appears to indicate a payback period of 10 years, measured from the opening of the toll road.

A table of discounted of cash flows for this project is set out in Figure 3.3. All taxation has been ignored for simplicity. After taking financial advice the owner has chosen an annual discounting rate of 5 per cent (based on the return that the owner could have expected from some other reasonably safe form of investment). The tabulation indicates that this project will have to be operated not for 10 years after start up, but for over 15 years if the notional costs of capital are added to the operating costs.

Project authorization

We live in a world that is volatile, uncertain, complex and ambiguous, which is summed up in today's jargon by the acronym VUCA. Unless a formal authorization and rejection process is followed there is grave danger that money might be wasted

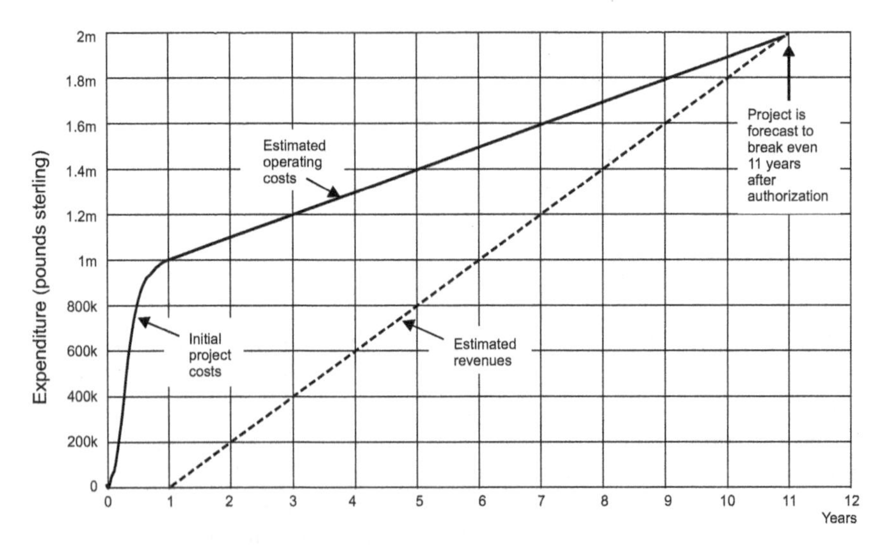

Figure 3.2 Simple break-even graph for the toll road project.

on 'wish lists' and vanity projects that could be left unfinished or otherwise fail disastrously to reach their intended objectives. This uncertainty and high risk can make projects exciting when compared with business-as-usual activities but clearly serious time and money can be lost if wrong decisions are made.

A common route for new projects to enter an organization is when sales engineers are successful in persuading a potential customer to consider placing an order. However, a prudent company will not accept an order and commit to project start-up without first making some checks on the potential project purchaser. Of course, if the potential customer has ordered before and satisfied all the previous financial and technical obligations (paid the bills on time and approved drawings without causing delays or problems) a new order will probably be attractive and accepted. Some companies consider these matters at high-level meetings, when additional factors such as current capacity and confidence in being able to fulfil the order will rank high.

Project authorization procedures include notifying all divisions and departments of the organization that the new project has become live. Essential project information must be included in the authorization document including, not least, the name of the client (if external), the unique project description and identifier, authorized budgets, delivery schedule and so on.

Small projects are often authorized by means of a works order, which is signed by a senior manager and circulated to all relevant departmental managers plus other functional departments (such as finance). Figure 3.4 shows a works order format suitable for use in a factory, and Figure 3.5 is a format that was used successfully by a mining engineering company for small and larger capital projects.

Year	Cash outflows (a)	Cash inflows (b)	Net cash flows (b-a)	Discount factor at 5%	Discounted annual cash flows	Net present value
0	£(1 000 000)	0	£(1000 000)	1.000	£(1 000 000)	£(1 000 000)
1	(100 000)	200 000	100 00	0.952	95 200	(904 800)
2	(100 000)	200 000	100 000	0.907	90 700	(814 100)
3	(100 000)	200 000	100 000	0.864	86 400	(727 700)
4	(100 000)	200 000	100 000	0.823	82 300	(645 400)
5	(100 000)	200 000	100 000	0.784	78 400	(567 000)
6	(100 000)	200 000	100 000	0.746	74 600	(492 400)
7	(100 000)	200 000	100 000	0.711	71 100	(485 300)
8	(100 000)	200 000	100 000	0.677	67 700	(417 600)
9	(100 000)	200 000	100 000	0.645	64 500	(353 100)
10	(100 000)	200 000	100 000	0.614	61 400	(291 700)
11	(100 000)	200 000	100 000	0.585	58 500	(233 200)
12	(100 000)	200 000	100 000	0.557	55 700	(177 500)
13	(100 000)	200 000	100 000	0.530	53 000	(124 500)
14	(100 000)	200 000	100 000	0.505	50 500	(74 000)
15	(100 000)	200 000	100 000	0.481	48 100	(25 900)
16	(100 000)	200 000	100 000	0.458	45 800	19 900

Figure 3.3 Discounted cash flow tabulation for the toll road project.

Works order	**Project number**

Customer	**Delivery address (if different)**

Project title

Documents defining this project	**Number**	**Rev.**

Budget summary	***Hours***
Engineering design	
Design after issue	
Works	
Assembly	
Final testing	
Installation	
Commissioning	
Materials, services and expenses	£

Schedule summary Project start and finish dates are firm. Others subject to detailed planning.

	Start	***finish***
Design and drawing		
Purchasing		
Manufacturing		
Assembly		
Final test		
Install and commission		
Overall project dates		

Commercial summary	**Contract type and total price**
Sales engineer:	
Sales reference:	
Customer's order No.	

Notes/limitations

Authorization (subject to any limitations listed above)

Project manager: Project authorized by:

Distribution

Project manager	Chief engineer	Works director	Materials manager	Quality manager	Accounts manager	Office manager
☐	☐	☐	☐	☐	☐	☐

Figure 3.4 Works order example for a manufacturing project.

Project authorization by charter and contract

Some organizations prefer to authorize their projects by means of a charter and contract procedure. The process can be used for significant internal investment or management projects and for projects carried out for customers and clients. There are

PROJECT AUTHORIZATION

Client ______________________________

Scope of work ______________________________

____

____

____

____

Source documents ______________________________

____

____

Project number (to be entered by accounts department) [| | | | |]

Project title (for computer reports) [| | | | | | | | | | | | | | | | | |]

Project manager (name) ____________________ Staff number [| | |]

Project engineer (name) ____________________ Staff number [| | |]

Project start date (enter as 01-JAN-14) [| - | | | - |]

Target finish date (enter as 01-JAN-14) [| - | | | - |]

Contract type:

Reimbursable ☐ Lump sum ☐ Other (specify) ______________________________

Estimate of man-hours

Standard cost rate	1	2	3	4	5	6	7	8
Man-hour totals								

Notes:

Authorization (1) *Authorization (2)*

Project manager		Marketing		Contracts dept.		Purchasing			
Project engineer		Central registry		Cost/planning		Accounts dept.			

Figure 3.5 Project authorization form used by a mining engineering company.

no doubt many variations of the charter and contract procedure in use in different organizations and many do not use it at all. However, the following paragraphs outline the process.

Project charter

The project charter (sometimes known instead as a project initiation document or PID) defines the project and its expected outcomes. It is widely used for internal management change projects that seek to improve business performance by organizational change, updating IT systems, performing company acquisitions and mergers, and so forth.

Where the company is the owner (of a product or facility) or the ultimate client, it will also use a project charter to authorize its capital projects for extending or improving its products and assets, or for adding new facilities. Capital projects tend to be complex and often involve the owner or ultimate client in placing commercial contracts for other organizations to perform the work.

For internal projects the charter summarizes key details from the business case. This charter must be concise but accurate and leave no ambiguity about what the project is intended to deliver. Some organizations will ensure that the project charter is updated as the project proceeds so that it remains valid and conforms to changes made throughout the project lifecycle. However, companies that wisely have rigid systems for managing project changes will keep change records that supplement the original charter and so will eventually record the 'as-built' or 'as delivered' condition of the project.

Preparing a project charter can be very time-consuming, adding costs without value. Some skill is needed to achieve brevity while recording all the essential details from the business plan or project specification. One danger is that those who join a project later in its lifecycle might confuse the project charter with the original business case, and so miss or misconstrue important information.

A typical project charter or PID will include information under the following headings:

- project title and number;
- project authorization signature;
- name of the project manager;
- names and titles of other key personnel;
- project scope and objectives;
- expected deliverables;
- summary, describing the project as envisaged and defined at start-up (the project baseline);
- summary of estimated costs or authorized budgets;
- sponsorship and principal stakeholders;
- project management methodology or governance;
- reporting requirements;
- record of authorized changes including a summary of their effects on this charter;
- issue date and revision number for this document.

Project contract

A project contract in this context of charter and contract is a document that forms a legal and binding contract between two organizations or parties that defines the scope of work to be performed by the contractor and the terms and conditions of financial settlement. The contractor performing the work is one party to the contract and the owner or ultimate client is the other party. While both parties will have their own representatives managing the project from their own perspectives, the contract recognizes the owner or ultimate client representative as the project manager.

The contract document will define the expected project deliverables and give detailed budgets. Key milestones should be identified, particularly giving any dates that have already been agreed. The contract will also list the project manager's targets, responsibilities and limits of authority. In short, the contract will define and quantify the project with as much detail as is known at this early stage.

Project registration and numbering

Once a new project of any kind has entered an organization it has to be formally 'entered into the system' so that all the necessary accounting, planning, progressing and other administrative procedures can be put in place.

One of the first steps in initiating a project after authorization is to allocate a project identification number which, depending on the procedures of the particular company, will be used henceforth as a basis for drawing numbers, cost codes, purchase orders and other important documentation. Project numbers are usually reserved in blocks of numbers where for example, the first two or three digits could be associated with one of the organization's business or operational units or a particularly valued customer. The remaining digits are allocated sequentially. An example of a project register page is shown in Figure 3.6.

The project register is a very important document for many reasons, both during the project lifecycle and long after the project has been delivered. Its purpose is, of course, not only to allocate numbers. For current projects the register lists all

Project register Issue date:

Project number	Project title	Project sponsor	Project manager	Customer	Date opened	Comments Special restrictions	Date closed

Figure 3.6 Essential elements of a project register.

authorized work within the organization against which work time may legitimately be booked by staff on timesheets and against which the costs of project materials and expenses may be charged. However, even when all the projects on a register page have been closed the register lives on in the company's archives.

Whenever it is necessary to retrieve information about a project (current or past) the project register is usually the best, often the only, starting place to begin the search. In most management information systems or archives the project number will provide a gateway to finding the various document files and project data. But historical searches often start with only a vague recollection of the project description, or the name of the customer, or the approximate project dates. So these data can be used as search keywords. Each entry in the project register should therefore include the following data:

- project title;
- project number;
- start date and (eventually) closure date;
- project sponsor;
- project manager responsible;
- the customer;
- and possibly, the customer's order number or letter reference.

A well-kept register will enable any project to be identified even when only one or two of the associated data items listed above are known. Register information for past projects should be kept in a secure but accessible archive.

Conclusion

If the procedures outlined in this chapter result in a new project being authorized, that can be a matter for rejoicing, especially in a company that was previously short of work. Now is the time to announce the new project to participating staff. One way for engendering enthusiasm from the project start is to organize a 'kick-off meeting' where the project manager, project staff and senior management can gather to ensure the new project gets off to a good start.

4
CONTROL PRINCIPLES

Dennis Lock

Projects from ancient history that still survive relatively intact today after thousands of years were managed without the benefit of professional organizations, recognized qualifications, textbooks, advice from 'scientific management' studies, critical path networks or even paper drawing sheets. There were no office copiers and printers, radio or telephonic communications, computers, mechanized construction equipment, plastic materials, satellite navigation, health and safety rules, air transport: none of those advantages were available to our ancestors. Yet they managed amazing feats of engineering and construction. Conversely now we have many modern advantages but manage to 'achieve' some spectacular project and company failures. This chapter introduces a few basic rules and principles for getting things done on time, within budget and to the satisfaction of all the stakeholders.

Control cycles

People engaged in manufacturing management know that there is a logical pattern of activities that should observed in order to discover and correct any failure in a repetitive process. That pattern is known as the plan-do-check-act cycle (or the Deming cycle), one version of which is shown in Figure 4.1. The principle is obvious from the illustration, as follows:

1 **Plan** to carry out a project task.
2 **Do** the work.
3 **Check** the result.
4 **Act** to correct any deviation from plan (nonconformity) as soon as possible.

That procedure is appropriate for making small objects in batches, where the total cycle time is short (perhaps measurable in minutes or hours). Strictly, one other element should be added to the cycle, which is to learn from the experience, so that the process becomes plan-do-check-act-learn.

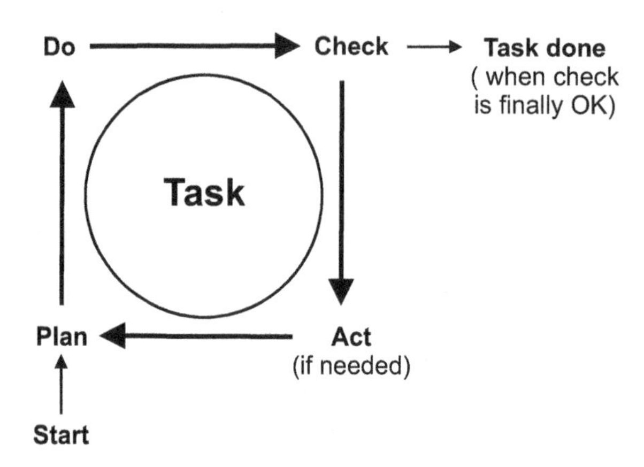

Figure 4.1 One representation of the plan-do-check-act control loop.

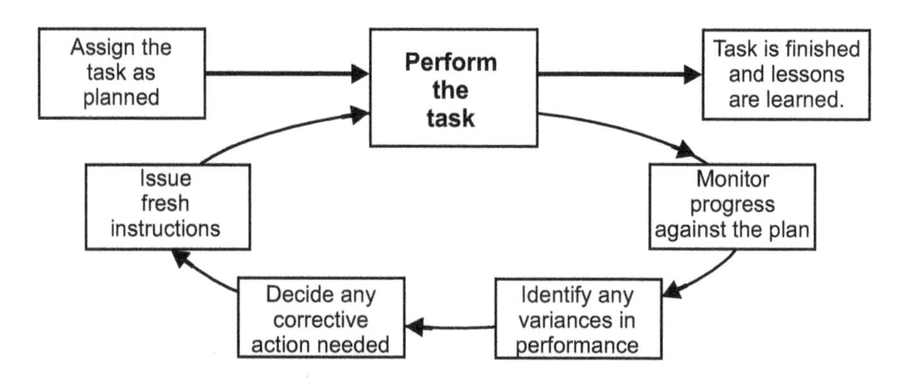

Figure 4.2 A more proactive view of a plan-do-check-act control loop.

Project managers do not usually have the luxury of being able to make trial runs or test batches. In projects the 'do' part of the plan-do-check cycle is normally longer than a few minutes or hours and might be weeks, months or more, so that the activity must be monitored continuously, allowing any necessary corrective action to be taken in good time. Project work is non-repetitive, which means that it has to be 'right first time' and having to restart a task would be a serious failure. We can and must still use a control loop, but have to be alert and proactive all the time, monitoring every task from start to finish while it is in progress (Figure 4.2). Then it should be possible to detect any fault or poor performance early enough for effective action to be taken and for the desired result to be achieved.

The control loop principle should be applied not only at the single task level, but at all other levels including work on the entire project. In projects, not only the work itself has to be measured, but the time taken and the costs incurred are clearly equally important. Some of the chapters in this handbook deal with these subjects in considerable detail, but here in this chapter I want to introduce and describe a few principles that need to be understood if we are to plan, monitor, control and achieve successfully.

Principles of planning and scheduling

Planning methods have evolved over many years, beginning with simple lists of target dates and then progressing to charts. Bar charts are often called Gantt charts after the early industrial scientist Henry Gantt, but his charts were really more concerned with planning activities in manufacturing facilities than with capital projects. In project management, before the 1950s we relied on charts drawn on paper, chalked on blackboards or constructed as wall charts. Some wall-mounted charts used quite ingenious proprietary kits, one of which used horizontal 6mm wide Lego-like strips that could be cut to timescale length, each strip representing one task. I confess to having used such methods myself, with some success.

Manual charting methods were fine for very small projects, difficult for larger projects and impossible for big projects subject to changes. None of these charting methods could show clearly how interdependencies between different tasks might affect the schedule and there was nothing to indicate which tasks claimed the highest priority for the allocation of limited resources. Then critical path networks came to our rescue. The earliest example of which I am aware was the link and circle method, dating from around 1950, developed by the long-dead and largely forgotten English Electric Company, which bore some similarity to the precedence networks that are in common use today. In those days project workers did not have the advantage of readily available computers and everything had to be calculated using only brain power or cumbersome mechanical difference engines. When digital computers first became available they were large mainframe machines needing special accommodation and typically costing hundreds of thousands of pounds.

The general availability of small-sized, affordable computers and software capable of calculating complex project schedules now provide project managers with powerful scheduling tools that can take available resources into account, looking across a whole organization and all projects that compete for those resources. Further, the results can be sorted, edited and printed so that managers receive detailed plans and reports relating only to the project tasks for which they are directly responsible and are able to control.

You will find critical path planning methods described in Part IV of this handbook, but I need to list a few principles of scheduling and progress control here, as follows:

1. All tasks needed for project completion should be included in the plan.
2. Dependencies between the starts and finishes of different tasks must be apparent in the plan and obeyed. Thus no task should be scheduled to start until all preceding tasks have been finished.
3. Some people have difficulty in deciding the level of detail to be shown in the plan but I apply the rule that no single task should extend beyond the direct

control of one manager. If the plan shows a task that moves from the control of one manager to another during its duration, that task should be split into two separate tasks (one for each manager to control).

4 It is preferable if all planned tasks, their estimated durations and the levels and types of resources needed can be agreed by the managers who will ultimately be responsible for controlling the work. That usually means organizing one or more planning meetings at the beginning of every project. The project manager should clearly be present at such meetings but it will be found most useful if an experienced planning engineer asks all the pertinent questions and develops the draft plan on a wall board that everyone can see. The planning engineer might be a member or leader of a PMO.
5 Scheduling must take account of the resources available, or which can be made available.
6 Scheduling must also take account of all other projects in the organization that make claims on common resources.
7 Departmental managers must receive regular advance reports that list the project activities that their departments will be expected to perform during each following period. Reporting periods might, for example, be four weeks or calendar months.
8 When the project is in progress every task must be monitored and the manager responsible must be proactive in taking immediate action to rectify any fault or delay.
9 The software used to perform the calculations, schedule resources, edit, sort and report the schedule data and performance results must be up to the job and chosen with care. The IT department is not usually the best placed company department to make that choice, but they must of course be kept informed and consulted. It is probable that the person or group of people best able to choose and recommend the most appropriate software will be found in the PMO. The most popular and universally available software is fine for small projects, but for any project involving more than a hundred or so tasks, and for all multiple projects, programmes and portfolios it cannot be recommended – at least not by me.
10 Organizations like to keep a baseline plan for every project, which is the initial plan for the project before it undergoes the ravages or changes and missed milestones. That is one measure by which divergence from intentions and ultimate success in project performance can be judged.
11 Companies that undertake similar projects repetitively find it useful to keep and refer to plans of previous projects so that lessons can be learned and advantage can be taken of previous experience.

Controlling progress

Controlling progress against the schedule requires that each manager arranges for regular checks to be made, ensuring that tasks which are scheduled to be in progress are actually taking place, and also looking at the rate of progress and the amount of work achieved to make certain that no task will run late.

Here is a useful spot-check to ensure that the actual rate of working is generally in line with the planned progress. Let's take design engineering as an example and suppose that you are the project manager. Have a look at the project schedule and note how many designers should be engaged on tasks for your project today. Suppose that number should be 15 people. Now go to the design office and ask its manager to point out the designers who are actually working on your project design tasks. You might get two unpleasant answers. The first unpleasantness can come from the design manager, who resents your intrusion. Then, when the answer is forthcoming, you could find that only six people are actually at work on your project. That will give you early warning that your project will run late unless immediate action is taken to inject more effort.

If your company regularly outsources work such as design or special processes, it is advisable to appoint someone (perhaps known as a project liaison engineer) whose role is to engage with those external organizations, taking fresh work out to them, checking on progress (and sometimes quality) and collecting finished work. If well managed, that process can help to build good relationships with the external providers, resulting in better service and fewer project delays.

Individuals and supervisors often tend to be optimistic when reporting progress. One danger is that tasks are reported as finished when actually a small amount of work or 'tidying up' still remains to be done. That might not always be serious, but it can happen in such cases that something goes wrong with a task at the last minute so that it has to be corrected or even restarted. Here are two examples from my own experience.

Example 1. A miniature high voltage resin-encapsulated electrical transformer for a prototype guided weapons project was reported as finished, but it subsequently failed on final testing and work had to begin again on a replacement.

Example 2. In order to achieve our departmental monthly revenue targets I allowed an invoice to be issued to a customer when the relevant equipment was still undergoing final inspection and testing. The equipment failed its test and we had to cancel the customer invoice with a grovelling apology. Delivery was delayed for a month.

In project management, the vital question that must be asked when any task is reported as finished is 'Can the next task or tasks on the schedule be started now?'

Immediate action orders

If a critical task does fail, clearly the unexpected additional work in putting things right becomes super-critical. Then special and decisive action must be taken to get things back on track, even if that means spending money on additional resources or special methods. One such method uses immediate action orders. These are described in Lock (2013), but their essential features are:

- An immediate action order takes precedence over all other work. If the urgent item needs the use of a machine or facilities currently engaged on another job, that other work must be interrupted, and your urgent item has to take immediate priority.

- A progress clerk or chaser is assigned to follow the job from beginning to end, actually staying with the work, armed with a route card that is signed in and out of each responsible work centre with a time and date stamp.
- Costs are virtually ignored: if a part that you need is in Scotland and your project is in Cornwall, then get it by the fastest method possible, even if someone has to get on an aircraft and collect it.
- An immediate action order must be authorized by a very senior manager, possibly at board level.
- Because of the disruptive effects on other work, only one immediate action order may be in place in the organization at any one time.

Those measures may seem drastic and costly, but they are very effective. Always remember the maxim TIME IS MONEY (from the short pamphlet *Advice to a Young Tradesman*, written by Benjamin Franklin in 1748).

Example 1 above tells of a miniature transformer that failed during testing. That prototype had taken six weeks of elapsed time in the factory from the issue of drawings to testing. An immediate action order was issued to produce a replacement transformer. A progress clerk was assigned to the job of following all stages through from redesign to final testing. The immediate action order, printed on orange striped paper and signed by the general manager, was carried into the design office and given to the design manager. A designer was immediately assigned to decide why the unit had failed (which turned out to be the creation of air bubbles during epoxy resin encapsulation). Fresh drawings were issued and carried to the production manager, who immediately arranged for the casting mould to be modified and for a new transformer to be manufactured. The initial (failed) transformer had taken six weeks of elapsed time to produce. With the immediate action order in force, the (successful) replacement was finished in three days.

Controlling project costs

Project costs usually comprise the costs of materials and bought-out services *plus* the costs of direct labour for all project tasks from design to delivery *plus* the organization's indirect costs (overheads). Part V of this book deals in considerable detail with the nature of project costs and cost control, so it is not necessary to go into much detail in this chapter. So here I just want to give some basic tips to project managers and staff on costs and cost control.

1. You must adhere to the project specification and business plan, prevent scope creep and not allow unauthorized project changes to happen.
2. Always remember that there is a direct relationship between project costs and progress against schedule. If your project runs late it will almost certainly overrun its budget, even converting an intended profit into a substantial loss.

3. Establish good communications and rapport with your organization's financial management, including the cost and management accountant. Gain an insight into your organization's cost accounting procedures.
4. Because materials and bought-out services typically account for a high proportion of costs in most projects, establish good relations with your organization's purchasing department and ensure that wherever possible your project purchases take advantage of any advantageous pricing and contract terms available to your organization.
5. Ensure that individual members of your project staff do not (deliberately or otherwise) bypass purchase enquiry and order procedures by entering into unauthorized contractual commitments with the suppliers of goods and services. Binding commitments can easily be made (unwittingly or otherwise) during meetings or online communication between project staff and suppliers' representatives.

References and further reading

Lock, D. (2013), *Project Management*, 10th edn, Farnham: Gower.
Turner, J.R. (ed.) (2007), *Gower Handbook of Project Management*, 4th edn, Aldershot: Gower.

5

ESSENTIAL CODING STRUCTURES

Dennis Lock

The world of industry and commerce could not function without numbers and codes. Everything in our daily lives is governed in some way by times, dates, numbers, accounts codes and so on. This dependency on codes is especially true for project controls, and this chapter will examine some important aspects of this subject.

An introduction to codes

A code is a shorthand and precise way for describing a piece of information or project data. A code might be allocated to identify an entire project, or the smallest item of data needed to manage a project. Codes can also define objects or materials. The following are some uses of codes:

- A code can act as an identifier for an object or a piece of data. We use such codes in our daily lives when we purchase goods using catalogue numbers.
- A code can act as an address, so that it locates an item within its physical or organizational context. Post codes and zip codes are familiar examples.
- Codes can describe the physical properties of materials and objects, such as a grade of steel.
- Departments in an organization can be given identifying codes.
- Codes can be used to denote the personal skills of people – for example designers, bricklayers, electricians and so forth.

In fact, without codes no database could function and all the benefits of modern management and work control systems would be lost to us.

Codes for calendar dates

Although codes, their meanings and their relationships can be complex, I want to begin with the apparently simple topic of calendar dates. Think of a project team in the UK beginning their project on 1 December 2022, a date which they would commonly express (code) as 1/12/22. But when dealing with some organizations overseas (particularly in the US) 1/12/22 would signify not 1 December 2022, but 12 January 2022. Clearly there is much room for misunderstanding and error here. Competent project management software should give the user options on the arrangement for inputting and reporting calendar dates. The sensible choice is to write all project dates in the form DDMMMYY, so that 1 December 2022 would be written 01DEC22.

Coding consistency throughout an organization

I was once engaged as external project management consultant to a manufacturing organization. There were 10 departments in this company that could produce design drawings and each of these departments used its own system of drawing and part numbers. So there were no fewer than 10 different coding systems in existence for drawing and part numbers within the same organization. That exceptional case (which we did manage to improve considerably) illustrates the need for conformity of codes throughout the organization.

Take, for example, the code of accounts that is usually operated through the company's cost and management accounting system. The person, department or committee responsible for designing project management coding systems should try to ensure that codes used in project costing and estimating are consistent with the code of accounts as far as possible.

A standard coding system that can be used again and again, for project after project has many advantages. For example, it will be easier for future engineering and project managers to access records of past projects to assist them in such things as comparative cost estimating. Also, from the design and engineering point of view, it will be easier to access past designs that can be used again on future projects (a process which has been called retained engineering). However, the ideal of completely standardized coding systems throughout an organization can be very difficult to achieve.

Coding becomes more complicated when different companies working on the same project use their own individual systems. This problem of non-compatibility is revealed, for example, in every project where a client's organization and the project company each needs to keep records and copies of the same drawings and specifications on file for the project. That problem can be solved by numbering each document with two numbers, one for using the customer's system and the other in the project contractor's system. At one time that would have been very difficult to do, but now we can let the search and find capabilities of the computer taken the strain.

Organizational breakdown structures and codes

Department codes

All project control data requires coding so that each data item can be identified with the department responsible. Then reports and data searches can be made so that each departmental manager can find or be given project data that they need to know for controlling work, costs and progress within their department. A typical way of achieving this is to assign a code to every department in the organization, the organizational breakdown commonly being referred to as the organizational breakdown structure (OBS).

Departmental codes can typically comprise two characters (alpha, numeric or a mix of both). These codes will later be embedded in (or otherwise associated with) more comprehensive codes for each project task or cost element. Such arrangements are not dependent on the organization shape, so that task forces, team and matrix structures can all have their departments or groups coded in the same way. Even non-project departments in the organization (such as central administration, purchasing, HRM and finance) should have their own codes because their managers will need connection, communication and selective reports from the organization's central database.

Staff codes within the organization

It is customary and necessary that each person within the permanent project organization should have a staff code. This is usually a three or four digit number. It is used to identify the staff member on timesheets, and also for indexing in HRM records and so on. Staff code numbers are typically allocated when each member of staff joins the organization.

Every staff code should be associated directly with a standard cost code for the member of staff. This applies to staff at all levels, whether or not they are engaged on direct project work or not. Standard cost codes signify the cost at which time recorded on timesheets will be charged to the job code. It is important to keep this system as simple as possible and to have as few standard cost levels as possible. For example, in one large company where I was a manager the following simple system was found to be adequate:

Grade 1: Company directors and one or two other very senior managers and specialists.
Grade 2: Departmental managers.
Grade 3: Senior design and engineering staff.
Grade 4: Design staff including group leaders and checkers.
Grade 5: Lower grade design staff.
Grade 6: All clerical staff and (surprisingly) non-project administrative staff including (for example) staff in the contracts department.

In the company from which this example was taken, some staff privileges were closely associated with these grades so that (for example) those in grades 1 and 2 were given company cars.

When entering timesheet data into the system, each person needs to know and write only their individual staff number because the system database can associate this with their department and relevant standard cost rate.

Resource codes

Resource codes are used particularly when scheduling and levelling project or whole organization resources by computer. A typical arrangement will allow a two-digit code that identifies each resource. Resources most likely to be scheduled are people, usually defined by a resource type or skill for this purpose. However, other resources such as bulk materials might be scheduled and thus need codes.

Project identifier codes

Every project in the company's programme or portfolio of projects will need its own identifying code. All information in the company's database pertaining to each project can then be pinned to this code. Ideally this code should identify the project and the project customer or client. A useful and relatively simple method is to allocate a two or three letter or code that identifies the client and then add the last two digits of the year in which the project was authorized. Then, a two- or three-digit serial number can be added to identify the particular project for the customer. For example, the third project to be authorized for the Lox Wonderful Widgets Company in 2021 would be authorized and opened with the project identifier LW2103.

Internal company projects

Most companies carry out internal projects, which are funded not by external clients but from the companies' own cash reserves or other means (such as bank loans). These projects need the same controls as projects for external clients but of course will not receive contractual payments and will not directly result in a profit or loss. Internal change projects come into this category.

These internal projects can be coded and managed in just the same way as those for external clients, but the internal project number will clearly identify the project as internal. For example, the project identifier for all internal projects carried out by the engineering division of one company were identified by having ED as the first two letters in the project number code. The costs for such projects are usually all indirect (overheads) and the accountants will therefore cost all work using only prime costs (without adding overhead cost rates). Cost accounting methods are introduced in Chapter 9.

Project register

Project identifier code numbers for all projects (external and internal) are best allocated from (or with reference to) a register kept and managed by a PMO or a registry. These codes will be found very useful – even essential - not only for current project controls but also for in future years, when information from closed projects might need to be retrieved to assist in post project problems, lessons learned, legal issues and so on.

Coded work breakdown structure

A work breakdown structure (to which I shall refer by its usual abbreviation WBS) is vital for controlling all but the very simplest project. Stephen Devaux wrote the following: 'If I could wish but one thing for every project, it would be a comprehensive and detailed WBS. The lack of a good WBS probably results in more inefficiency, schedule slippage, and cost overruns on projects than any other single cause. When a consultant is brought in to perform the role of "project doctor", invariably there has been no WBS developed. No one knows what work has been done, nor what work needs to be done. The first thing to do is assemble the planning team and teach them how to create a WBS' (Devaux, 1999).

For tiny projects the WBS is little more than a list of tasks. In larger projects the WBS can be a complex layered structure containing thousands of tasks, each with its unique identifying code. The WBS is like a pyramid, with the entire task (the project), at the apex. All project work then breaks down through a series of levels until the lowest level comprises the many simplest tasks and components for the project.

The composition of every code in the WBS should be designed so that it signifies the relationship of each WBS element with tasks at higher levels in the WBS. This can be illustrated by the following example, for which I have chosen a large capital project to create a copper mining facility.

WBS example for a copper mine development project

The fictional example here is for a project being carried out by a mining engineering company (MEC) to design and build a copper mining facility for an international mine owner and operator called Lox Minerals. This project is the first to be authorized for this client in the year 2020, and is coded Project LM2001. The project breaks down into several major components as follows:

01: The mine shaft complex, complete with winding gear.
02: Concentrator plant, comprising conveyors, ore bins, crushers, separation plant (flotation tanks) and tailings disposal.
03: Smelter, complete with ovens and associated and fuel bins and so on.
04: Refinery, comprising electrolytic tanks, controls and the associated building.
05: Warehousing.
06: Site roads and vehicle parks.
07: Accommodation and messing.
08: Hospital.
09: Airstrip.
10: Railhead,

and so on.

Figure 5.1 shows how MEC might set out and code some of the upper WBS levels for this large capital project. For simplicity, this diagram shows only a small selection from the total WBS.

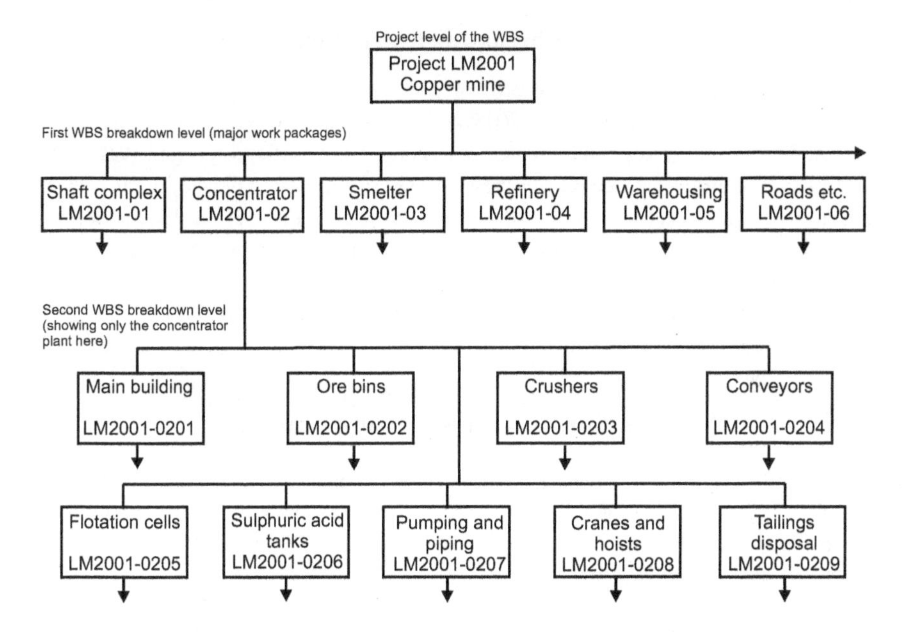

Figure 5.1 Top levels of the WBS for a project to establish a new copper mine.

The top level is the project itself, and the next level below shows major work packages from the first breakdown level (six items are shown here, taken from the above list). Each item at this first level breaks down into second level items, and I have shown how that might be done for the concentrator plant. The breakdown for this project would continue to several lower levels, resulting in thousands of individual tasks at the bottom level.

For every item at each level in the WBS the code not only identifies the item (for use in costing, design engineering, work allocation in manufacture or construction and so on) but it also acts as an address, from which the position of every item in the WBS at any level can be related to its place in the design and fulfilment of the entire project. Take, for example, the item crushers LM2001-0203, which occurs at the second level of the WBS in Figure 5.1. Apart from identifying the crushers, this code breaks down to convey the following information:

- LM identifies the project client as Lox Minerals;
- 2001 informs us that this project was the first to be authorized for this client in the year 2020;
- 02 positions this task in the concentrator plant;
- 03 identifies the task as pertaining to the crushers in the concentrator plant.

Lower levels of breakdown will follow. In this example the next level down might be used to distinguish between fine and coarse crushers, and the code level then descends further to identify separate small components.

Cost breakdown structure

The cost breakdown structure (CBS) for a project requires input for each task or other cost item built around the WBS and the associated OBS code. This is best explained by a matrix diagram. Figure 5.2 is an example of a combined WBS, CBS and OBS. The project is necessarily very simple (a lawnmower development project) because a large project would be impossible to depict in these pages. In practice it is never necessary to draw the CBS in a diagram because the codes fall out automatically from the central database.

Particularly useful illustrations of combined work, organization and CBS are given in Harrison and Lock (2004).

Avoiding unnecessary complication

A task which spans more than one department, which is at a low level in the WBS and which also needs cost and departmental codes could result in a very long number. Long numbers can be associated with clerical errors and cause discontent among those who have to use them.

However, even long composite codes can be made simple for each user. For example, bar coded documents could be one solution. The central database can also remove much of this problem. For example staff numbers will automatically be associated with their relevant department codes. A person from the electrical

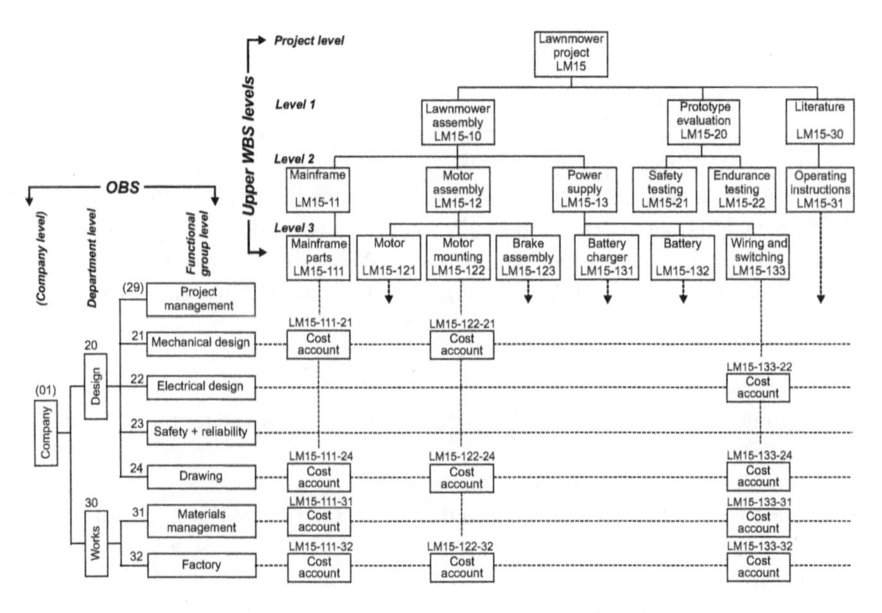

Figure 5.2 Combined WBS, OBS and CBS for a lawnmower development project.
Source: Taken from Lock (2013).

department need not normally enter a separate skills code because it can be assumed that all staff in that department will be involved in electrical tasks.

Part numbers

In projects such as manufacturing, aerospace, defence equipment and so on it is important for many purposes to allocate code numbers to the whole product and its components. Initially these can derive from the WBS for the design, which should break the project prototype and its subsequent batches down logically into main assemblies, subassemblies and so on. The process of parts identification and coding should be familiar to any car owner seeking a replacement component, for which the make, model and chassis number will usually be required in order to identify the correct part.

Consider the result of a project to build a prototype aircraft. After much testing and several test flights this aircraft will go into production. For each tiny component, subassembly and the main assembly it is important to allocate a code so that the correct replacements can be made during maintenance. So the WBS can be set out and structured to show all the components and assemblies with their part numbers.

When the aircraft goes into production, some of the parts and assemblies will have undergone one or more modifications. There are some rules that must be applied here to ensure reliability and safety of the product, achieved through the use of serial numbers.

Serial numbers

Every component should carry not only its part number, but also a serial number from which it can be related to a particular design modification state. An important rule here is that if a component design is modified to the extent that it is no longer interchangeable with corresponding components from the past, the part number itself must be changed (not just its batch or serial number). For example, suppose that part number 123–5956 refers to a spacer bush made from brass and that subsequently this component is changed by design to a nylon bush that is electrically insulating, the new bush must be given a new part number.

Build schedules and traceability

Over the course of time, products such as cars and aircraft undergo many modifications to combat unreliability or to satisfy particular customer orders. So we arrive at the concept of build schedules. A build schedule is compiled each time a new batch of the product is made. It lists the part number, drawing issue and description of every component and assembly, the list being structured in tree fashion, resembling the original WBS.

In many cases it is necessary to provide each component with a nameplate or to stamp it to indicate its part number and modification status. Also, build schedules should allow the source of supply for components to be traceable (by naming the manufacturer and the production batch number used during assembly). The reason for this is to allow traceability, the process by which engineers and accident investigators can trace the design status and source of any component that has been found to cause a malfunction or (worse) an accident. Traceability, combined with product build schedules, allows an airline or an automobile manufacturer to identify all aircraft or vehicles currently in use that contain the faulty or unreliable part, thus allowing urgent preventative and maintenance action.

References and further reading

Devaux, S. (1999), *Total Project Control*, New York: Wiley.
Harrison, F. and Lock, D. (2004), *Advanced Project Management: A Structured Approach*, 4th edn, Farnham: Gower.
Lock, D. (2013), *Project Management*, 10th edn, Farnham: Gower.

PART II

Project organization

6

ORGANIZATION STRUCTURES

Dennis Lock

Many different company organization structures exist in a wide range of complexities and configurations. Organizations can be simple, based on one location or (at the other extreme) large international groups. However, for the purposes of project management and controls it is usually possible to categorize an organization according to one of the examples described in this chapter or in Chapter 7. Some of the views expressed here are personal, based on experience.

Introduction to organizations and management control

Understanding the principles of organization structures and their effects on command and communications is important and essential for the effective implementation of project controls. In most cases a project manager is appointed to an existing organization, or to a newly formed organization that has been already been prescribed by senior management or (Heaven help us, by management consultants or, far worse, by a government department).

First, it is necessary to be able to understand how organizations are depicted in charts (sometimes called organigrams or organograms) and to appreciate the limitations of such charts. Figure 6.1 is an organization chart that is fairly typical for a small manufacturing enterprise. Each box represents a function of some kind and the vertical levels of the boxes depend on the status of the person or people occupying the position. Even at this early stage in the discussion a point has been reached where simply distributing a new organization chart can cause resentment and disputes, because people might perceive themselves to have a higher organizational status than that depicted in the chart. Recognizing this potential problem, many organization charts bear the legend 'levels do not necessarily indicate status' (which will usually fail to appease those who feel aggrieved).

At the summit of the chart is the board of directors. Nomenclature changes with the passing years and at the time of writing this would be called the 'C-suite'. Directors now tend to carry the title executive officer so, for example, a former

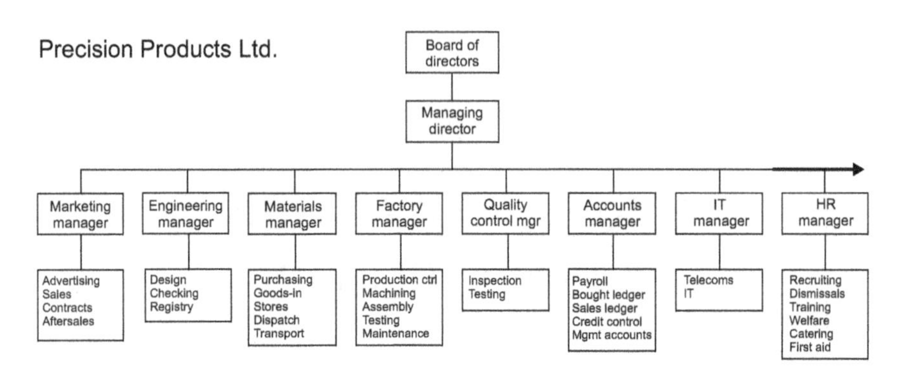

Figure 6.1 Possible organization chart for a small manufacturing enterprise.

managing director might be known as the CEO (chief executive officer) and the financial director might be known as the CFO (chief financial officer).

The vertical lines in an organization chart are lines of command (looking downwards) and lines of reporting or feedback (looking upwards). But most organizations also contain less formal lines of communication between people, which can be horizontal or diagonal and are sometimes shown as dotted lines. No organization can function effectively without both formal and informal communications. The chart in Figure 6.1 is fairly representative of a small to medium manufacturing enterprise and, for our purposes, please suppose that this company manufactures small components but also accepts orders to manufacture items to drawings and specifications from customers. Given competent management, staff and facilities, there is no reason apparent from the organization chart why this company should not be able to carry out those tasks. For the sake of argument let's suppose that this company has the registered name of 'Precision Products Limited', which abbreviates to PPL.

Management span of control

The study of organizations is interesting and fertile ground for debate. For instance, the managing director of PPL has at least eight subordinate managers reporting directly to her. Any one of these might approach her at any time to ask for advice or permission on one of many topics or problems. If more managers were to be added to those reporting directly, the managing director's time available to deal with each individual manager would clearly be diluted. A large number of direct subordinates leads to a 'wide organization'. Introducing more layers of junior managers below would cause the organization to become more vertical, lessening the load on each manager but lengthening communication and reporting channels and (importantly) increasing the company's overhead salary burden. This subject comes within an argument called the 'span of control' and there was once (perhaps still is) the 'rule of five', which suggested that no manager should have more than five direct subordinates. That is arguable, and it's time to leave that debate and move on to consider how PPL might manage its first project.

Coordinated project matrix organizations

PPL could become involved in their first project (and therefore project management) either because of an internal change project or through accepting a customer's order that includes a larger element of development, design and testing than that usually experienced. The organization shown in Figure 6.1 should be able to cope with the company's usual range of orders which flow seamlessly from sales orders, through manufacturing processes and then to dispatch and invoicing. But when hit by a more complex project, there is no person in that company who can logically 'own responsibility' for seeing the project pass seamlessly through all the stages of its lifecycle.

Recognizing this difficulty leads some companies to appoint a project coordinator. This person will usually have no executive authority and might report (for example) to the head of a design department. However, project coordinators must have some training and skills in project management. They should know how to produce a workable plan and be capable of following the progress of the project against that plan through all company departments until it is delivered on time and within budget. A project coordinator has to rely on the line managers to get things done, and can only prompt them or seek their support to keep the project on track. The coordinator must be a good communicator and will sometimes need to use tact and persuasion to get results. Coordinators might have the additional duty of drafting progress reports.

This is the weakest form of project management organization and is sometimes known as a coordinated functional matrix. Figure 6.2 shows the organization of PPL after the appointment of its project coordinator. The project coordinator might not necessarily report to the engineering manager as shown here. The role could be temporary. In a company that is conducting only one project in addition to its business-as-usual activities, the coordinator's role will become redundant when the project has been successfully delivered. But some senior project managers began their professional careers as coordinators.

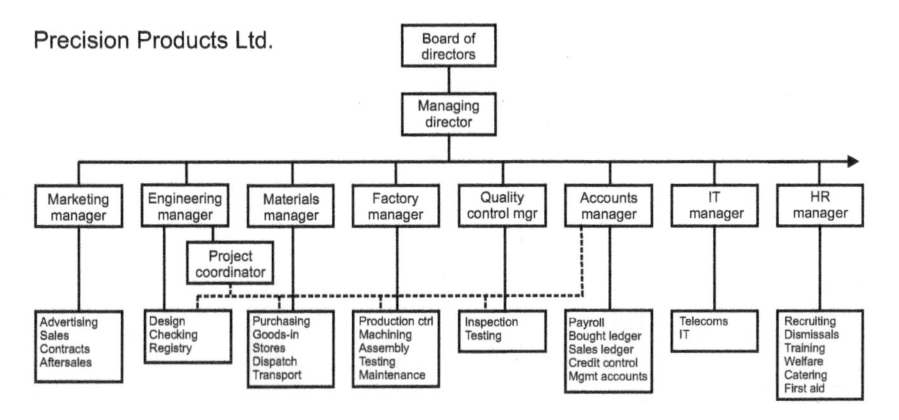

Figure 6.2 PPL has developed into a coordinated functional matrix. The project coordinator has no line authority but is responsible for project progress.

Balanced matrix organizations

Project organizations are often described as weak or strong, depending on the degree of authority enjoyed by the project manager. A coordinated matrix (Figure 6.2) is the weakest option. The balanced matrix is favoured by many, and an example is shown in Figure 6.3. This time the company chosen is engaged in the mining industry, but a similar arrangement can apply equally to all other industries.

In a balanced matrix project managers and managers of the functional groups share authority over the various group members to a large extent. Project managers can dictate priorities and sometimes even allocate work to individual staff. The functional managers ensure technical excellence within their particular specialities, are responsible for professional engineering standards and generally manage the career progressions of their staff.

A balanced matrix can be a breeding ground for disagreement and conflict. The managers of different projects might compete with each other and with the functional managers for the allocation of scarce resources. Individual project managers can find difficulty in achieving a team spirit, because no team actually exists. If the projects director grants more power to the project managers and allows them to override day-to-day decisions (such as work priorities), then the matrix is said to have more strength and it becomes a strong matrix. Conversely, if the balance of power moves to the functional managers, the organization becomes a weak matrix.

Disputes often arise between project managers and the functional managers on technical or priority issues. However, when compared with the team and task force organizations (described below) a matrix organization offers relative organizational stability because, although projects can come and go, the core organization remains practically unchanged. Individual group members can see long-term career paths ahead within their specialized functions. Technical excellence is more easily achieved because each group concentrates its technical talents, can maintain relevant standards

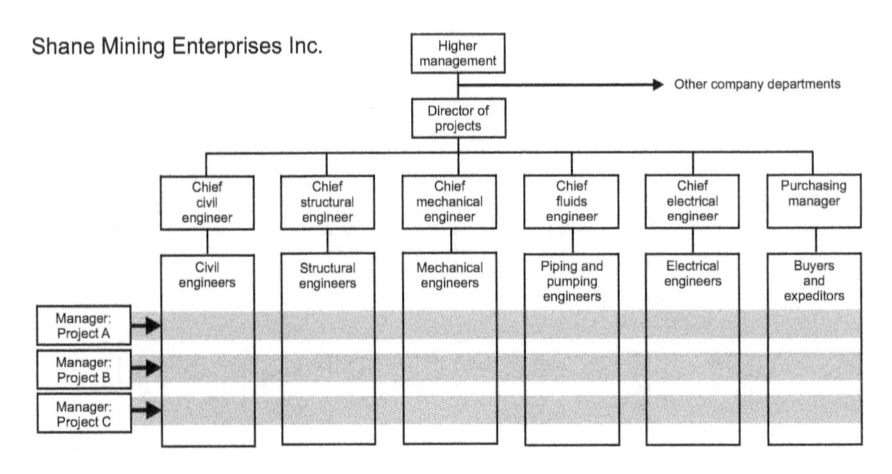

Figure 6.3 A balanced project matrix organization example. Project managers share authority with functional managers.

libraries, has the advantage of an experienced technical leader and so on. Individual members of a functional group can see a potential career path in their chosen profession, perhaps with chief engineer as the pinnacle. Although a mining company was chosen for this example, functional matrices are equally applicable to other industries, including manufacturing.

Team organizations

In a team organization each project has its own team, all reporting to the project manager. There is no conflict of command. Where a team actually exists it is far easier to generate a team spirit – until the end of the project comes into sight and the team members begin to wonder what will happen to them next. That end-of-project feeling can be a real demotivator and is discussed in Chapter 7. Project teams are sometimes also known as projectized organizations.

In a team the project manager has supreme command. Functional managers might allocate their staff to work in a team and can impose a degree of technical direction in the particular professional domain but the project manager has the 'casting vote' and decides who does what in the day-to-day allocation and progressing of tasks.

Task force organizations

A task force is a specially focused form of team. All the advantages and disadvantages of a team organization are more pronounced. A task force in the middle of an organization conducting other projects and business-as-usual can be a highly disruptive element. However, there is one indication where a project team is the clear choice and that is when a project is in distress or is otherwise in need of urgent care.

In a project team the members are taken from their usual functional departments and placed under the direct command of the project manager, who is given powers that could be described as dictatorial. Ideally the members of a dedicated project team should be located together physically. They should have a conference room or office dedicated to them, which they might call their 'project war room'.

Conclusion

In this introductory chapter on organizational structures the principal differences between team and matrix organizations were described. Chapter 7 continues this discussion. References and further reading for both Chapters 6 and 7 are given at the end of Chapter 7.

7
MORE COMPLEX ORGANIZATIONS

Shane Forth

This chapter continues the consideration of project organizations by looking at structures which are more complex than the clearly-recognizable team and matrix organizations described in Chapter 6.

Hybrid organizations

In the previous chapter team organizations and matrix organizations were described and discussed as alternative solutions. However, real life often provides examples of organizations in which these concepts of team and matrix are mixed or blurred. Figure 7.1 is an example (adapted from real life) of a mining engineering company that was based in Central London. This company suffered attention from some management consultants and, as a result of their recommendations, was reorganized as a balanced matrix, very much along the lines of the example shown in Figure 6.3 in the previous chapter.

A year or two after that reorganization, two projects were carried out for a mining company in Africa. Names have been changed, but otherwise the facts are as reported here. One project was undertaken by the London engineering company to reclaim a mineshaft and galleries that had, some years previously, been flooded with huge quantities of mud and waste that broke through from a surface tailings dam. The structural integrity of the shaft and galleries was not in doubt (all hewn from rock) so this project consisted almost entirely of difficult pumping operations. So the project manager and small team for that project were contained entirely within the fluids functional group, reporting to the chief fluids engineer.

At the same time a new project was authorized for the supply and installation of a large electrical transformer, complete with its associated switchgear and safety equipment. This was an all-electrical project, managed by a senior electrical engineer who reported to the chief electrical engineer. So this project had a team organization contained within the electrical engineering functional group.

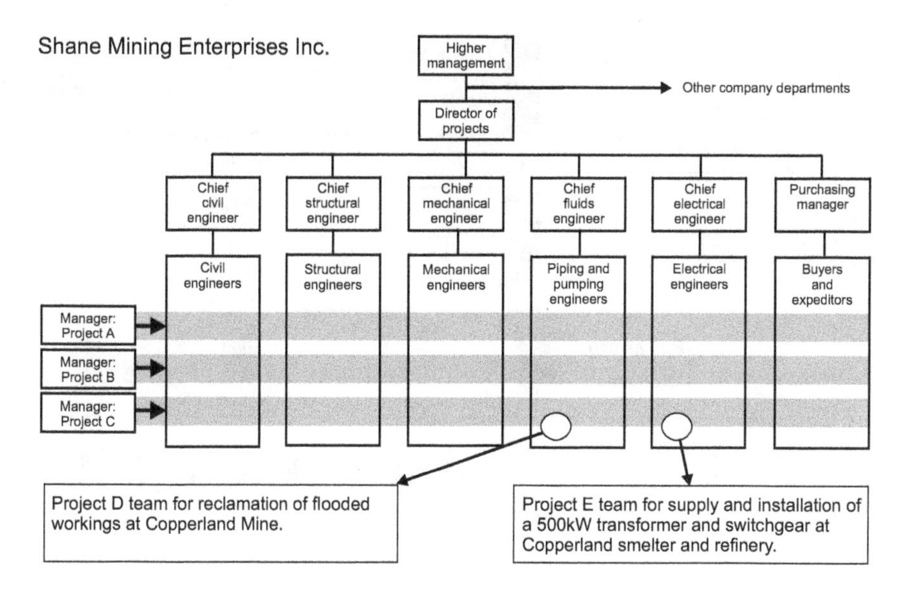

Figure 7.1 A hybrid project organization example. Project managers A, B and C share authority with the functional managers in a functional matrix. Project D manager is a fluids engineer and manages a small, dedicated team entirely within the fluids group. Project E manager is an electrical engineer leading a very small dedicated team within the electrical group for the purchase and installation of a transformer that will upgrade the existing facilities at the Copperland mine's smelter and refinery complex.

So here was a case of a company successfully operating a mix of different organizational concepts. Most projects were managed by project managers who relied on the functional groups for their resources. Those project managers shared power and authority with the chief engineers who managed the functional groups. That arrangement (known as a functional matrix or balanced matrix) allowed flexibility and worked well. But for the two specialized projects described above, each project manager was wholly in charge of a dedicated project team.

Project organizations sometimes have to be adapted so that they suit the special conditions and needs of the projects that they must fulfil.

Contract matrix organizations

In large projects the customer/supplier chains are sometimes found in a type of organization which is called a contract matrix. A contract matrix is illustrated in Figure 7.2, closely based on an actual project case. In this example the project owner (the customer or client) has engaged a managing contractor to design the project and manage all the works. The managing contractor's organization includes the design team, the purchasing department or purchasing agent and the senior project site manager. If the managing contractor had been one of the major construction organizations it would also have had all or most of the services needed to fulfil the project within

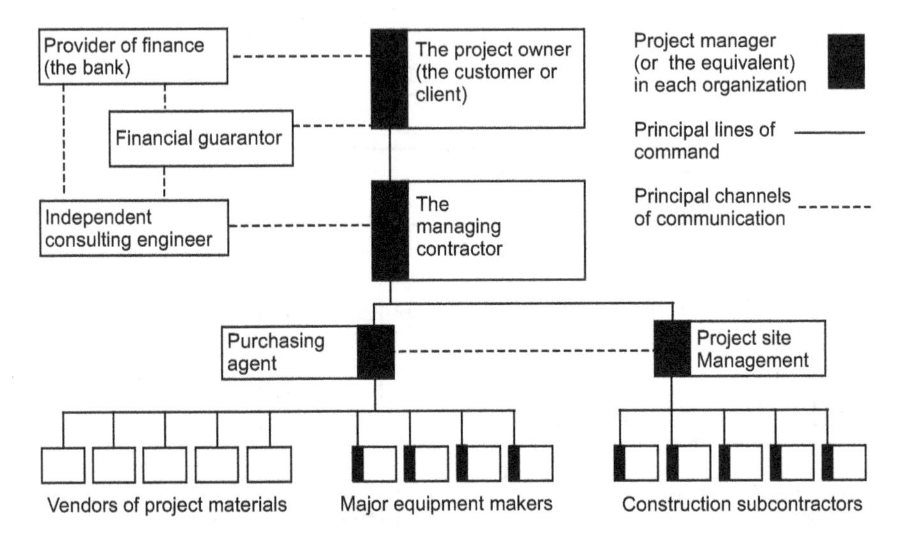

Figure 7.2 A contract matrix project organization.
Source: Used, with permission, from Lock (2013).

its own workforce and resources. However, in this case the managing contractor is really only an engineering consultancy and management organization, so a number of subcontractors had to be engaged for the actual work.

Most big projects will have at least two project managers; one employed by the organization with principal responsibility for carrying out the work and the other representing the client or project owner. The contract matrix organization described here includes several people who must practise project management skills. Every special equipment manufacturer, for example, needs a project manager to plan and control its share of the project.

The organization chart in Figure 7.2 shows that many of the companies involved in the project will have their own project managers (indicated here by the black rectangles) in addition to the principal project manager employed by the main contractor.

This project is being funded initially by a bank. The bank in this case has lent the funds on condition that the project owner finds a guarantor who is willing to underwrite a substantial part of the lending risk. In the UK, for example, the Export Credits Guarantee Department (ECGD) acts as guarantor for some projects carried out for overseas clients. Visit <www.ecgd.gov.uk> for more information about ECGD.

Neither the bank nor the guarantor has the technical knowledge and experience to understand the project details and they certainly cannot judge how the project is progressing in terms of the value created when compared with the various claims for payment. So the bank, particularly, will need an expert independent engineer who can understand how well the project is progressing, give regular unbiased reports and testify as to whether the series of payment claims to be paid from the loan funds are fair and reasonable. The expert engineer in the example of Figure 7.2 is not a single person, but a professional engineering organization. This organization, sometimes known simply as 'the engineer', can inspect progress and certify all significant claims for payments so that monies are only paid against work that has actually been performed correctly and in the quantities listed on the contractors' invoices. The engineer's fees are charged to the project owner.

One disadvantage of the contract matrix organization is that the companies involved are independent of each other and this does not encourage the collaborative approach needed to deliver a large project. So project success depends very much on the competence and capabilities of the managing contractor.

Joint venture and consortium organizations

For a major project or especially a megaproject (typically requiring capital expenditure measured in billions, rather than millions of pounds) the outlay and risk can be greater than a single organization is capable of bearing alone. Then a collaborative approach is needed between different companies, who can work together, combining their skills, capabilities and resources for the good of the project. This ensures that the risks are allocated to the participants appropriately and reduces the potential for project cost and schedule overruns, technical failure, reputational damage and serious financial consequences.

This collaborative approach requires a more complex project organization. That can be achieved by the member companies establishing a *joint venture* or *consortium* to provide a single integrated project management team. The terms 'joint venture' and 'consortium' are often regarded as having the same meaning, but there is an important difference. As a separate legal entity, a joint venture is a more formal organization for delivering the project (both from a financial and project execution perspective) than a consortium (in which each company retains a reasonable level of independence). However, the objective is the same for both a joint venture and a consortium: that is to form an organization capable of successfully delivering the project. For the remainder of this chapter I shall use the term joint venture (which is often abbreviated to the initials JV). A JV is a single entity where in some cases each partner contributes equity and receives a share of the revenue.

Example of a joint venture organization and integrated project management team

A JV will have a project team that reflects the contribution each member company makes to the delivery of the total project. It is not unusual for the participants to be based in separate locations, which can often be thousands of miles apart. Figure 7.3 shows one such example, which was for a large project in which construction took place in the Far East. In this case the JV partners were as follows:

- Company A: engineering and project management (based in the UK).
- Company B: the principal vendor (based in North America).
- Company C: construction and commissioning (based in the Far East).
- Company D: local design institute (also based in Far East).

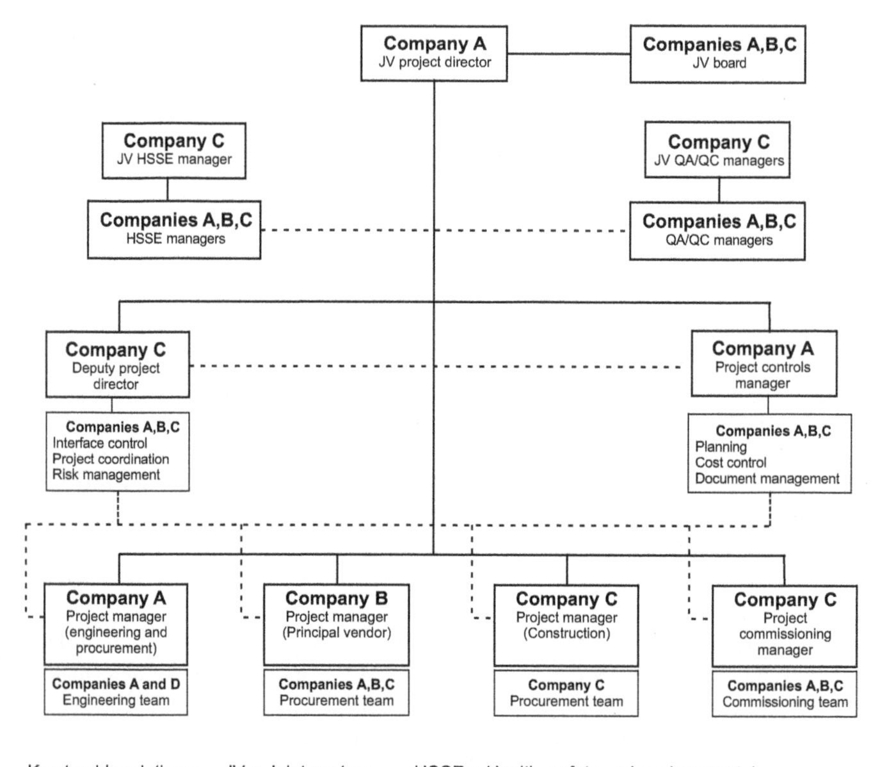

Figure 7.3 Example of a joint venture integrated project management team.

The organization for the management team of this JV is shown in Figure 7.3.

The JV board, which is sometimes known as the JV steering group, included an executive director from each of the JV member companies and was responsible for oversight of the project. Regular JV board meetings were held, typically on a monthly basis, with a more thorough and in-depth review carried out quarterly. These meetings were attended by the JV board members, the JV project director and other key members of the integrated JV project management team on an 'as required' basis.

The dynamic nature of a JV organization

The JV project management team shown in Figure 7.3 had a single project director from Company A, which was the leading party in the JV when the project engineering phase began. The deputy project director, who had an important interface management role, was from Company C, which was responsible for construction. I have seen several large projects where these two roles are reversed when the engineering work nears completion and the project moves into its construction phase. Other functional management roles in the integrated JV project team can change in a similar manner. This dynamic approach to the nature and shape of the project organization as the project progresses through its lifecycle phases is known as *situational leadership*.

Best athlete approach in joint venture organizations

Providing the commitment of the JV board remains strong and supportive and their objectives remain aligned, the likelihood of a successful project outcome for all concerned is significantly increased by the JV arrangement.

A major benefit in the integrated project team provided by a JV is that the resource pool available from the combined strength of the JV partners is greater than any one of them could provide independently. This enables a 'best athlete' approach in which the key roles in the integrated project team are filled by the best and most suitably qualified and experienced people available, rather than by people who happen to be available from only one of the JV partners. If a key member of the team should leave the project, searching through all the JV partners will improve the probability of a suitable replacement being found quickly.

Implications for project controls people and procedures in joint venture organizations

The project controls team should include the best people identified from the different JV partners. The team will report to the most suitably experienced project control manager available from the JV partners. Initially at least, planning engineers and cost engineers who were accustomed to working on smaller projects and had only ever reported to someone from within their own company might find this difficult. On the other hand they could regard their new roles in the larger, more prestigious project as an important career development opportunity.

The starting point for evolving a single suite of project control procedures, processes and software tools for managing the project might be to consider those that are available from the JV partner which has the lead responsibility for project controls (which in the case of this chapter is Company A). However, providing there is enough time available in the project mobilization period, it is better for the JV partners to arrive at a solution by reviewing all their management systems (including project controls). Then they can identify and combine the best practices and thus establish a single suite of management and processes and tools.

The extent of project controls reporting may be greater with a JV than those needed for a smaller, more straightforward project. This not just because the project is larger, more complex, or carries greater risk, but because there is often a requirement to produce an array of progress reports, monthly reports and so on: not only for the integrated JV project team but also for the individual companies.

Timesheet procedures can be more complicated in a JV organization. I have seen people having to enter timesheets for both the JV and their own company systems.

However, with the recent advent of business intelligence tools there is an opportunity to eliminate this and other issues.

Joint venture integrated project team organizations – what can go wrong?

Provided that the commitment of the directors from each of the partners that form the JV steering group remains strong and their objectives remain aligned, the likelihood for all concerned is a very successful project outcome, or at the very least less risk of a cost and schedule overrun. But a JV is not a panacea. Conflict will arise, for example, if the wrong people are appointed to the JV board. Should this happen, the essential collaborative principles will not be cascaded through the project director to the integrated project team.

For international projects, the different cultural values, language, beliefs and traditional behaviours of people from different countries, team building and ensuring good communication are other issues that must be considered to avoid conflict and misunderstandings within the project team. There is always a risk of the effort needed to achieve this being underestimated because project people tend to become task-focused at the expense of team and individual considerations.

At the beginning of my career I saw many major engineering and construction projects in the UK where relationships between clients and contractors were adversarial. Behaviour was so aggressive and immature that I thought I had returned to the environment of my school playground.

Later in my career (in the mid-1990s) I was involved in many JV organizations where cooperation within the project resulted in a success. I found that when I visited project construction sites I was no longer at risk of making a wasted journey because the people I was expecting to visit were unavailable (caught up in unscheduled meetings, earnestly discussing the latest conflicts of the day or busy helping the construction project manager to draft the latest letter as part of a war of words between the client and the contracting organization).

On the other hand, I was once involved in a consortium where three participating companies had people working in a single project office. The people who worked for my employer were actively discouraged from leaving the office at the end of the day with people from the other consortium companies, even though all were staying in the same hotel. Out-of-hours socializing was actively discouraged.

Today, alliancing, partnering and target cost contract arrangements are still evident, but my observations are that in recent years there has unfortunately been a return towards the adversarial behaviour that I saw earlier in my career.

References and further reading

Constructing Excellence (2004), *Partnering*.
Egan, J. (1998), *Rethinking Construction*, London: Department of Trade and Industry.
Kim, S. (2019), 'The Incheon Bridge Project', in Lock, D., and Reinhard, R., *The Handbook of Project Portfolio Management*, Abingdon: Routledge.

Latham, M. (1994), *Constructing the Team*, London: TSO.
Lock, D. (2013), *Project Management*, 10th edn, Farnham: Gower.
Mosey, D. (2019), *Collaborative Construction Procurement and Improved Value,* Hoboken, NJ: Wiley-Blackwell.
Roe, S. and Jenkins, J. (2003), *Partnering and Alliancing in Construction Projects*, London: Sweet & Maxwell.

8
THE PROJECT MANAGEMENT OFFICE

Shane Forth

Project management office (or PMO) is the name usually given to the project management support group in business organizations. PMOs are a relatively new concept, but since the year 2000 they have become more common in companies that regularly conduct projects. They should now be regarded as an essential part of the project management function.

What's in a name?

Because company organizations and their projects come in many shapes and sizes, it follows that their project management support functions (their PMOs) also have many differences in size, function and staffing. A common cry is that 'There is no one-size PMO which fits all', and that has certainly been my experience in engineering and construction projects. The key factor is that the PMO should be seen by the people it serves as meeting the needs of the business and helping to improve profitability.

The abbreviation PMO has become increasingly common but its meaning is not the same across all organizations. For example, in recent years I have come across programme offices, project offices, centres of excellence and project directorates. A more significant complication is that the letter 'P' in PMO can mean several different things. First there are the (sometimes subjective) differences between project management and programme management. More recent years have seen the emergence of portfolio management, which introduces a third P.

Different organizations have their own ideas about which of their endeavours represent a project, a programme or a portfolio. It is common for a large company to do work involving a mixture of all three, perhaps with a bias towards one or the other. At one time when a number of projects were in progress in the same organization we talked simply about multiprojects. Not surprisingly there is now often confusion about what the P in PMO stands for. For simplicity, I shall designate project, programme and portfolio management as the '3Ps' in the remainder of this chapter.

Some professional associations, such as those listed in Chapter 1, go to considerable trouble in their published 'bodies of knowledge' to explain the differences between the 3Ps. In practice, however different companies will interpret these variations in their own ways. Until fairly recently life was more straightforward, with all work other than business-as-usual referred to as projects. If further clarification was needed we would simply talk about multiprojects, internal projects and external projects.

Now that PMOs have become far more common it is interesting to compare the results from two studies, one carried out by Hobbs (2007) and the other by PMO Flashmob (2016) to discover which of the 3Ps PMOs most associate themselves with. The trends are readily apparent, as shown below:

Designation	*Perception*	
	2007 (Hobbs)	*2016 (Flashmob)*
PMO = project management office	59%	20%
PMO = programme management office	12%	50%
PMO = portfolio management office	0%	18%

These results show dramatic changes over the nine-year interval separating the two studies. PMOs have declined and their former dominance has been replaced with programme management PMOs. The perception of PMOs as portfolio management offices rose from zero in 2007 to 18 per cent in 2016. So portfolio management offices have become almost as common as PMOs. Who knows where these trends will end? However, because this handbook is focused on project management controls I shall give prominence in the remainder of this chapter to the PMO's role in controlling project management functions.

PMO organizational structures

The organizational structure, functions and roles of people in PMOs vary widely across different companies. At its simplest level a PMO can provide basic project administration and simple reporting. On the other hand, a large PMO might be part of a strong functional matrix, with the responsibility for governance and assurance of projects across the entire business. In some cases the PMO might even include the project managers.

A company might have a single PMO for the entire business. Rarely (but possibly) a PMO might even be temporary, its ephemeral existence lasting just long enough to support an 'only child' project. On the other hand, a sizeable business that manages projects from more than one office might have several PMOs. These PMOs could well be scattered across different locations, at different levels, and acting for specific regions, business units or service lines.

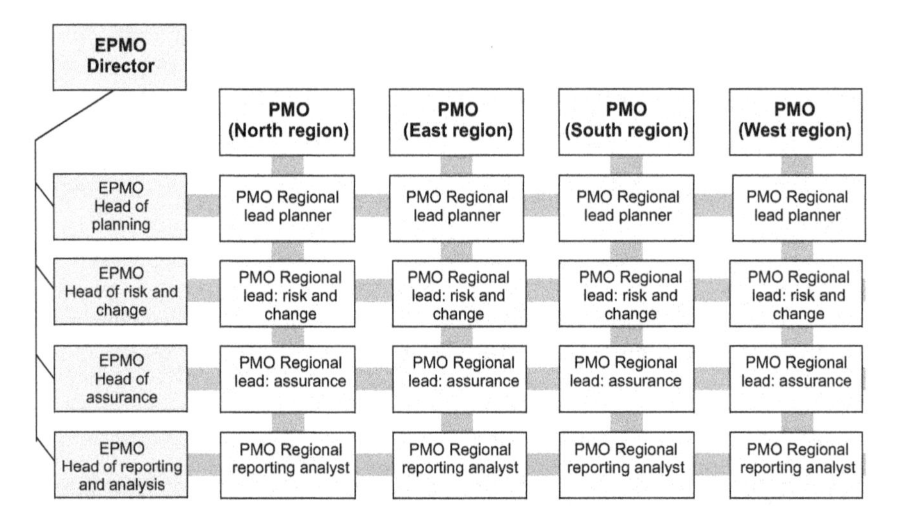

Figure 8.1 Example of a hub and spoke PMO organization.

Where multiple PMOs exist in an organization it is sensible and highly desirable to nominate one as the enterprise PMO (EPMO). That EPMO can then become the 'umbrella PMO' that sets the standards to be followed by all the other PMOs in the organization. The APM Body of Knowledge (2019) describes this as a 'hub and spoke' arrangement, which is similar to the functional matrix organization described in Chapter 7. Figure 8.1 is a simplified representation of an actual hub and spoke PMO arrangement from the UK infrastructure sector.

Different types of PMO

Some PMOs, whether they are stand-alone or part of a hub and spoke arrangement, might contain additional or very different functions from those indicated by the roles shown in Figure 8.1. For example, a PMO in an organization that is involved principally in developing IT software might include people with roles in business analysis, IT, systems development, solution architecture, data integration and deployment and so forth.

Professional organizations, consultants and businesses describe many different types of PMO, which can be categorized at four or five levels. Some of these are shown in Figure 8.2, although these are only a selection from many possibilities. Ten Six (2016), for example, offers two or three more descriptions for the five PMO types they have listed. From the professional organization's and consultant's viewpoints each level of the PMO provides different services. As the PMO progresses through the four or five levels each step is considered to represent an increase in maturity.

Other types of PMO are also appearing at an ever-increasing rate, based on the latest project management thinking. We now have agile PMOs, evolved PMOs,

	Professional organizations		*Consultants*	
	APM PMO SIG	*PMI (2013)*	*Project Solutions Inc (2014)*	*Ten Six (2016)*
1	Supportive PMO	Supportive PMO	Basic PMO	Business unit PMO
2	Operational PMO	Controlling PMO	Established PMO	Project PMO
3	Directive PMO	Directive PMO	Institutionalized PMO	Project support PMO
4	Centre of excellence		Strategic PMO	Enterprise EPMO
5			Best in class PMO	Centre of excellence

Figure 8.2 Some different PMO types and levels.

strategic PMOs, digital PMOs and more. All this can add to the confusion as people try to keep up with the changes.

Essential PMO services

Given the 3Ps, the four to five possible PMO levels and the different business needs, there is no standard set of essential PMO functions. In an attempt to understand what PMOs do, studies by Hobbs (2007) and PMO Flashmob (2016) identified over 20 different services that some PMOs provide. The same studies also ascertained which of these services were most important to PMOs and were provided most often (with the findings expressed as percentages). Results from these two studies are summarized and compared in Figure 8.3.

Attempts to draw definitive conclusions by comparing and contrasting the data displayed in Figure 8.3 only confirm the large number of services that PMOs might provide, and also that many of the services listed are provided by fewer than half of today's PMOs. This is further evidence that PMOs differ greatly in what they do. However, the top service provided is reporting and analysis of project status.

There are a number of similarities between the two lists in Figure 8.3. To some extent the sequence of importance given to the topics in each list may be following trends in the continuing evolution of project management and controls. In 2007 for example, I was working in a team developing a standard project management methodology (the second item on the Hobbs list). In 2016, assurance (which is closely related to governance) and project management tools were both services that were seen as important and close to top of the list.

More recently much is heard about benefits management, which relates principally to the outcomes of internal projects, programmes and portfolio management. This is a key topic in the Programme Management Framework - Managing Successful Programmes (MSP®). Benefits management has not yet become a prominent feature in PMOs but time will tell. The study by PMO Flashmob (2016) also

Hobbs (2007)

#	Service	%
1	Report project status to upper management.	83%
2	Develop and implement a standard methodology.	76%
3	Monitor and control project performance.	65%
4	Develop competency of personnel, including training.	65%
5	Implement and operate a project information system.	60%
6	Provide advice to upper management.	60%
7	Coordinate between projects.	59%
8	Develop and maintain a project scoreboard.	58%
9	Promote project management within the organization.	55%
10	Monitor and control performance of the PMO.	50%
11	Participate in strategic planning.	49%
12	Provide mentoring for project managers.	49%
13	Manage one or more portfolios.	49%
14	Identify, select and prioritize new projects.	48%
15	Manage project documentation archives.	48%
16	Manage one or more programmes.	48%
17	Conduct project audits.	45%
18	Manage customer interfaces.	45%
19	Provide a set of tools, with no effort to standardize.	42%
20	Execute specialized tasks for project managers.	42%
21	Allocate resources between projects.	40%
22	Conduct post-project reviews.	38%
23	Implement and manage a database of lessons learned.	34%
24	Implement and manage a risk database.	29%
25	Manage benefits.	28%
26	Conduct networking and environmental scanning.	25%
27	Recruit, select, evaluate and determine salaries for project managers.	22%

PMO Flashmob (2016)

#	Service	%
1	Reporting and analysis.	75%
2	Governance.	74%
3	Administration.	61%
4	Risk management.	61%
5	Communications.	58%
6	Project management tools.	57%
7	PMO management.	55%
8	Methodologies.	54%
9	Planning.	53%
10	Finance/budgets/cost.	50%
11	Change management and control.	50%
12	Issue management.	50%
13	Stakeholder management.	44%
14	Resource management.	39%
15	Knowledge management.	37%
16	Portfolio management.	34%
17	People management.	33%
18	Quality management.	24%
19	Benefits management.	24%
20	Configuration management.	20%
21	Business analysis.	19%
22	Supplier management.	18%

Figure 8.3 Essential PMO services identified by two different management studies.

looked separately at the three variants of PMOs that considered themselves as project, programme or portfolio PMOs and found that benefits management was the second-from-top service provided by portfolio management PMOs.

Over many years, I have been involved with a number of PMOs. These have had various names, been of different sizes and provided a variety of services. Figure 8.4 shows how PMOs have greatly developed and extended their range of services over the last 50 years or so.

Staffing a PMO

PMO sizes vary considerably. PMOs might be found containing anything between a single person and 300 people. However, according to Hobbs (2007), Project Management Solutions (2014) and PMO Flashmob (2016) the number most often lies somewhere between two and eight people. A PMO in a large company might start as a very small team resembling a supportive PMO but, if it becomes well established, moves up through the levels and improves its value to the business, the number of people will increase. However, owing to the challenges faced by a typical PMO, sometimes the opposite is true. Reduction in size or even abolition of a PMO can result, for example, if the company has a senior management change or is the victim of a merger and acquisition.

PMOs do not necessarily include project managers. Hobbs (2007) found a wide variation between different organizations, with 30 per cent of PMOs including all project managers and another 30 per cent having no project manager at all. In looking across a wide range of private and public sector groups, the Association for Project Management's (APM) PMO special interest group found that 75 per cent of a PMO's responsibilities are project controls, which agrees with my own experience.

There are many possible role titles for PMO staff, complicated further by the differences between the 3Ps. So, for example, we might now come across project planners, programme planners and portfolio planners. The following list originates from a study that I carried out for an organization which had too many different PMO job titles. This list makes allowance for the fact that a PMO can have many different functions or disciplines.

- PMO manager;
- PMO lead (with the specific function added here);
- PMO analyst;
- PMO administrator.

Conclusion: challenges for the PMO manager

If a PMO develops good communications with other company departments, such as design and engineering, HRM, finance and procurement, the PMO manager's task will be far easier. Every PMO depends on maintaining good communications and cooperative relationships with all stakeholders, both within and (to some extent) outside the company.

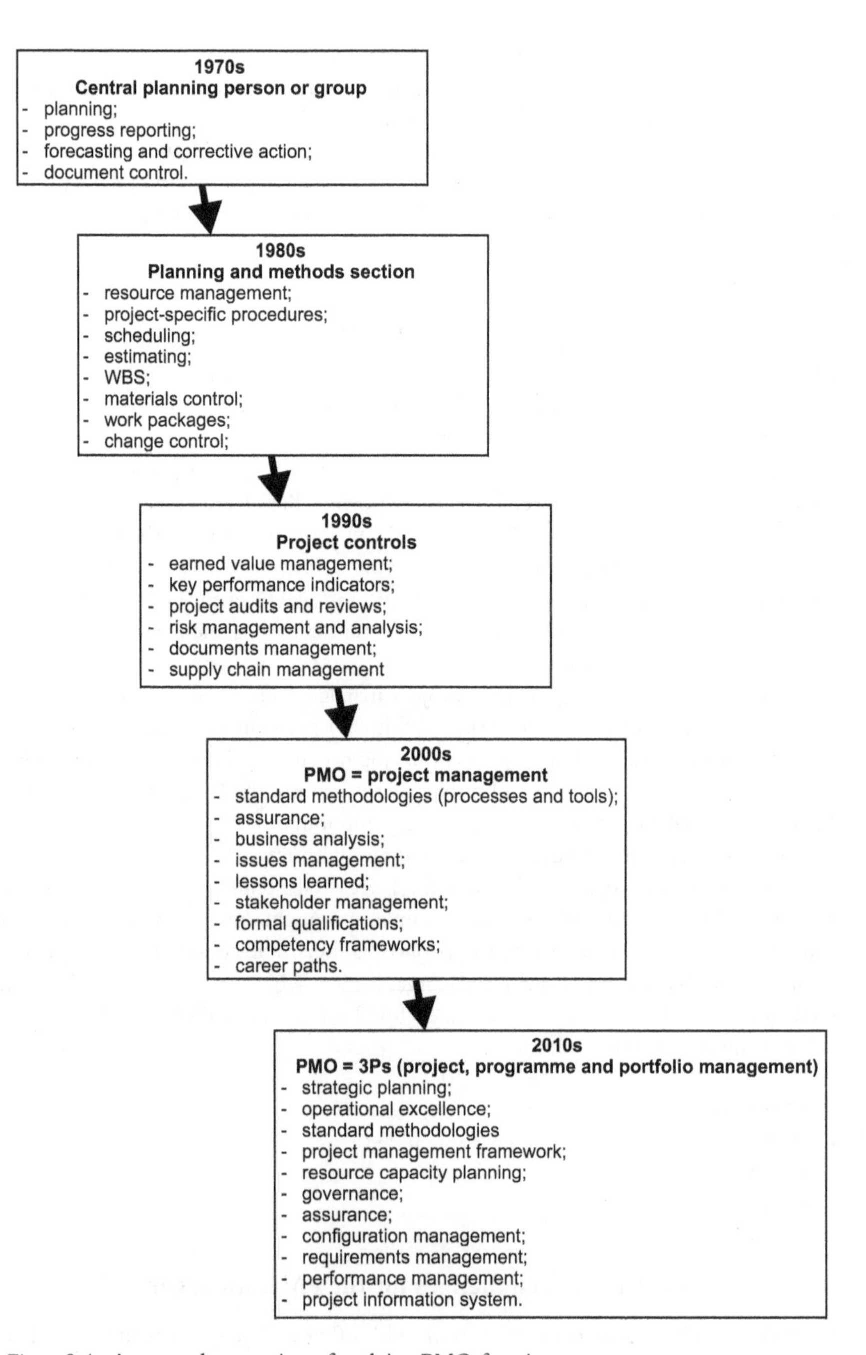

Figure 8.4 A personal perception of evolving PMO functions.

A PMO, especially when it first established, must not be too ambitious and try to achieve too much. It has to concentrate on the core project control functions. Usually all PMO costs are overhead expenses, and most organizations try to be efficient and keep their overhead costs as low as possible. Senior management, particularly, will always be looking at ways in which the company's overhead expenses might be reduced. When times are hard, a PMO can be a prime target for the downsizing axe. So a good PMO manager should always keep records that show (by quantified results) how the PMO contributes to the business. Evidence of this contribution, showing how the PMO contributes to profitability is useful in this context, as are letters of praise and congratulations from customers and clients.

From the journey through my own career, it seems to me that the PMO has been there all along in all but name and has simply continued to evolve alongside project management and controls. Going even further back, to the Polaris Missile Program, Admiral William F. Raborn was leader of the US Navy SPO Plans and Programs Division, which may have been one of the earliest PMOs (as we know them today.)

References and further reading

APM (2019), *APM Body of Knowledge*, 7th edn, Princes Risborough: Association for Project Management.

AXELOS (2013), *Portfolio, Programme and Project Offices*, Norwich: TSO.

Duggal, J.S. (2007), The Project, Programme or Portfolio Office, in Turner, J.R. (ed.), *Gower Handbook of Project Management*, 4th edn, Aldershot: Gower.

Giriadau, L. and Monaldi, E. (2015), 'PMO evolution from the origin to the future', Global Congress 2015 -EMEA, Newtown Square, PA: Project Management Institute.www.pmsolutions.com/reports/State_of_the_PMO_2014_Research_Report_FINAL.pdf (accessed 1June 2019).

Hobbs, B. (2007), *The Multi-Project PMO: A Global Analysis of the Current State of Practice,* Newtown Square, PA: Project Management Institute.

Managing Successful Programmes (MSP) at https://itgovernance.co.uk/msp

PMO Flashmob (2016), *PMO Benchmark Report 2016,* https://pmoflashmob.org/pmo-benchmark-report-2016-3/ (accessed 1st June 2019).

PMI (2013), *A Guide to the Project Management Body of Knowledge,* 5th edn, Newtown Square, PA: Project Management Institute.

Taylor, P. and Mead, R. (2016), *Delivering Successful PMOs: How to Deliver the Best Project Management Office for your Business*, Abingdon: Routledge.

Ten Six (2016), What Sort of PMO Are You?, https://tensix.com/2016/06/what-sort-of-pmo-are-you/ (accessed 1June 2019).

Wellingtone (2018), *The State of Project Management Annual Survey 2018*, Windsor: Wellingtone Limited.

PART III

Cost control

9
INTRODUCTION TO COST ACCOUNTING

Dennis Lock

Cost control is clearly important in all projects. Consideration of project costs is one of the three elements in the familiar triangle of objectives. A project that overruns its budget can cripple or destroy a company. So it is vital that project managers have at least a rudimentary understanding of how the costs of their projects are incurred, recorded, analysed and reported in their organization.

The finance department

All project managers need to keep abreast of expenditure compared with estimates and budgets. That means establishing good and regular communication with the department responsible for collecting and reporting project costs, which might be called the finance department or the accounts department. That department will comprise several sections, each of which performs a particular function such as one of the following:

- bought ledger (looking at invoices and costs from the suppliers of goods and services);
- sales ledger (invoices submitted to customers and project clients);
- timesheets (submitted by employees). These record and allow analysis of how all the people employed spend their time (whether they are at work on specified jobs, off sick, on holiday or otherwise);
- credit control (particularly chasing up customers and clients who pay late or not at all);
- payroll (payment of salaries and wages and analysis of those costs);
- management accounting (which produces accounts statements for internal use and for statutory purposes among other things);
- taxation.

In a very small company some of the above functions will be performed by only one or two accounts clerks or they might even be combined. The department will probably be headed up by a cost and management accountant who will report to the CFO. That describes an arrangement that I have seen in several organizations but no doubt there are many variants.

Costs

A professional accountant reading this chapter would probably object to the heading of this section. It is a general rule that one should never simply talk about 'costs', but should always specify what kind of costs is intended. Costs can be divided and specified in different ways, but the principal distinction for project managers is whether they are direct costs or indirect costs.

Direct costs are generally those costs that can be identified and recorded as being directly associated with labour or materials that can be attributed fairly, accurately and honestly to a sold product or project.

Indirect costs include all the costs of running and managing the company and its fixed facilities and premises, plus the wages and salaries of people (such as top managers and cleaners) who do no work that can be charged directly to jobs or customers. That's the simple story, but life is usually slightly more complicated.

Pricing and costs

The way in which the prices charged to customers and clients are decided depends on many factors and there is considerable variation between different companies and the different markets in which they operate. In the haute couture fashion industry, for example, the production costs of garments might be marked up several times to arrive at the price sold. Fortunately this handbook is concerned only with commercial and industrial projects, so the explanation of costs and pricing can be simpler here.

The best way to introduce the subject of how project costs are related to the price charged is to consider a small project in which a manufacturer designs and makes a special product for a purchaser. This is an order involving special manufacture, assembly, design and testing, so it can be considered as a project even though it is very tiny. The company might use a formula such as the following to arrive at the price invoiced to the customer:

Cost of direct labour:	£ 10 000
Cost of direct materials:	£ 6 000
Direct expenses:	£ 500
Prime cost:	£ 16 500
Plus overhead costs at 50%:	£ 8 250

Factory cost:	£ 24 750
Plus 50% mark up for profit:	£ 12 375
Price charged to customer:	£ 37 125

Some companies might add a further very small percentage cost element to the price for 'general and administrative expenses'. General and administrative expenses are incurred, for example, when the project company is part of a larger group and needs to recoup charges made by the group head office. So the actual price charged for the above project could be (say) £38,000 or £40,000.

Pricing is not always as straightforward as this example suggests. When competition is fierce there is a limit to the profit that can be expected. On occasions a company might want to offer work at a 'come and buy me' price in order to 'capture' a customer and attract future business. However, these issues are usually beyond the project manager's remit and only costs (not pricing) will be discussed in the remainder of this chapter.

Direct labour costs

Consider all the people engaged on a typical project. These will include managers and supervisors, engineers and designers for different purposes, factory or construction workers, inspectors and more. The salaries of senior managers and many office workers will be accounted for in the company's overhead costs, so we can usually (but not quite always) eliminate those from consideration of direct labour costs. Let's just consider the design department. Suppose it employs 100 people. Some of those will be senior engineers, all earning different salaries according to factors such as their qualifications, length of service, personal achievement and so forth. When the cost estimators are working out the cost of a future project they can have no idea which of those senior engineers will actually be working on the project or what the actual salary costs will be. So the company accountants will work out an average hourly rate for all its senior engineers and that rate can be used with confidence henceforth, not only in advance by the cost estimators but also for working out the actual costs after the work has been performed. This same averaging process can be used for all staff types and is the major component of a very common accounting method known as standard costing.

Now suppose that you are the person who has to decide the standard costs to be charged to project work for all the different jobs and grades of labour in the company. That would seem to be a very difficult task. But it could be made very much simpler if the company decided to limit the number of different standard grades and cost rates. This is not the place to tell readers how they should go about these calculations and indeed most readers will not be involved in the detail. You only need to know what the standard cost rates are in your own organization, so that these can be applied in project cost estimates and subsequent job costing.

Potentially there could be a significant number of different standard cost rates but remember the aim must be to keep the system as simple as possible, consistent with estimating and recording costs that bear a close relationship with the very detailed facts. Here is how one engineering company in which I was a manager solved this problem very successfully and simply. We slotted all our staff into one of six categories, as follows:

- Grade 1 staff: Company directors.
- Grade 2 staff: Departmental managers and project managers.
- Grade 3 staff: Project engineers and other senior engineers.
- Grade 4 staff: Drawing office group leaders and other supervisory staff.
- Grade 5 staff: Detailing draughtsmen and checkers.
- Grade 6 staff: All staff not included above, including clerical and secretarial staff.

So here was a company employing hundreds of people (ranging from maintenance workers up to directors) that was able to carry out all its estimating and costing successfully with only six standard cost labour grades. The rule, as in so many other areas of management and control, is to keep it simple. Incidentally, various privileges were associated with these grades. For example, all staff in grades 1 and 2 were allocated a company car.

Timesheets

Every member of staff in a company that works on projects will usually be expected to submit a weekly timesheet. Timesheets data can be written on forms or entered digitally directly into a computer system. A timesheet will usually be headed with the person's name and department (or departmental manager's name). A staff code number is usually assigned to each person. Then, perhaps to the nearest half hour, each employee will be expected to list for each day the time spent on each job. For project work, the job code will usually bear a close relationship with the project number and the relevant WBS code.

Supervision of timesheet entries is highly desirable, and where a client is being invoiced on a cost plus basis for time actually spent on their project, supervision and audit are essential (and will probably be insisted upon by the client). Figure 9.1 is an example of a timesheet for handwritten entries. There is more explanation of timesheets in Chapter 13.

Indirect timesheet bookings by direct workers

All staff who usually work directly on projects are often known by accountants as direct workers. But even direct workers occasionally spend time that cannot be charged directly to a project task (such as when they are on leave, sick, on a training course and so forth). So the company will provide 'job numbers' against which direct workers can allocate their unproductive (indirect) time. The costs of those unproductive hours will eventually find their way into the company's overhead costs.

Timesheet

Name: ____________________ Staff number: ____________

Department: For week ending:

Job number	Monday		Tuesday		Wednesday		Thursday		Friday		Weekend	
	Normal	O'time	Normal	O'time	Normal	O'time	Normal	O'time	Normal	O'time	Sat	Sun

Enter times to nearest half-hour. For holidays use 0096 sickness 0097; special leave 0098; waiting time 0099

Signed: ____________________ Approved: ____________________

Figure 9.1 Example of a weekly timesheet.

Direct timesheet bookings by indirect workers

It sometimes happens that a member of staff who usually spends all their time on indirect tasks will spend time that can legitimately be charged to a project task (and, for cost plus projects, ultimately to the client). For that reason all staff, including those normally regarded as indirect workers, should be expected to submit weekly timesheets. That should really even include the top managers.

Temporary and agency workers

All employers will need to employ people from agencies on a temporary basis from time to time. Although clerical and secretarial duties are usually considered the most common form of work provided by agency workers, other more senior tasks such as design are often performed by agencies. Those agency design tasks might be carried out in house, or in the premises of the agencies.

Agencies will submit their own timesheets to support their claims for payment, charging agreed standard hourly rates for their staff. However, those agency timesheets will be of no use to the cost and management accountant attempting to assess project costs unless each agency worker enters data in the same format as that required by permanent staff. This usually means that every agency worker should really be asked also to fill in an individual weekly timesheet, in the same way as the company's permanent workers.

Where paper forms are used as timesheets, there is no reason why the timesheet design should not be similar to that for permanent staff, but it is good practice to print the agency staff timesheets in red or some other colour that distinguishes them.

Timesheet analysis and reporting

The accounts department will analyse all the data entered on weekly timesheets. That analysis should allow the accountants to report the labour costs spent periodically on the project to the relevant departmental managers. Then the project manager and the departmental managers will be able to compare actual labour times against the original estimates and budgets.

Sometimes timesheet analysis can take weeks, so that the reports back to departmental managers are not made in sufficient time for overspending on job budgets to become apparent. So steps must always be taken to ensure quick timesheet analysis and reporting. If the company is operating earned value analysis (EVA) that is another reason why timesheet analysis by the accountants should be conducted efficiently and urgently. Departmental and project managers should be able to expect analysis of their labour costs within a week or 10 days. Some accounts departments can take longer than that, in which case the project manager has cause for complaint.

Accounting for project materials costs

In the same way as standard costs exist for labour, so standard costing can apply to materials and components that are routinely kept in stock. But there is nothing routine about projects and project materials are often ordered specially, and not supplied from regular stores stocks. So standard costing for materials is not always applicable to projects.

The earliest indication of actual materials costs will come not from the accounts department but from the purchasing department. In this case, unlike labour costs, the earliest information comes from the time when the purchase orders are placed. So, just as the project manager needs good communications with the accounts department for labour costs, so the project manager is heavily dependent on regular and good communications with the purchasing department for early warning of materials costs. Then the PM can establish records of committed purchasing costs for comparison against budget. This same argument applies to specialist services supplied for the project, particularly when the orders for these are placed by the purchasing department.

Credit control

One important function of an accounts department is to issue invoices to customers and clients and then, most importantly, follow them up if payments are not forthcoming. Some project contracts run on very tight margins, and non-payment of invoices can cause the contractor serious cash flow difficulties. So the accountants will keep the project manager aware of any payment defaulter and if necessary the project manager can apply pressure on that defaulter. The following case is real, but it happened long ago, the companies no longer exist, and all the human players are now if not actually dead, extremely wrinkled.

Company A (let's call them Red Aviation) was a supplier of military aircraft to HM Government in the UK. Company B, in which I worked, had a contract with

Red Aviation. So we were a subcontractor. Our task was to design, test, manufacture and supply one prototype plus six ATEs. In case you are wondering, ATE stands for automatic test equipment. Each ATE was constructed on a large trailer that could be towed by a tractor to the vicinity of a military aircraft, upon which it could perform checks and diagnoses in a matter of hours, where previously people would have needed two weeks.

My company shipped the prototype and it proved successful. So manufacture of the six further ATEs was authorized and put in hand. But when the time came to ship the first of these to Company A, we were still owed a very large amount of money for the prototype. So my company refused to ship the first new ATE. When HM Government began to ask why the ATE was not in place on the airfield, we explained the reason, the government department came down like a ton of bricks on Red Aviation and we got our money and shipped the ATE.

Sometimes bad payers need to be taught a lesson but the injured party is not always in a strong enough position to taken punitive action. There are many cases where small businesses have collapsed because their main customers have been very late in settling the accounts. So, project controllers, don't let your subcontractors delay their payments. With the cooperation of your friendly accounts department, make sure that all claims for payment are made on time and, where necessary, followed up.

Conclusion

The project manager and the PMO need to have a good understanding of how costs are collected and reported in their organization. For that reason alone the project manager needs to establish and maintain good communications with both the accounts and purchasing departments. The accounts department can also have a vital role to play in credit control and cash flow.

References and further reading

Callahan, K.R., Stetz, G. and Brooks, L.M. (2011), *Project Management Accounting*, 2nd edn, Hoboken, NJ: Wiley.

Holm, L. (2019), *Cost Accounting and Financial Management for Construction Project Managers*, Abingdon: Routledge.

10
INTRODUCTION TO COST ESTIMATING

Dennis Lock

It's no fun being a project cost estimator. There's no glamour in the job. When the project is successfully finished and everyone at the celebration party is congratulating each other, the cost estimator takes a back seat. But if the project ends in a catastrophic overspend, suddenly the cost estimator is centre stage for all the wrong reasons.

Relevance of cost estimating to project controls

Cost estimates are an essential factor in pricing external commercial projects and in the business plans of internal investment projects. For commercial projects, the actual project contract price may be arrived at after complex considerations depending on many things, but the project cost estimates will always be a key factor. When margins are tight and the competition is fierce, your company can price more keenly with less risk and greater confidence if you have a good cost estimator. If your organization practises EVA, detailed estimates will be needed for evaluating the budget costs of work performed.

Unlike manufacturing tasks, many project tasks are difficult to define and, therefore, more difficult for the cost estimator. However, without cost estimates there can be no contract pricing and no sensible project budgets. Budgets are a key factor in cost control. So initial cost estimating is a key component of budgetary control and project pricing.

Reliability and accuracy of project cost estimates

Entirely accurate cost estimates happen sometimes by chance, but all estimates rely to some extent on judgement and it is really never correct to talk about the accuracy of estimates. At the beginning of any project there are always too many unknowns, and too many risks lying ahead for any prediction of costs and times to be guaranteed as accurate from the start. But given a well-defined project with a good, detailed, complete WBS, the cost estimator has a good chance of being able to come close to estimating what the project *should* cost. What the project actually *will* cost by the time it is finished depends on many things that are beyond the cost estimator's control and (in some cases) also beyond the project manager's control.

So, rather than talking about the accuracy of cost estimates, it is more practicable and sensible to talk about the degree of confidence that can be placed in them. That degree of confidence will depend on many things but most of all upon the following factors, which are not listed here in any particular order of importance:

1 Competence of the cost estimator.
2 How well and in how much detail the project can be defined.
3 The degree of risk pertaining to the new project and its physical environment.
4 The quality and detail in archived records of costs from comparable completed projects.
5 The economic climate, particularly relating to cost inflation, international currency exchange rates and so on.

Classifying estimates according to their reliability

The reliability of any cost estimate will obviously be at most risk of error when a project begins, with many unknown quantities lying ahead. It is always recommended for the purposes of reporting and control that the cost estimate of any project should be updated at stages throughout the lifecycle. If that practice is followed, for example when applying EVA, you will have a project cost estimate that gradually becomes nearer to the actual final cost until the day when the project is delivered and the updated cost estimate and actual records meet. No initial estimate can ever be claimed as 100 per cent accurate or (in other words having an error of zero per cent) but here is a suggestion, based on experience, for labelling estimates according to the degree of confidence that they probably deserve.

1 Ballpark estimates. When the project is just a gleam in the project manager's eye, very little detail will be available. Unfortunately, this situation often applies when it is necessary to quote for a fixed price project when time is short, competition is fierce and you desperately need the work. Ballpark estimates look at the entire project, considering it 'top-down' without delving into all the possible detail (which would be impossible in any case because the details are unknown). However some experienced managers, using a mix of fortune telling and good judgement, can get within ±25 per cent of the actual project cost.
2 Comparative cost estimating is a common method that is often used as a basis for initial budgeting and pricing. It consists of comparing as much of the newly proposed project as possible with similar projects (or bits of projects) that have been performed in the past and for which you have access to accurate historical records. Of course there is no guarantee that even a complete repetition of a previous project will cost the same as its predecessor. However, experience is a great teacher and a comparative cost estimate for a project that is subsequently well managed and free of unexpected disasters might come within ±15 per cent

of the final project cost. Given some actual project examples (such as London's Crossrail transport project), many project managers would give their lives for that kind of accuracy.

3 Feasibility estimates rely on the availability of considerable advanced design detail. That might be thought of as an impossible condition when bidding for (say) a new building construction project. However, I have been in the position of representing the client of a construction project and some of the bidders had clearly performed much of the essential layout drawing before submitting their tenders. My experience suggests that an error factor of ±10 per cent should be applied to a feasibility estimate.
4 Definitive estimates depend on a considerable amount of project work already having been finished. So, clearly they are not available when the project begins, and can usually have no place is setting the original estimates, budgets and project price. Every company that carries out EVA performs definitive estimates when it compares the actual cost of work (ACWP) with the estimated or budget cost of work performed (BCWP). Cost estimates made in this way can probably earn the label of definitive estimates when their value comes within ±5 per cent of the finished project costs.

Competence of the cost estimator

Cost estimators may be occupied full time in that role, or they might also perform other duties. They will probably have a relevant qualification in cost engineering, accountancy or more general project management. It is essential that the cost estimator has a thorough understanding of the company and the way in which its accounting systems operate. But if you ask four different people to forecast the probable cost of a given task you might well get four different answers. If you are a project manager or senior manager in an organization you will in time come to learn something about the reliability of the forecasts from different individuals. Long ago I learned to classify estimators' abilities and place them in one of the following four categories:

1 Optimistic estimators, whose cost and time estimates are usually too low.
2 Pessimistic estimators, whose estimates are usually too high.
3 Accurate estimators, whose estimates can always be trusted and used with complete confidence (rare).
4 Unreliable estimators who produce random results, sometimes good, sometimes optimistic and sometimes pessimistic.

You can use your personal knowledge of individuals from the above four categories in the following ways. If your estimator is consistently either optimistic or pessimistic, you can apply correction factors to their estimates. For example, I used commonly to add 50 per cent to estimates that I considered to be unduly optimistic. Pessimistic estimators are rarer than the optimists, but their figures can be factored downwards once you have established the trend. So, in time, you can learn to evaluate each individual and judge what factor to apply to that person's time and cost estimates.

If you have accurate estimators, count your blessings and pay them well. Those in the fourth category, who produce random results, should be moved to another role (job seeker is one that comes readily to mind).

Project definition

Ideally, your project should be very well-defined (especially in scope) before any attempt is made to estimate its costs. It is to be hoped that your organization has experience of similar projects from the past. You will need a work breakdown, although that will probably not be available in any great detail at the project proposal stage. Large new capital projects are often subjected to a feasibility study, and that will probably result in a rudimentary work breakdown and a summary of the risks to be expected.

One very great danger in estimating for any new project is to forget some important factor and omit it from the estimates. That can spell disaster, whether the project is internal and you are developing a business plan or you are bidding to a potential customer for designing, building, selling and commissioning a project against competition. Thus it is important to supplement the early WBS with one or more checklists, just to be certain that nothing of significance has been forgotten. For a new company, with no previous experience, compiling checklists can be difficult but most project organizations have records of past projects that can provide not only checklists but also performance figures against which new estimates can be calculated or compared.

When a proposed project is to be conducted in a foreign country, many other uncertainties come into play. The national government could be unstable, there might be risks from terrorists or aggressive neighbouring countries or there could be local conditions that will add costs to your project. Some parts of the world are particularly prone to violent weather or earthquakes and volcanic activity. Many of my former colleagues worked on projects in undeveloped countries and subsequently suffered from chronic or even fatal illnesses. Figure 10.1 is part of a very comprehensive checklist once used by a mining company when considering any development project in virgin territory.

Documenting the project cost estimate

Cost estimates have to be tabulated and recorded so that they can be kept safely for subsequent reference. The total cost estimate record should be numbered to correspond with the project number (if known) or some other identifier such as a proposal number or sales reference. Where several estimates have to be made, each with reference to a specific proposed project case, then every resulting estimate has to be labelled so that it can clearly be associated with the relevant case.

Estimates should not be made freehand on blank pages, but must be set out according to a standard format that encourages clerical discipline, avoids ambiguity and eventually assists historical comparisons. Figure 10.2 is a form suitable for estimating the labour costs of individual tasks in a project. This example assumes that the company uses standard costing, as explained in Chapter 9. It is most suited to

Project site checklist
Name of proposed project:
Location:
Health considerations
- Medical facilities, doctors, hospital;
- Endemic diseases;

Physical conditions
- Seismic situation;
- Altitude;
- Temperature range;
- Rainfall or other precipitation;
- Humidity;
- Wind force and direction;
- Dust/sand;
- Barometric pressure;
- Soil investigation and foundation requirements.

Utilities available
- Electrical power;
- Drinking water;
- Other water;
- Sewerage;
- Other services.

Transport
- Existing roads;
- Access difficulties such as low bridges, tunnels, weight restrictions;
- Nearest airstrip;
- Nearest commercial airport;
- Nearest suitable railpoint;
- Nearest seaport;
- Local transport and insurance arrangements.

Local human resources available
- Professional;
- Skilled;
- Unskilled.

Existing site accommodation arrangements
- Offices:
- Secure storage;
- Hotels for VIPs,
- Short stay visitors;
- Expatriate managers and engineers;
- Artisans;
- Married quarters;
- Catering.

Figure 10.1 Part of a checklist for a mining project in new territory.

Source: A more detailed version of a similar checklist appears as figure 3.2 in Lock (2013).

COST ESTIMATE

Project or sales proposal number:

Estimate for: Compiled by:

Estimate number:
Date:
Sheet of

WBS Code	Task description	Estimated labour times and costs by standard grade												Estimated direct labour cost
		Grade 1		Grade 2		Grade 3		Grade 4		Grade 5		Grade 6		
		Hours	£	Hours	£	Hours	£	Hours	£	Hours	£	Hours	£	

Figure 10.2 Suggested format for estimating task labour costs.

ESTIMATE FOR PURCHASED EQUIPMENT AND MATERIALS

Project or sales proposal number:

Estimate for: Compiled by:

Estimate number:
Date:
Sheet of

WBS Code	Description	Spec ref No.	Proposed supplier (code).	Quantity required	Quoted F.O.B. cost	Quoted currency	Exchange rate used	F.O.B. cost sterling	Shipping mode proposed	Shipping weight (tonnes)	Freight cost	Taxes/ duties payable	Estimated total time (weeks)	Estimated delivered cost

Figure 10.3 Suggested format for estimating task material costs.

manufacturing projects but with small amendments, particularly to the materials section, can be used for most projects. Figure 10.3 illustrates the design of a form intended for the purpose of recording the estimated costs of materials in a large project where international shipments are involved. These are only examples and every company will have its own preferences and different versions. Other documents relevant to cost estimating are:

- drawing schedules, which list all expected project drawings and thus provide an input for estimating design and drawing costs, and;
- purchase schedules (sometimes known as purchase control schedules) which list every significant item of equipment, along with details such as the source of supply, means of transport and so on. Beware that bulk materials are not always included in purchase schedules but must of course not be forgotten.

Drawing and purchase schedules should be started before the project is authorized and when it is at the proposal stage. These records have to be kept up to date as the project proceeds through design to execution and handover. After project completion drawing and purchase schedules form part of the essential archived 'as-built' records. There is more information on drawing and purchase control schedules in Lock (2013).

Overhead costs

When all the project tasks and materials costs have been estimated, it is time to consider the overhead (indirect) costs and add those to the cost estimate. In most cases overhead costs will only be added to the labour costs, but some companies will add a small additional percentage amount to the estimated costs of materials and equipment (often called a materials handling charge). Materials handling charges are usually in the order of 5 per cent.

The overhead costs added to direct labour costs can vary enormously from one company to the next. The contractor that operates a lean, efficient organization will have lower overheads, and thus be far more competitive than a company with high overheads. Some clients ask for details of how the contractor arrived at the quoted price, and I have known cases where clients have demanded to see how the overhead costs are calculated and what they contain.

Below the line costs

For large projects with high risk it is usual to add some additional costs to the estimate as a cushion against risk and other eventualities. The most common below the line costs are explained below.

Contingency allowance

A contingency allowance is added, as its name suggests, to provide a cushion against possible unforeseen costs, breakages, events or mishaps that were not included as tasks in the original estimate. These are costs that cannot usually be passed on to the client, so if they occur they will reduce profitability or, in the worst case, could cause a loss. The amount of the contingency allowance will depend on factors such as the degree of confidence that can be placed in the project definition, the quality of the estimates, the novelty of the project, risks associated with the project and its environment, and so on. The amount of contingency allowance that can be added will depend quite a lot on the degree of competition faced by the bidder. A common contingency allowance is 5 per cent.

Cost escalation

If a project is planned to last for more than a year, and if national cost inflation is significantly high, then the cost of a task performed near the end of the project would be slightly higher than the cost of the same task performed earlier because of intervening rises in the prices of labour and materials.

So some estimates for projects with a delayed start or with durations of more than a year or so might have a below-the-line amount to cover expected inflation.

However, it is more common and preferable for cost escalation to be agreed as one of the terms and conditions of the contract. So, project contracts might contain an escalation clause that will allow the contractor to increase the price of the project by agreement with the client, should cost inflation become significant.

References and further reading

Lock, D. (2013), *Project Management*, 10th edn, Farnham: Gower.
The main further reading list for project cost estimating can be found at the end of Chapter 11.

11

COST ESTIMATING FOR CONSTRUCTION

Shane Forth

This chapter continues the important subject of project cost estimating, with particular reference to construction projects.

Optimism and risk

Capital construction projects, especially those in the public sector, often run late and exceed their original budgets by eye-watering amounts of money. At the time of writing, the UK's HS2 and London's Crossrail projects are two examples. Cost estimating can only suggest what a project should cost. Who knew, for example, that Crossrail would be delayed and overspent partly because tunnelling revealed burial sites, with all those ancestors having to be removed and reburied elsewhere with appropriate dignity and respect? Cost estimates can only predict what the project should cost if everything goes according to plan. It cannot allow for risks such as finding a small pond full of rare protected newts bang in the middle of the route for a new eight-lane highway that your team is supposed to be constructing.

However, risk only means uncertainty, with outcomes differing from expectations. Thus (as described in Chapter 23) by no means all risk outcomes are bad. Risk outcomes can be positive and beneficial, perhaps even presenting opportunities. I have attended many meetings with project teams where the intention has been to identify project risks that need to be managed. People at these meetings usually pay scant (if any) regard to positive opportunities. The discussions tend to be focused entirely on the gloom and doom prophesies of negative risks (threats). It seems that almost everyone becomes very pessimist when it comes to identifying risks.

Curiously, when it comes to estimating costs, whether for single tasks for an entire project, the opposite is true. People become very optimistic about what can be achieved in a short time, with cost projections that are well understated. This applies to some of the same people who are so pessimistic when they are talking about project risks. Thus impossible or highly unlikely project timescales and budgets are announced and accepted as realistic targets. The following cases illustrate some common reasons for producing optimistic estimates.

Case example 1

A government minister introduces a report to parliament in connection with a proposed public sector project. The report holds nothing back in explaining the tremendous benefits to be expected for the environment, for the business community and for the public at large. The estimated project costs are confidently given as £1 billion, with a timescale to completion of five years. What parliament and the public do not know is that the first draft of this report estimated the costs at £2 billion with a time-to-completion estimate of seven years (and even those figures were optimistic).

The minister, for political or other reasons not connected with reality, sends that first report back for redrafting, with instructions that the time and cost predictions have to be reduced. Otherwise the chances of getting parliamentary approval would be minimal. Such ventures are sometimes known as vanity projects. Vanity projects are too common.

Case example 2

A project investor or client in the private sector will be far more likely to agree to a proposed project if the estimated costs are low by comparison with the predicted financial benefits and if the timescale will allow those benefits to be realized as early as possible. So the project contractor, who is bidding competitively for the project against other potential contractors, will be tempted to bid low for the project and promise an unachievable early delivery. Unfortunately, if that bidder does win the contract, the project will begin with euphoria and end with dismay (and some people seeking new jobs).

Case example 3

A planning engineer took meticulous care in preparing a project schedule with a total predicted duration of 34 months, even though very similar previous projects had needed 36 months. Yet the engineer was asked by his manager why he had 'stretched the schedule'. The customer wanted completion in only 32 months. So when the estimated times in the schedule were revised to satisfy the client, failure to deliver on time was inevitable. From my experience optimistic promises of project completion times is common and 80 per cent of projects stand no chance of being delivered precisely on time.

Case example 4

In many contracting companies it is common practice for senior management to review project price bids before they are submitted in proposals to potential clients. In one such company these meetings were known as mark-up meetings, because the purpose was to add a mark up to the cost estimates to arrive at a bid price. However, these meetings would have been better described as mark-down meetings because the total cost estimate would often be reduced without justification (hoping that the lower costs could be achieved). So the prices bid for these projects were certainly competitive, but too low to give any chance of profitability. The problem of unwarranted optimism persists and is well recognized. A number of reports reveal that warnings of project failure owing to over-optimism are repeatedly ignored (see for example EY, 2014).

The importance of project scope definition

The cost estimator can only make realistic estimates if the project has been defined as accurately as possible. Put another way, the project requirements must be clear. Prints of a popular cartoon adorned many office noticeboards some years ago, showing a garden swing, suspended from the limb of a large tree by two ropes. Successive images in the cartoon showed the swing components placed in various ridiculous juxtapositions, captioned variously as 'How the project leader understood it', 'How the business consultant described it' and so on. Being unsure of copyright issues, we have included our own unique version here (Figure 11.1).

These simple pictures illustrate an important lesson: inadequate or inaccurate specification by the customer can result in the wrong project being delivered at the wrong time and at the wrong price.

Foreign currency exchange rates

Project goods and services are often sourced from overseas, which poses the following questions when compiling a project cost estimate:

- Which currency should be used?
- How should we deal with exchange rate changes?

This problem can be exacerbated in international projects, where project labour and supplies typically come from different national sources. The cost estimating solution often adopted is to choose the most common project currency (which for UK based projects would normally be sterling). Then rates and quotations from overseas suppliers and contractors have to be converted into that base currency using the exchange rates prevailing when the cost estimates are carried out (sometimes known as 'money of the day').

The terms and conditions of contracts with external clients for international projects should specify whether or not parts of the total project price can be renegotiated should one or more exchange rates change significantly during the project lifecycle. Also, the tender document and resulting contract should specify the exchange rates that applied when arriving at the quoted project prices.

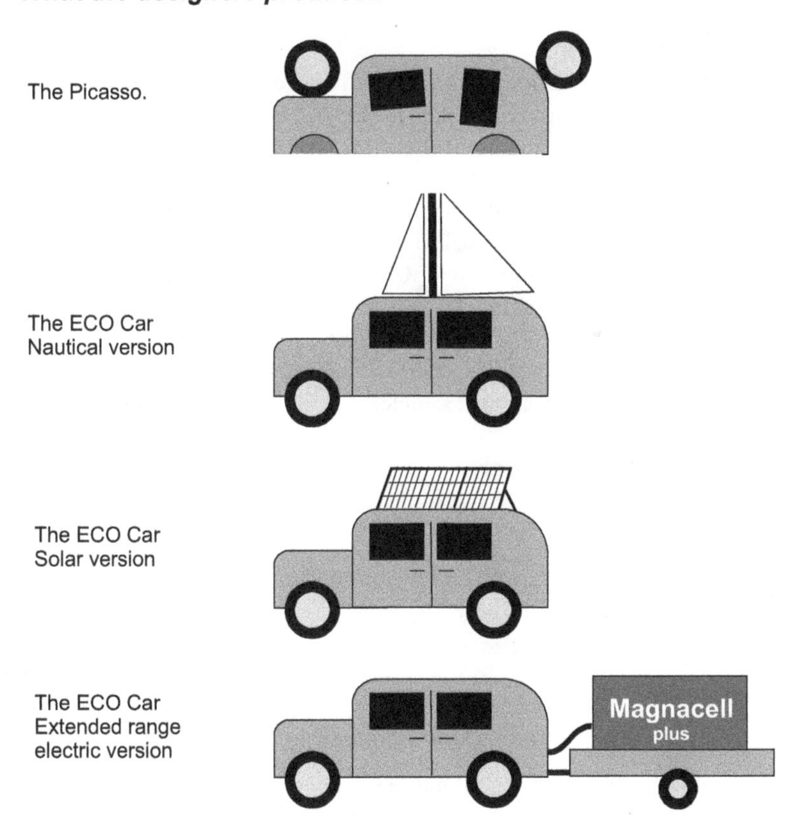

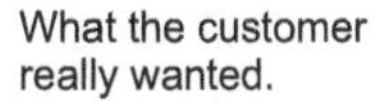

Figure 11.1 Costs and project satisfaction depend on providing an accurate project specification.

Lang and Hand factors

The top-down and ballpark estimates described in Chapter 10 use what are known as parametric or factoring methods. These are based on mathematical relationships (known as cost engineering relationships) between the different components. Of course these early estimates should be treated with great caution and given a 'health warning'. The over-optimism mentioned earlier in this chapter must be avoided. Factoring methods are used, for example, for process plant equipment estimates. The factors convert the costs of equipment manufacture and supply into their installed cost and, ultimately, into the completed project cost.

Lang factors

Lang factors originated in 1947. They are derived from project experience and are used as follows:

1 A numeric factor is associated with a given type of process plant.
2 The costs of manufacture and supply for the required process plant are multiplied by the relevant factor to arrive at the total installed cost of the equipment.

The formula is: TEC × L = TIC, where TEC is the total equipment cost, L is the Lang factor and TIC is the total installed cost.

Suppose that the early estimated TEC for a fluids processing plant was £20m. Now suppose that the Lang factor for that kind of plant is known from experience to be 4.7. Then the TIC for that processing plant can be predicted as £(20 × 4.7)m, which is £94m. This total installed cost will include all the project costs associated with that item of plant such as (for example):

- equipment purchase and supply;
- installation;
- project management;
- associated design;
- construction management;
- site establishment;
- and so on.

Hand factors

Hand factors date from the 1960s and are somewhat similar to the Lang factors described above.

Hand factors are associated more specifically and directly with the different components of process plant equipment, such as columns, heat exchangers, vessels, pumps and so on. Hand factors can vary between 2.0 and 3.5, depending on the type of equipment.

When the relevant Hand factors are applied to each type of manufactured process plant equipment the results are likely to be only about half the TIC (where the Lang factor would have given a 100 per cent result). The reason here is that Hand factors exclude all the other associated project costs (listed above). Thus those other associated project costs will have to be estimated separately when using Hand factors.

Predicting labour hours: use of norms at composite and elemental levels

There are many reasons why confidence in the estimate of construction labour hours (that are then costed using the rates of pay for each trade) is important. This component typically represents around a third of the total installed project cost. At the time of project sanction or final investment decision no main construction contract will have been awarded, so the cost estimate might lack any expert input.

The prediction of construction labour hours uses different methods, each of which requires greater effort and preparation time, consistent with an increasing level of project scope definition and through which the level of confidence in the estimate should increase.

For the early ballpark or comparative estimates, the prediction of construction labour hours is derived from the estimated cost of construction divided by an all-inclusive labour hours rate for installation. This is continually improved as project scope definition matures and is increasingly based on bills of material or parts lists defined that quantify the components for the different construction installation work, to which base norms are applied for the number of labour hours for different work operations or quantities installed. These base norms (or output rates) are then adjusted for project-specific considerations such as the properties and location of the site. They are available as published information, or can be derived in-house from historical data and used to calculate and predict the construction labour hours required for specified construction activities or discrete operations.

In the early stages of project scope definition on completion of basic or preliminary design, bulk material quantities will typically be high-level and composite in nature. These might include, for example, total metres of pipe or tonnes of steel. At this time, detailed component quantities by commodity code, size and specification will not yet be available because they will require working (or approved-for-construction) drawings. Therefore, the output rates are high-level and known as composite 'norms' because they do not fully consider size and complexity. At best, composite 'norms' may (using pipework as the example) be by size range (for example, 6 to 12 inch pipe diameter) within which the total lengths of pipe are estimated by an experienced individual very approximately from the early information available (such as layout drawings and plot plans or early 2D or 3D CAD modelling).

Gradually, as the project definition matures and the working drawings become available, material take-offs provide detailed quantities for the components (by size, specification and so forth) to which more detailed, component level elemental 'norms' are applied in order to predict construction labour hours with increasingly greater confidence (at least in theory). Once project execution gets underway, construction labour hours can be predicted by measuring performance and productivity and the use of EVA.

International location factors

When estimating the costs of international projects, labour productivity and cost rates must reflect the country and region in which the work is to be undertaken. Costs and productivity can vary widely from one country to another. Base norms must therefore be determined and used only for the country and region originally intended. When used elsewhere those norms must be factored to suit the particular location. This is done by first establishing base norms associated with the country for which the estimates were originally made, and giving those base norms a location factor of 1.00.

Here is an example. Suppose that a project is to be constructed in Houston Texas and that all the project estimates were made originally with that location in mind. So no factoring will be needed. All the estimated labour times and costs can be used as they were originally calculated. Now suppose that an identical project is to be constructed in Loxlandia, a tropical country where, because of local factors, every construction task can be expected to need three times the number of labour hours estimated for Houston, Texas.

If the base norms for Houston are used for estimating the Loxlandia project they will have to be multiplied by a location factor of 3.00. However, although the Loxlandia project will need three times as many labour hours as the Houston project this does not necessarily mean that the project is not viable. The labour pay rates in Loxlandia might be considerably less than those in Houston, Texas - even to the extent that the total labour costs could be lower, despite the three-fold increase in construction labour hours.

References and further reading

Dysert, L.R. (2003). 'Sharpen Your Cost Estimating Skills'. *Cost Engineering*, 45(6), 22–30.

ECIA (n.d.). *Databank of Existing Norms*. London: Engineering Construction Industry Association.

EY (2014). *Spotlight on Oil and Gas Megaprojects*. London: EYGM.

Flyvbjerg, B., Bruzelius, N. and Rothengatter, W. (2003). *Megaprojects and Risk: An Anatomy of Ambition*. Cambridge: Cambridge University Press.

Locatelli, G. and Mancini, M. (2012). 'Looking back to see the future: building nuclear power plants in Europe', *Construction Management and Economics*, 30(8), 623–637.

Lovallo, D. and Kahneman, D. (2003). 'Delusions of Success How Optimism Undermines Executives' Decisions'. *Harvard Business Review*, 81(7), 56–63.

Marks, T. (2012). *20:20 Project Management How to Deliver on Time, on Budget and on Spec*. London: Kogan Page.

Merrow, E.W. (2012). 'Oil and Gas Industry Megaprojects Our Recent Track Record', *Oil and Gas Facilities*, April, 38–42.

OECD and NEA (2016). *Costs of Nuclear Decommissioning*, Paris: OECD/NEA Publishing.

Spon (an imprint of Taylor & Francis) publish estimating and pricing guides for use in construction projects.

12
COST ESTIMATING ACCURACY

Shane Forth

Cost estimating is one of the key foundations of project controls, which is one reason why so many chapters in this handbook have been devoted to this subject. Now it is time to reflect on the accuracy that can be expected from cost estimates.

The myth of accuracy in cost estimates

Suppose that you are a project manager and your estimated total project costs were £10m. Now your project is finished and everyone is at a party celebrating its hand-over to a satisfied client. Your telephone buzzes and you see on its little screen that you have an email to read from your company's accountant. So you open your email and read the following message. 'Hi there – just thought you would like to know that the total recorded costs for your project were £10m'. You blink your eyes in disbelief and look at the email again. There's no mistake. It's not the effects of the champagne. Your recorded project costs really did come in exactly on budget and estimate. You envisage promotion.

Now here's the snag. Read that email once more and it does not say that the project costs were £10m. It read 'the total *recorded* costs'. That means the costs entered in the project account files after analysing possibly thousands of materials costs, invoices and labour timesheet entries. Many of those timesheet entries could have been inaccurate. Some cost items might have been erroneously charged to other projects, or your reported project costs could include amounts that had nothing to do with your project. Recording the actual costs of any large project is an exercise that is prone to many mistakes and inaccuracies. What all this means is that the true final cost of some projects might never be known.

This is not to imply that all recorded project costs are very inaccurate or unreliable. Simply remember that project costing is not always an exact undertaking. Records of actual project costs are very important for a number of reasons. In the context of estimating costs for future project costs, records of recent completed projects include cost data that will help to estimate the costs of comparable future projects or, at least parts of those projects.

Degree of confidence in cost estimates

In Chapter 10 four suggested classifications of cost estimates were listed. These depended to a very large extent on when the estimate was made in respect to the project lifecycle. Clearly the greatest potential error occurs when least is known about the project, which usually means at the time of the project proposal or (for an internal project) when making the business case. The project cannot be defined in all of its aspects in those early days and it's most likely that only ballpark or (at best) comparative estimates can be made.

It is good (and recommended) practice to review the project estimates at regular intervals during the lifecycle of any project and compare them on an item-by-item and total project basis. This process will entail replacing more and more of the estimated costs with the actual recorded costs as the project moves forward. So the project estimate accuracy should improve gradually as the project work progresses and the estimated costs are replaced with actual recorded costs. This process of gradually increasing confidence (and reducing error margins) is well known. It is often represented graphically in a diagram known as the estimating funnel (Figure 12.1). This is a useful concept but several observations need to be made about some of the assumptions made here.

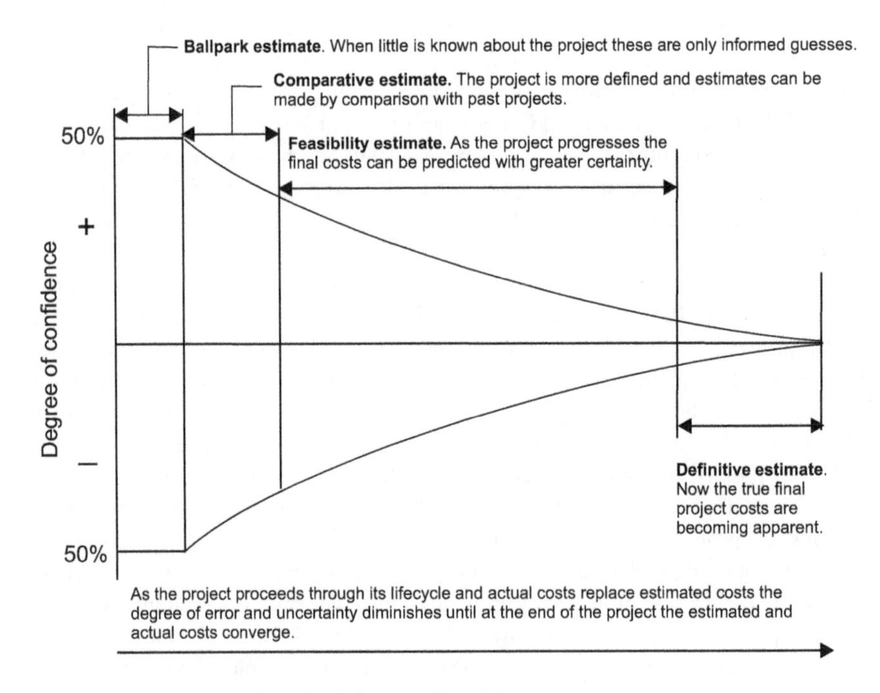

Figure 12.1 Estimating funnel and degree of confidence.

First, the estimating errors might not be equidistant about the horizontal timeline. The experience of many people suggests that initial estimates tend to be optimistic (understating the project costs). This implies that the funnel in the diagram should really be slewed anticlockwise about the central horizontal dividing line, giving greater prominence to the negative errors.

Different organizations will have their own ideas about what constitutes feasibility and definitive estimates. For example, some companies use feasibility estimates very early in the project, perhaps as part of making a business case before project authorization, when the estimated project costs are very relevant in deciding whether or not to go ahead with the project. So the experience of some readers might (correctly in their cases) indicate that the spacing of the vertical divisions in the estimating funnel relative to the project lifecycle should be quite different from those shown in Figure 12.1.

Now here's a really depressing thought. Although cost accountants are, by and large, very competent and likeable people, they can produce cost reports that (through no fault of their own) are less accurate than one would want. There are many reasons for this. The highest degree of accuracy can be expected when the entire organization is working on only one project. Then all the organization's direct costs can confidently be attributed to the project. However, when there is a mix of projects in the same organization, all at different stages of progress but all using common resources, then the risk of assigning costs to the wrong projects is increased. For example, a person filling in a timesheet at the end of a week has to remember how time was apportioned between the various cost coded tasks and, human memory is by no means infallible.

The upshot of all this is that the true costs of some projects might never be known, so the accuracy of the original estimates cannot be verified. That's a depressing thought, and it bodes ill for any estimator who attempts to produce a comparative cost estimate for a new project, when the recorded costs for the completed sample project are not reliable.

Rating the project definition

The Project Definition Rating Index (PDRI) is a useful tool from the Construction Industry Institute www.construction-institute.org. Separate PDRI tools are available for industrial, building and infrastructure projects. A PDRI provides some objectivity and sense-checking when assessing project definition and is aimed at early (ballpark and comparative) cost estimates. A detailed checklist of 70 project definition and design deliverables is provided and each deliverable has to be assessed for completeness using the following definition scale:

1 (best) complete definition;
2 minor deficiencies in the definition;
3 some deficiencies;
4 major deficiences;
5 incomplete or poor definition.

Not all deliverables have equal value in terms of their contribution to defining the scope of the project, so each deliverable must be weighted differently before the above scores are associated with it. The total score for all the deliverables can be anywhere between 70–1000. However, the lowest total score will clearly give the better rating. Project scores lower than 200 can give reasonable confidence that the final total installed cost (TIC) of the project will be between 3 and 5 per cent lower than the cost estimate. By contrast, a higher PDRI score (say above 200) will suggest less confidence in the cost estimate. In other words, the higher the PDRI score, the greater is the risk that the final cost of the project will exceed estimates and budgets.

A PDRI assessment can be conducted in various ways. Workshops are very productive in encouraging discussion among the project or bid team to improve the quality of the estimate and to determine systematically the range of accuracy that has been achieved. That is better than taking the purely subjective approach, which could be driven by the over-optimism mentioned in Chapter 11.

Estimate score sheet

A few years ago, people in the company for which I worked were routinely asked to quote fixed prices for major international engineering and construction projects. The problem was that the project definition was often insufficient to allow a cost estimate in which a high degree of confidence could be placed. However, the company desperately needed the work. Invitations to tender (ITTs) that had more favourable terms than being restricted to a fixed price quotation were rare. Like many other companies engaged in projects, this organization implemented a formal regime of senior management meetings to examine each ITT as part of its corporate risk management approach. The outcome of each meeting could be one of the following:

1 decide to bid for the project;
2 decide not to bid;
3 go back to the potential client and ask for some clarification.

On many occasions the decision on whether or not to bid had to be made knowing that some risk had to be accepted. Otherwise the company would not have priced any work at all, at least not on major projects, and the business would no longer have had a viable order book. Under this regime, in time (as more cost estimates and proposals were processed) an apparent confidence and even optimism in the cost estimates emerged in most of the people involved. That optimism was helped to some extent knowing that each estimate included below-the-line allowances for management risk and contingencies.

This optimism was not however felt by the cost estimators. Eventually, they came to me to express their concern that they were repeatedly being asked to produce definitive cost estimates (accurate to within plus 10 or minus 5 per cent). That, in their opinion, was simply not possible. So, when they did as they had been asked, they had little or no confidence in the cost estimates that they produced.

After reflecting upon their problem I created what I called an 'estimate score sheet' to help them explain each situation to senior management. This was intended for all estimates designated as class 1 definitive estimates (accurate to within plus 10 or minus 5 per cent) or class 2 budget estimates (with a reduced declared accuracy of between plus 15 or minus 10 per cent). The purpose of this approach was to:

- establish a confidence level (designated as high, medium or low) for each estimate;
- confirm that the confidence level was appropriate to the designated class of estimate (which should be high for class 1 or medium for class 2 estimates);
- improve determination of the amount of allowances and contingencies to be added to the estimate (and where to add these);
- decide other actions needed to improve confidence and finalize the estimate.

The completeness of definition was scored separately by the estimator for each high-level element of the estimate for offsites and for process within engineering design, procurement and construction. The definition scale described earlier was used (ranging from 1 for complete definition to 5 for incomplete or no definition). The definition scores for each element were then added and divided by the number of applicable elements to give a definition score (average estimate confidence level) of between 1 and 5 for both offsites and process (see the example in Figure 12.2). Remember that the lowest score corresponds to the highest level of definition.

The following confidence level ranges were used to establish whether the overall confidence level against the designated class of estimate for each of the offsites and process was high, medium or low. An example of this is shown at the head of the completed estimate score sheet (Figure 12.3), which refers to a class 1 definitive estimate. Here it can be seen that the offsites estimate average score was 2.3, suggesting a medium confidence level. For the process tasks the overall confidence level was 3.7, and that suggests very low confidence. The completed estimate score sheet for this example is shown in Figure 12.3.

Reference class forecasting

The idea behind reference class forecasting is to overcome the optimism bias during the early stages of project definition and improve the degree of confidence in the cost estimate and project schedule. It comes from Daniel Kahneman, an Israeli-American psychologist and economist and Amos Tversky, a cognitive and mathematical

	High	*Medium*	*Low*
Class 1, definitive (+10% to −5%)	< 1.5	1.5 to 2.5	> 1.5
Class 2, budget (+15% to −10%)	< 2.5	2.5 to 3.5	> 2.5

Figure 12.2 Confidence levels for different classes of estimates.

ESTIMATE SCORE SHEET

PROJECT TITLE: Lox Castle **ENQUIRY REF:** 9999

EXPECTED ESTIMATE CLASS: (Customer): Class 1 Definitive (+10 to -5%)

CONFIDENCE LEVEL RANGES:

	High	Medium	Low
Class 1, definitive (+10% to -5%)	< 1.5	1.5 to 2.5	> 1.5
Class 2, budget (+15% to -10%)	< 2.5	2.5 to 3.5	> 2.5

OFFSITES:

Definition score aggregate:	58
Confidence level (definition score average)	2.3

	Definition score
A Engineering design (Level of detail of material take-off)	
I Civil	2
ii Mechanical equipment	1
iii Insulation of equipment	1
iv Instrumentation	2
v Electrical	5
vi Pipework	1
vii Pipework insulation	1
viii Structural	3
ix Ancillary equipment	2
B Procurement (Reliability of quotations received)	
I Civil	2
ii Mechanical equipment	2
iii Insulation of equipment	2
iv Instrumentation	2
v Electrical	5
vi Pipework	2
vii Pipework insulation	n/a
viii Structural	3
ix Ancillary equipment	2
C Construction (Quality and level of confidence in subcontractor quotations)	
I Civil	2
ii Mechanical equipment	2
iii Insulation of equipment	2
iv Instrumentation	2
v Electrical	5
vi Pipework	2
vii Pipework insulation	n/a
viii Structural	3
ix Ancillary equipment	2

PROCESS

Definition score aggregate:	88
Confidence level (definition score average)	3.7

	Definition score
A Engineering design (Level of detail of material take-off)	
I Civil	3
ii Mechanical equipment	5
iii Insulation of equipment	5
iv Instrumentation	3
v Electrical	5
vi Pipework	3
vii Pipework insulation	3
viii Structural	3
B Procurement (Reliability of quotations received)	
I Civil	3
ii Mechanical equipment	5
iii Insulation of equipment	5
iv Instrumentation	3
v Electrical	5
vi Pipework	3
vii Pipework insulation	3
viii Structural	3
C Construction (Quality and level of confidence in subcontractor quotations)	
I Civil	3
ii Mechanical equipment	3
iii Insulation of equipment	5
iv Instrumentation	3
v Electrical	5
vi Pipework	3
vii Pipework insulation	3
viii Structural	3

COMMENTS (required for any item with a definition level of 4 or 5)

Offsites:

A (Bv), C(v), Distributed control system (DCS) is not defined.

Process:

A (ii), Few specifications issued or no information received.
A (iii) No insulation specification.
B (ii) Mechanical equipment specs based only on basic info.
B (iii) Insulation of equipment quotations based on John Brown's specifications.
C (iii) Insulation of equipment was estimated.
A (v), B (v), C (v) Distributed control system (DCS) not defined.

Figure 12.3 A completed estimate score sheet.

psychologist. Kahneman also has a lot to say about how the tendency towards optimism can frustrate any attempt to produce a cost estimate that has a reasonable degree of confidence. See Lovallo and Kahneman (2003).

Reference class forecasting is simple in practice and involves three steps:

1. Get information from similar past projects both within and outside the company.
2. Establish the outcomes for metrics such as:
   - cost at completion;
   - construction timescale;
   - construction labour hours.
3. Compare the new project to the information gathered in step 2, to establish its position and predict the outcome.

One of the benefits of reference class forecasting is that it encourages companies to consider information beyond that obtained from their own projects. The issue here, of course, is that the results from using this approach might mean that the cost estimate for the new project will not be acceptable to the organization or its customer. However, at least the level of confidence will be identified as suitably low and decisions can be made accordingly.

Documenting the estimate basis

Documenting the basis upon which the cost estimate has been prepared is important to avoid misunderstandings between clients, contractors and other project stakeholders about what is, and what is not included in the cost estimate. The document prepared for this purpose is typically known as the basis of estimate (BOE). The contents of this document should include:

- A brief description of the project.
- The class of estimate.
- A high-level breakdown of the estimate summarized form the estimate detail.
- Explanation of how the estimate was developed, including key reference data used.
- The WBS used.
- The main companies and organizations involved in the project, such as client, regulatory bodies, main contractor, subcontractors and supply chain.
- Work and responsibility split between the organizations involved in the project.
- Important assumptions, constraints and exclusions.
- The source of the rates, norms and quotes used in compiling the different cost elements (both direct and indirect).
- The different levels of confidence in the various work elements in the estimate.
- The level and nature of management and risk contingency and allowances.
- Estimate currencies(s) and project reporting implications (including use of exchange rates, conversion factors and "project currency").
- Identification of work required to convert the estimate into the project control budget.
- Intentions regarding further development of the estimate during the life of the project.

Meetings during cost estimating

During development of the cost estimate for a major project a series of formal meetings should be held. These might include for example:

- A kick-off-meeting led by the project or bid manager to introduce the people who will be involved in preparing and supporting the preparation of the estimate to the project and to establish an estimate preparation strategy.
- A mid-tender review also led by the project or bid manager. This meeting should ensure that the project is introduced to the people involved in preparing and supporting the compilation of the estimate. The meeting should also ensure that the estimate and outline project execution plan are aligned and that all the information needed for preparing the estimate is available. The scope of the project must be made as clear and free from ambiguity as possible.
- Meetings with principal subcontractors and suppliers.
- Risk review. This meeting might be facilitated by the project manager, but should really be chaired by someone who was not involved in preparing the estimate.
- Price fixing. This meeting, or series of meetings, takes place when the cost estimates and outline delivery programme have been prepared (or, at least, drafted). The estimating and proposals team would usually be present. One or more separate meetings at executive or board level will finalize the bid price. This procedure will depend on several factors and will vary greatly from one company to another.

Reviews and checks

The need for an independent review of a cost estimate within an organization will depend on company governance procedures. A 'deep-dive' review of the basis of estimate document will usually be required. This will entail meetings and interviews with the project or bid manager, estimators and other people who were involved in preparing the estimate. An independent review provides a fresh pair of eyes in looking at the estimate and provides recommendation that can result in improved confidence.

Checklists

Very early in my career I assisted a much more experienced estimator to compile the cost estimate and develop the contract programme for a project for designing and building a research centre in the north east of England. Following the contract award I attended a project kick-off meeting with the client, whose project manager was something of a bully. At this meeting my contract programme was reviewed and it turned out that it did not include the installation of a high pressure steam line. That was a very significant omission. The client's project manager became extremely angry, tore up my project schedule and threw it into a bin. This was traumatic for me and caused a short-term loss of confidence. This could have been avoided if my company had used a checklist to make sure that no significant item was omitted from the estimate.

The level of detail in a checklist should be based on templates for different project types. A reliable checklist should be organized within company standard cost types or a code of accounts and (if possible) a high-level WBS, so that lessons are learned over time and captured across the organization.

References and further reading

CII (2009). *Project Definition Rating Index.* Austin, TX: Construction Industry Institute.

Dysert, L.R. (2003). 'Sharpen Your Cost Estimating Skills'. *Cost Engineering*, 45(6), 22–30.

ECI (2000). *Cost Estimating Value Enhancement Practice.* Loughborough: European Construction Institute.

ECI (2002). *The Fast Track Manual.* Loughborough: European Construction Industry Institute.

ECIA (N.D.). *Databank of Existing Norms.* London: Engineering Construction Industry Association.

Ernst & Young (2014). *Spotlight on Oil and Gas Megaprojects.*

Flyvbjerg, B., Bruzelius, N. and Rothengatter, W. (2003). *Megaprojects and Risk: An Anatomy of Ambition.* Cambridge: Cambridge University Press.

Lovallo, D. and Kahneman, D. (2003). 'Delusions of Success: How Optimism Undermines Executives' Decisions'. *Harvard Business Review*, 81(7), 56–63.

13

PROJECT COST ACCOUNTING AND CONTROL 1

Dennis Lock

The subject of cost accounting was introduced in Chapter 9, which should be read before delving into this chapter and Chapter 14.

Introduction

When we talk about controlling project costs it has to be recognized that other project controls play a very important part in preventing budgetary excesses. For example, one very high risk to cost budgets is scope creep and the most important safeguard against that is to have an effective change control procedure. Above all, it must always be remembered that a project that finishes late will almost certainly fail to meet its cost budgets.

For most of the project lifecycle, project costs will continue to accrue simply because of the overhead costs associated with the project. That applies to all projects, whether they are internal or conducted for an external customer. When project funding is dependent on payments from a client or customer, the timing of invoice submissions will very often depend on the achievement of project milestones or on the independently certified value of work done. So, for many reasons, controlling progress successfully is a huge step towards controlling project costs.

An important topic often neglected in discussions about project costs is purchasing. The costs of purchased goods, equipment and services typically account for well over half the direct costs of a project. Thus efficient purchasing and contract control is an essential component of cost control.

Cost control interfaces

An effective project manager or PMO will ensure that regular communications are established between project control staff and other departments in the organization. For cost control that applies particularly to the accounts and purchasing departments.

An efficient project management team will build up a good rapport with the accounts and purchasing managers. The cost engineering function in the PMO will ensure that the project manager is kept aware of costs spent and committed in relation to budgets.

On medium and large projects a very useful step is to have one or more purchasing officers seconded from the purchasing department to the project management area, so that they can operate alongside the project staff. These purchasing officers must still report to their purchasing manager, but embedding them in the project organization will provide far better day-to-day communication and control than relying only on formal periodical interdepartmental reports.

Cash flow

Cash is the lifeblood of an organization. Without cash in the bank a company cannot operate. Cash is needed to pay for the provision and upkeep of premises, pay staff wages and salaries and settle bills from the suppliers of goods and services on time. Failure to pay suppliers, for example, will inevitably cause them to give priority to other companies who do pay on time. Suppliers might even withhold the release of materials needed for projects and that can only make matters worse, leading to a terrifying downwards spiral towards financial collapse.

There have been many well publicized cases where famous names have disappeared from the lists of active companies simply because those companies ran out of cash. Maintaining a positive cash flow is vital. Every senior person in the project organization must understand the principles of managing cash flow.

Cash flow has several important elements, all of which are related to time. These are as follows:

1. Cash outflows. These are the expenses incurred in running the business, paying employees' salaries and wages and settling invoices from external organizations (such as the suppliers of goods and services). Cash outflows comprise both direct costs and indirect costs (overheads).
2. Cash inflows. These typically result from payments received from customers and investments. These payments and all other incoming receipts represent cash flowing into the organization (positive cash flows).
3. Net cash flows. Net cash flows are the difference between cash inflows and cash outflows. These differences can either be measured within a period (such as a month or quarter) or cumulatively.

In a profitable company the cash inflows should normally exceed the outflows, resulting in profits that can be distributed as shareholders' dividends or retained as cash reserves (after payment of taxes).

Scheduling cash flows

Cash flow scheduling for a project requires the existence of a project schedule (preferably derived from the network diagram after resource scheduling). Because cash is a project resource, it can be scheduled using the principles of resource scheduling. Project scheduling software can be used for that purpose (perhaps needing a little ingenuity). Chapter 18 (scheduling project resources) describes the process of scheduling cash flows briefly. Figure 18.2 summarizes the net cash flow scheduling process and Figure 18.3 shows the principles of setting out a cash outflow schedule. These procedures will now be described here in greater detail.

Cash inflows

Using a grid format such as that shown in Figure 13.1, put each expected cash inflow amount into its planned time period. The source of project cash inflows will typically be payments from a client. Milestones on the project schedule will provide the clue for timing these payments because invoices and other interim claims for payment usually depend on the certified achievement of milestones. Companies beginning new projects for external clients will often ensure an initial boost to cash inflows by having those clients pay a cash deposit when contracts are signed.

For projects where there is no external client (such as in-house change projects) other financial sources will have to be found to produce the cash inflows. One possible source of funding for these internal projects might be cash reserves (typically arising from profits retained from earlier projects). Funding can in some fortunate cases come from financial grants (for example when a member of a group of companies is given a grant from group headquarters or when there are cash incentives from a government department). For large in-house projects funds can occasionally be generated by issuing new shares, either on the stock market or to private investors.

Cash outflows

Cash outflow scheduling can only take place after several preparatory steps have been performed in a logical sequence. The following method is typical and recommended.

1. Obtain a coded WBS for the project.
2. Estimate the direct costs associated with every WBS element.
3. Add the company's general overhead burden to obtain the total estimated cost of each WBS element. The overhead burden is typically added as a percentage of the direct labour costs. The amount of that percentage applicable to the entire project will be determined by the company accountant. The overhead burden added to labour costs is intended to pay for the general operating expenses of the company (accommodation, higher management and administration) and for the non-productive time of people when they are sick or on leave or training. Note that in some companies a charge (which might be 5 per cent) is also added to the direct costs of every planned materials purchase to allow for the costs of documentation and materials handling.
4. Refer to the project schedule and obtain the planned times for all the WBS elements.

5 Using all the above information put every item of expenditure into its relevant time period, using the same grid used for the inflows. For items spanning more than one time period, costs should be apportioned across the relevant periods.
6 Add the amounts in each period column to obtain the total predicted cash outflow for that period.
7 Add the period outflow totals from the beginning to end of the schedule (horizontally). The grand total in the extreme right hand column should equal the total predicted cash outflows (cost estimate) for the project.

Scheduling the amounts and expected timings of all those costs as described above will produce the cash outflow schedule.

Net cash flows

When all the predicted cash outflows have been added to the cash inflow schedule, the net cash flows can be determined by subtracting the outflows from the inflows. Again this should be done both for every period and cumulatively.

Example

Figure 13.1 is an example of a net cash flow schedule for a fairly large construction project. Using the convention applied by accountants, figures placed in brackets are negative values. Thus, in this example, the cumulative cash flows shown in the bottom two rows give a warning that the project is expected to run into temporary debt during some of the quarterly periods. Figure 13.1 has been adapted from Chapter 17 of Lock (2015).

Levels of authority for project expenditure

The levels of authority allowed for committing a company to expenditure vary widely from one company to another but no-one in a project team should be left in any doubt as to how much they are authorized to commit their company to expenditure through purchases or the signing of contracts. Good practice will limit the number of people in an organization who are allowed to put their signature to documents such as petty cash vouchers, service contracts and purchase orders.

With orders so easy to place online, the risk of unauthorized purchases is greater. There is also a risk of orders being placed unintentionally during telephone calls, which a potential supplier might record as evidence. Such accidental verbal commitments are so easy to make when design engineers are talking to potential suppliers to investigate details of their products, prices and delivery times.

Ideally all orders for project equipment and supplies should be submitted on the organization's standard purchase order forms and they should be signed by the purchasing manager. However, an organization might decide to allow other nominated people to sign purchase orders or project contract documents for limited amounts, because too much restriction can cause unwanted and unnecessary delays. Very small

PROJECTS UNLIMITED LTD					Loxylene Chemical Plant for Lox Chemical Company								Project number P21900 Issue date November 2022					
Cost item	*Quarterly periods - all figures are £1,000s*																	
	2021				*2022*				*2023*				*2024*				*2025*	*Budget*
Cash inflows	1	2	3	4	1	2	3	4	1	2	3	4	1	2	3	4	1	
From HQ	50				150													200
From Client	10		50	100	200	500	1500	1000	1000	1750	1000	1000	1000	3000	1000	1000	1000	15110
Total Infloows	60		50	100	350	500	1500	1000	1000	1750	1000	1000	1000	3000	1000	1000	1000	15310
Cash Outflows																		
Engineering	14	25	59	80	85	63	43	23	12	11	9	6	7	10	14	14		475
Purchasing		5	45	5	550	310	745	295	750	665	215	457	2242	76	2	470		6832
Construction				17	35	97	245	393	436	654	382	241	186	45	30			2761
Contigency					10	20	25	25	30	30	35	35	45	45	50	50		400
Escalation					35	29	74	59	110	136	70	88	322	30	32	82		1070
Total Outflows	14	30	104	102	715	519	1132	795	1338	1496	711	827	2802	206	128	619		11538
Net Cash Flows																		
Periodic	46	(30)	(54)	(2)	(365)	(19)	368	205	(338)	254	289	173	(1802)	2794	872	38	1000	
Cumulative	46	16	(38)	(40)	(405)	(424)	(56)	149	(189)	65	354	527	(1275)	1519	23941	2772	3772	3772

Figure 13.1 Quarterly net cash flow schedule for a construction project.

Source: Adapted from Lock (2015).

purchases are sometimes made by people who shop for cash, and then reclaim their expenses by means of petty cash vouchers, but that arrangement can be abused and lead to fraud if care is not taken.

Because of differences between the policies adopted by various organizations it is not feasible to say here who should be able to sign for what, or for what amount. But what can be said for certain is that, within any organization, clear rules should be established so that every member of staff knows what they are allowed to sign, and the limit of their spending authority. For most members of staff that limit must be zero.

When setting limits of financial authority, senior management should be aware that project managers in particular should be trusted with the authority to spend or commit money, provided that it is for project materials or work that is included in the authorized project budget. If allowable spending limits are set too high (or not at all) there are obvious potential risks but there must be a level of trust between senior management and project managers. The time and work of project managers should not be wasted in expecting them to seek permission for every tiny item of expenditure.

In my own case, for many years I had no upper expenditure limit and could sign purchase orders and contracts for any amount, the only proviso being that for any capital item over today's equivalent of £100,000 I should inform the financial director *afterwards* in order that the purchase could be reported to the board of directors and recorded in the minutes of board meetings. That arrangement worked very well, but it only worked because of the mutual trust established over a number of years. That level of trust made my management task much easier, but in the wrong hands it could have led to disaster.

So, to sum up, there is no clear recommendation here on how much personal authority should be allowed to commit the company to expenditure because circumstances vary so much from one organization to another. However, within every company it should be made clear where the authority lies for signing petty cash vouchers, purchase orders and other contracts that will commit the company to expenditure. If a project manager's limits are set too low, then their time will be wasted in continually having to seek permissions. My own view is that a project manager who has served without trouble in an organization for over (say) a year should be allowed to authorize all expenditure within the project budgets. When expenditure would cause budgets to be exceeded, then it is not a bad idea to insist that the project manager should seek permission from higher management and explain why the over-budget costs are necessary.

Timesheet management

The usually accepted method for collecting and analysing all labour costs depends on the submission of weekly timesheets to the accounts department. Some companies might exempt senior executives from having to compile timesheets. At the extreme lower levels, general maintenance and janitorial staff might also be excused but, although those people might never work on projects their attendance times are needed to record their hours worked (signing or clocking in and out of work is one method).

For project controls the main concern must be for recording and analysing the time spent by people who work on project tasks with reasonable accuracy. Timesheet data might be collected using paper forms (an example of which was shown in Figure 9.1) or, increasingly commonly, from direct data entries by individuals into the company's server.

Timesheet entries are not always accurate and reliable. Imagine a member of a project team who tries at the end of a week to remember which tasks he or she was engaged on during the previous Monday morning. So people should be encouraged to note the relevant times and task numbers at the end of each day. Daily direct data entries should reduce the risk of errors.

Task numbers for timesheet entries can usually be derived from the project WBS codes (coding methods were explained in Chapter 5). However, codes have to be kept as short and simple as possible to reduce the risk of errors. The person who designs a coding system that requires every person to enter a 16 character code for every job or task might feel very proud of that system, but the worker who has spent all day on top of scaffolding in the driving rain will be less impressed and in most cases such people will simply refuse to comply.

Timesheet accuracy is important for many reasons. For project control in general, inaccurate time recording would prevent true comparison between estimated (or budgeted) times and actual times. For internal projects and projects conducted for fixed price contracts, accurately recorded times (and their extension into actual costs) are important for at least two reasons, which are:

1 To monitor actual costs against budget, so revealing over- or under-expenditure at task level.
2 To provide archived records that can be referred to when estimating the hours needed for similar tasks on future projects.

When projects are being conducted for external clients on a cost-plus basis, client billing amounts are usually directly proportional to the hours booked on timesheets. So then there is an ethical requirement to ensure timesheet accuracy and avoid clients being over- or undercharged. Indeed, clients of cost-plus projects will sometimes need to be assured of the timesheet checks and authorizations in place.

Timesheet analysis

Each departmental manager should check (and preferably countersign or initial) every timesheet for staff in that department. A typical arrangement in a paper system was that departmental managers would submit their department's timesheets to the accounts department at the end of every week. Then, during the following week, the cost accountants would analyse those timesheets and ascertain the total hours booked to each WBS code. Those hours were then 'extended' into labour costs, using the appropriate standard cost rates (standard cost rates were explained in Chapter 9). That analysis had to be reported back to the departmental managers.

With digital systems and direct timesheet data entry the analysis process is considerably speeded up, but there is then an increased need for checks and supervision (which can be achieved by sampling rather than by checking all entries).

Conclusion

This chapter introduced aspects of project cost and cash flow accounting that are relevant to project controls. All those concerned with project controls should be acquainted with the accounting methods used in their organization. This important subject is continued in Chapter 14.

Reference

Lock, D. (2015), *Project Management*, 10th edn, Farnham: Gower.
More references and further reading sources can be found at the end of Chapter 14.

14

PROJECT COST ACCOUNTING AND CONTROL 2

Dennis Lock

This chapter concludes the discussion of cost accounting functions that are especially applicable to projects. For those who are new to this subject, Chapters 9 and 12 should be read first.

Overhead costs

Overhead cost recovery is a complex function that can give cost and management accountants some difficulty. It is not the easiest topic to explain, but all those at senior level in project management must grasp the principles and understand how project performance can affect not only project costs, but also the overhead operating costs of the project management organization.

First, consider all the costs incurred by a company that is conducting projects for external customers. This company might also be using its resources to produce manufactured goods for sale or to provide non-project customer services (such work is often described as business as usual). However, for simplicity, I shall take the example of an organization that is completely engaged in projects. All that project work and the resulting invoices sent to clients will result in cash inflows that are intended to pay for all the operating costs incurred by the company, with a margin left over for profit.

All the costs incurred from conducting project activities (labour, materials and expenses) can be attributed and charged directly to projects. Those costs are thus known as direct costs. Because the rate at which these costs are incurred will vary in proportion to the day-to-day level of project activity, direct costs are also known as variable costs.

Costs arising from running the business, such as providing the management framework and paying senior managers' salaries, paying for the maintenance of premises, paying business rates and so on are all overhead costs. Payments for certain staff benefits and facilities (such as welfare, leisure and catering) are also overhead costs.

Because overhead costs vary very little or not at all on a day-to-day basis regardless of project activity they are also known as fixed costs.

Cost engineers from the PMO should be able to assist the company's accountants by supplying the estimated direct project costs for labour, materials and expenses over the forthcoming period (which might be for six months or a year), based on everything running according to the project time schedule. Managers of administrative and non-project departments (such as HRM, marketing, facilities and premises management) can be asked to supply their estimates of indirect costs for the same calculation period.

Now the cost accountants have estimates for all the direct costs and all the indirect costs. Taking the example of a medium-sized company, suppose that the results of these predictions for a given period are as follows:

Estimated direct costs (resulting from all projects)	£
Direct labour	10 000 000
Materials	5 000 000
Prime cost	15 000 000
Estimated total indirect costs (fixed costs) for the period	5 000 000
Total estimated cost of sales	20 000 000

Ignoring the subsequent mark up for profit, the question arises 'How can this company account for and recover the £5m expected indirect (overhead or fixed) costs for this period?' The solution adopted by cost and management accountants is to add the overhead costs as a burden on the direct labour costs. In this example this overhead burden amounts to 50 per cent of the direct labour costs. So, when the cost estimators compile their project cost estimates they must add 50 per cent to the estimated direct labour costs as an overhead recovery allowance.

Although the materials estimate might include a small burden to allow for handling and storage, traditionally all the company's overhead costs will be recovered by the burden added to the direct labour costs.

Overhead over and under recovery

Please refer to the above figures and consider the following three different possibilities.

Case 1. The PMO and the project management function are all performing well. All project activities are running exactly on schedule and the original cost estimates are all proving to be accurate. Invoices are being issued to customers and clients at the expected times for the planned amounts.

Case 2. Project performance is better than forecast. Total costs are being incurred under budget levels, progress is ahead of schedule and invoices can be issued to customers and clients earlier than planned.

Case 3. Project performance is not good, with some work delayed or repeated, so that invoices to clients are issued late.

When the conditions are as described in Case 1, provided that clients pay promptly the cash inflows should be sufficient to pay for all the company's costs, plus a margin for profit.

In Case 2, again assuming that clients pay on time, the company's financial position will be better than forecast, with more money coming in than scheduled to cover the direct and overhead costs. In this case, the overhead costs will have been over-recovered. This is a satisfactory outcome although the company's sales team might argue that prices could have been set lower, allowing them to take advantage of better competitiveness against the company's business rivals.

Case 3 is an example of a company heading for financial trouble. Cash outflows to pay for fixed overhead costs occur as scheduled but the compensating inflows are late, leading to less money in the bank than planned. In serious cases of overhead under-recovery bank overdrafts can result, in the worst circumstances leading to uncontrollable and increasing debt.

Cost reporting

The responsibility for collecting, analysing and reporting project costs will usually fall upon the PMO. One or more cost engineers will perform the analysis, relying on cost data from the purchasing and accounts departments. The project manager (who bears responsibility for project performance) should issue the reports to senior managers in the organization. For fixed price contracts, reports to clients should normally be confined to reporting progress and, indeed, the company will usually wish to keep its costs confidential. For cost-plus contracts the situation is different and the project organization will probably have a duty of care to keep the client informed of costs incurred and the relationship of those costs compared with agreed budgets and progress.

The frequency of cost reports will depend to some extent on the size and duration of the project. For megaprojects lasting several years, cost reports might be issued at quarterly intervals but for many smaller projects monthly reporting is more appropriate. Although much of the responsibility for recording costs will fall upon the accounts department, it will usually be appropriate for cost reports to be issued under the signature of the project manager, because it is he or she who bears the responsibility for project performance. In some cases a more senior manager (such as a company director) might wish to add the final signature.

Cost reports should ideally always be combined with progress reports. Simply reporting project costs alone means little if the relevant state of progress is not given. So here I should really be describing cost and progress reports. Senior managers and other stakeholders receiving the reports will want to know:

- Is the project running on time and, if not, where do the problems lie and what corrective measures are being taken?
- What are the cumulative costs to date and how do these compare with the budget?
- Have there been any authorized changes since the last report and what will the effect of those changes be on costs and progress? This part of the report will also

have to describe whether or not the changes were requested by (and thus chargeable to) the client.
- Is the project still predicted to finish within the currently authorized budget?
- If an overspend is predicted, where will this occur and for what reason? This section of the report might directly or indirectly implicate who is to blame for the overspend and could have unfortunate consequences for the person or people concerned.

Level of detail in cost and progress reports

There might sometimes be a question in the project manager's mind about the level of detail for cost and progress data included in reports. The first consideration when answering that question is to consider the intended recipient of the report. Generally speaking reports to clients should be filtered so that information that should be kept confidential for commercial reasons is not included, although in cost-plus contracts the client will expect to be kept informed in considerable detail of how effectively the money is being spent.

Now consider reports intended only for consumption within the project company. The management scientist Elliott Jaques (1917–2003) devised a theory which he called the timespan of discretion (see Chapter 31 in Lock and Scott, 2013). The core of that theory was that when deciding the level of remuneration for executives and other staff, pay should be gauged in relation to the time taken for the results of their management decisions or other work to take effect and become apparent. Thus, taking extremes, the CEO of a large corporation would deserve a far higher salary than a machine operator because if the CEO made a wrong decision it might take years for the effects on the business to become apparent, but if the machine operator made a mistake the need to scrap the job and begin again would usually be obvious immediately. So, looking at Elliott Jaques' theory in reverse, there is a clue here on the level of detail that should be included in project cost and progress reports, namely that departmental managers need to see more detail in their reports, but senior executives need only be told the bigger picture. If the electrical work on a new building is running a week late and 5 per cent over budget, the manager of the electrical department must be told, but that is hardly an item to include in a monthly or quarterly report to the CEO.

Debtors

In the environment of project management, debtors are customers and client companies who owe money to your company for work that you can prove you have done or goods that you have supplied. Serious debtors pay their bills late, or as only a part of the amounts requested on invoices. Some do not even pay at all. There are companies which have a reputation for being 'bad payers' and they will delay payment for as long they can as part of their culture. A company that habitually pays late will find that advantageous to its cash flow (because cash is retained in their bank instead of being transferred to your own company's bank).

It is important to deal with debtors promptly for several reasons, which include the following:

- The simple honour of ensuring that they meet their contractual obligations.
- To gather cash in as soon as it becomes due, and so improve your own company's cash flow position.
- Failure of a contractor to pay might be an indication of inability to pay. That could mean that the contractor is in serious financial trouble. If a committed contractor goes out of business, that could leave you (as project manager) and your company with a king-sized headache.

If your project has been finished and delivered to a debtor customer, you have several possible actions open to you, none of which is pleasant and any of which might rebound on your own company's reputation. Options include the following:

- Refuse to deliver any items required as part of the project contract until the debt has been settled. Even after a project has been completed and handed over, there might be some work (such as the provision of operating and maintenance instructions or service under warranty) that might be bargaining points.
- Legal action. This is definitely only to be considered as a last resort because courts and lawyers are expensive and, if the action is unsuccessful, your debts and cash flow will be far worse than if you had done nothing.
- Threaten bad publicity. Pass the word around other companies and in the press. But this is a dangerous road to travel because of libel laws.

Close out

When all the project's cost and progress data have been safely gathered in, a final report should be written. A copy of that report should be placed in the company's archives. An essential part of a final project report is a detailed account of all project estimated costs (coded by the WBS) together with the actual amounts recorded. A company's capital is not only financial. It also includes its retained knowledge and 'lessons learned'. With due diligence any mistakes made should not be repeated on future projects.

References and further reading

Drury, C. (2018), *Management Accounting for Business*, 10th edn, Andover: Cengage.

Holm, L. (2018), *Cost Accounting and Financial Management for Construction Project Management*, Abingdon: Routledge.

Lock, D. and Scott, L. (eds) (2013), *Gower Handbook of People in Project Management*, Farnham: Gower.

PART IV

Scheduling

15

BASIC PLANNING METHODS

Dennis Lock

This chapter is the first of three that describe the principal planning and scheduling methods available to project managers and those working in PMOs. Here the subject is planning with simple charts. Following chapters will describe planning and scheduling with network diagrams. Plans for all but the very smallest projects take up large areas of paper when they are sketched or plotted. Clearly such plans cannot be shown in the pages of this handbook, so all the examples will be of very small projects, yet large enough to demonstrate the essential features of each described method.

Identifying and listing the project tasks

In order to identify project tasks, perhaps it is necessary first to understand what is meant by a 'task'. It might be thought that any job which requires the actions of people is a task. However, when planning all the activities in a project we also have to take account of anything that will cost money or take up time. Thus when planning a project, 'tasks' can include things such as purchasing materials or waiting for paint to dry. Tasks usually (but by no means always) take up time, add costs to the project and require resources. Some actions that occupy practically no time or cost include things such as the approval of a design strategy, yet such things are important and must be included in plans as tasks because they can delay progress if not planned and done on time.

The concept of gathering a few relevant people together so that they can make a long list of all the tasks before a project starts might sound sensible and feasible, but that's not the best approach (although it is necessary to produce such a list for the initial cost estimates). So, although project planning might begin with a list of perceived tasks, a team discussion about the expected plan and work sequence will itself reveal tasks that might otherwise have been forgotten and left out of the plan. So a good way of identifying and listing all the project tasks is to gather a team together of experienced project participants (they may be senior engineers or department managers) to develop an initial project plan. The planning method used can be any of those described in this and the following chapter.

Developing a detailed project plan can be a kind of brainstorming session, during which project tasks come to light that would otherwise have been forgotten until too late. Clearly any task that is forgotten and not included in the plan could risk late project completion, not to mention an omission from the cost estimates and budgets.

Estimating task durations

Although not always possible to arrange, it is preferable for the duration of every project task to be estimated by the manager (or senior delegate) from the department that will eventually perform the task). Then we establish a preferred principle that the person who makes each estimate will ultimately be responsible for managing and completing the task on time and within budget. If a task would be performed better using more than one worker, then that resource information has to be recorded along with the duration estimate (five days, two electricians for example).

People vary considerably in their estimating capabilities. Project managers will in time come to learn which estimators tend towards optimism, pessimism or accuracy. Experienced project managers will even learn to make allowances for people and 'adjust' their estimates accordingly (a practice that I followed myself liberally and shamelessly in my junior days). A big problem is found with managers whose estimates exhibit no constant bias, but range randomly from optimism to pessimism (in which cases it is best to look for another estimator).

Task duration estimates are for 'elapsed time', so that when they are applied to all the planned tasks a project schedule can be developed that signifies when each task should begin and end. Clearly that process will ultimately predict the duration of the entire project. Some people recommend that tasks should be chosen at random when estimating durations, arguing that if consecutive tasks are considered the estimators might (subconsciously or otherwise) become aware of the cumulative estimates towards project completion and thus be influenced.

Elementary planning methods

Whenever any project is planned it is clearly necessary to record that plan and make it available to all who will be expected to follow it. A plan is not only the result of deliberation, calculation and consideration of everything that needs to be done in a project: it is also a means for communicating the project manager's intentions to everyone involved in the project and subsequently for assessing progress towards successful project completion on time.

Without a plan there can be no control.

Task lists and action lists

The very simplest planning method possible is just to list all the tasks that can be imagined for project completion. If we were to consider Christmas shopping as a project, then all that would be needed is such a list. However, many project plans do begin as task lists.

One very elementary method for listing tasks that should be performed is seen in the action plans that are often included when the minutes of meetings are distributed. Those 'plans' simply list all the actions agreed during the meeting and name the person or people responsible for taking each action, along with the required completion date. When the next meeting is called, the previous action list is reviewed and anyone who has failed in any way to perform the actions required can expect some form of chastisement or retribution.

Although project task lists will always be needed for such purposes as cost estimating, it is usually only when the work is planned in detail that all necessary tasks can be identified.

Gantt charts

For a project involving more than a few simple tasks, something more than a simple action list is needed. We need a method that sets out all the forthcoming project tasks clearly, showing their start and finish times. Then every task can be monitored, allowing corrective action to be taken in time where needed to speed things up and avoid late project completion. Where consecutive tasks are dependent on each other, those dependencies need to be shown in the plan and observed in practice.

Henry Gantt, one of several early industrial management researchers working in the US, developed the use of tables and charts for scheduling and progressing manufacturing and assembly work in factories around the year 1900. His name lives on in the horizontal bar charts described in this chapter.

Gantt charts have the great advantage that they can readily be understood at first sight without any need for special training. Anyone familiar with an office holiday chart will be able to understand a Gantt chart with no difficulty. The principle is simple. All significant known jobs are listed down a vertical column at the extreme left-hand edge of the chart. The chart is divided horizontally into vertical columns, representing successive equal time periods (such as weeks). Every task is plotted as a horizontal bar along its named row, and positioned from right to left across its appropriate period column or columns. Such early charts were chalked on blackboards, drawn on paper, or assembled on wallboards using proprietary kits that allowed editing and adjustment. The bars in those kits could be colour coded to indicate a particular property of the task (for example to show that a task required a particular resource or had high priority).

Gantt charts drawn on paper or assembled on boards have limited practicability for projects because they become too complicated to understand (requiring very large boards or sheets of paper) when the project has more than about 100 tasks or extends over a long period. The biggest disadvantage will be experienced when, owing to progress or a change in the project requirements, the chart has to be amended, redrawn or reconstructed. However, I shall now introduce a very small domestic project for which planning using a Gantt chart would be entirely suitable.

Gantt chart example for a garden workshop project

A man has decided to install a workshop in the grounds of his home, for which he can enlist the help of his teenage son. They will buy a flat-packed timber building with one central door and two windows, providing at least 200m^2 of usable internal floor space (which is a fairly big workshop that will cost several thousand pounds). They will choose their building with care from suppliers' catalogues and Internet searches, knowing that it will have to be erected on a level, flat, strong concrete base. Failure to provide that firm base would negate the supplier's guarantee of good quality and workmanship. For the purposes of this project example I have ignored electrical wiring and power supply. Let's suppose that an extension cable can be run from the house until a more permanent arrangement can be made.

Simple drainage into the surrounding earth is all that is needed, but the pitched timber roof boards will have to be covered with mineralized felt (supplied with the kit), with rainwater draining into gutters and downpipes. The flat-pack will come with the door furniture already fitted (latch, lock and hinges), but the windows will have to be glazed, using the pre-cut glass and beading strips supplied. Figure 15.1 is the task list for this simple project and Figure 15.2 shows the corresponding Gantt chart.

Although this plan is extremely simple, it includes features that are commonly used in Gantt charts for much larger projects. Every task has an identifying code and a short but adequate description. The duration estimates in this case are given in calendar days, and those same units are used throughout the plan. Simple resource codes (F and S) identify the resources needed for each task and the bars on the chart are shaded accordingly. Unshaded (white) bars require no resources from this family, so they represent only the predicted passage of time.

Calendar for the garden workshop project

The plan in Figure 15.2 shows sequential day numbers. In practice the project manager (the father in this case) would have to decide and specify the project calendar and show dates at the head of the columns. Does a project week contain 5, 6 or 7 working days? Here I have assumed that the father is retired from other full time work, and the project can take place while the son is enjoying his summer holiday. So the tasks can be performed with no time out for holidays or weekends.

In project schedules it is preferable to write dates in all documents in the form DDMMMYY (day, month, year). Thus the fifth day of July 2022 would be written 05JUL22. Then there can be no confusion in international projects where some nations express their written dates differently.

Linked Gantt charts

The garden workshop project is clearly a very simple example, and it was easy to construct the chart so that no task was planned to start before all its preceding tasks had

Task ID	*Task description*	*Estimated Duration*	*Resource*
01	Choose the site and decide the overall floor plan	1 day	1F
02	Research catalogues and the Internet	1 day	1F
03	Get quotes	5 days	-
04	Choose supplier and order the building	1day	1F
05	Supplier makes and delivers the building	28 days	-
06	Clear temporary space for delivery of the building kit	1 day	1F + 1S
07	Clear and level the chosen workshop site	3 days	1F + 1S
08	Get timber for concrete base frame	1 day	-
09	Make and level the concrete base frame	1 day	1F + 1S
10	Get cement and ballast	1 day	-
11	Mix, pour and level the concrete base	2 days	1F + 1S
12	Allow concrete base to cure	2 days	-
13	Unpack and set wooden bearers and floor in place	1 day	1F + 1S
14	Erect the four walls (including door)	1 day	1F + 1S
15	Fit roof supports and panels	1 day	1F + 1S
16	Fit gutters and downpipes	1 day	1F + 1S
17	Fit roofing felt	2 days	1F + 1S
18	Glaze the windows	1 day	1F
19	Green waste disposal	1 day	1F + 1S

Figure 15.1 Task list for a garden workshop project.

Notes: 1 All duration estimates have been rounded up to the nearest whole day. 2 The resource codes are F for father and S for son. 3 A few tasks (such as hiring a concrete mixer) have been omitted for clarity.

been finished. When a chart is drawn for a larger project it is less easy to remember all the finish-start relationships between tasks and there is some danger of showing tasks beginning before all their dependent preceding tasks have been finished. That risk clearly increases when a larger number of tasks have to be planned, and greater still if a change is introduced to the project. An early way of preventing such mistakes was to place vertical link lines to highlight the inter-task dependencies, as shown in Figure 15.3. However, all those difficulties disappear when we use critical path diagrams, which are introduced in Chapter 16.

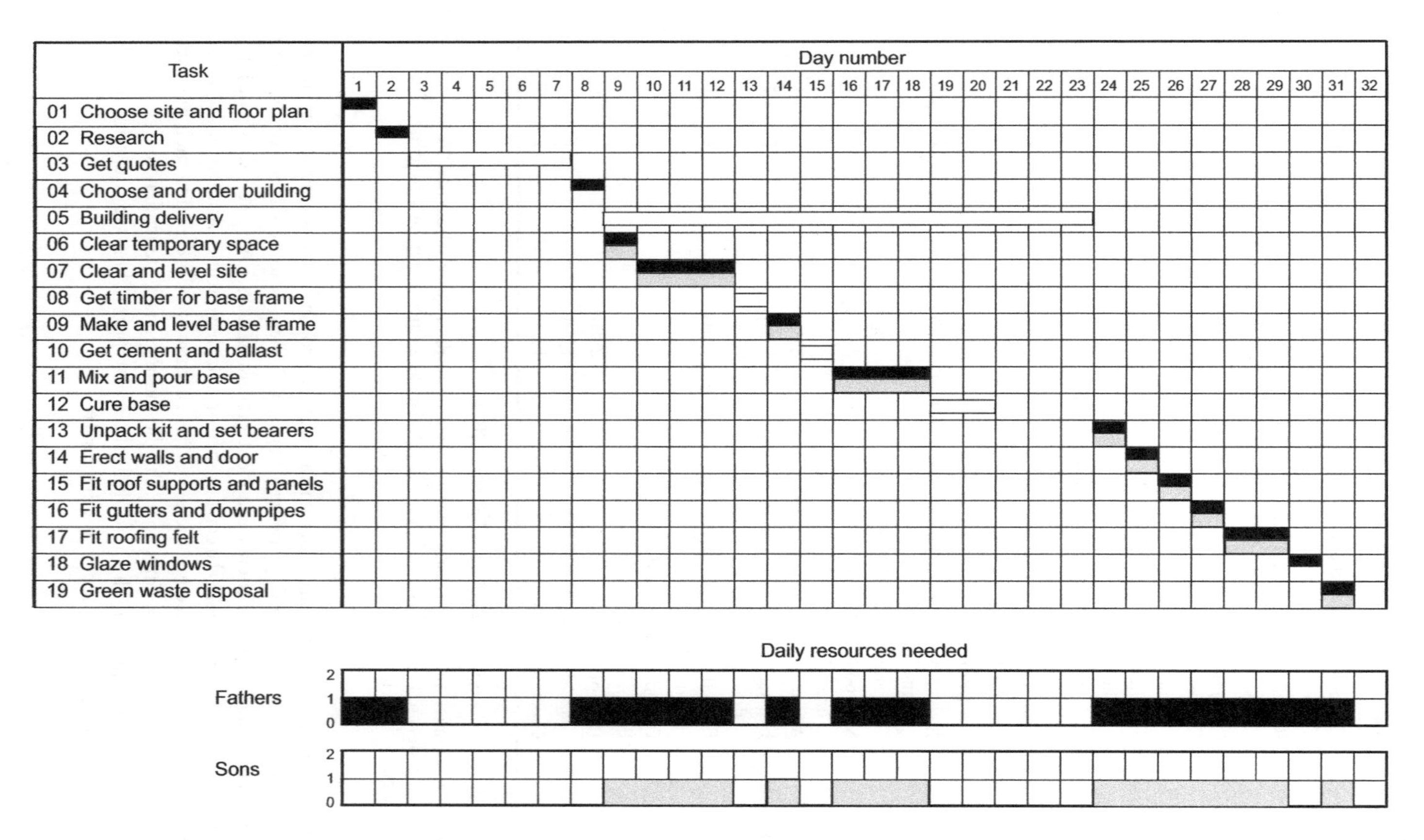

Figure 15.2 Gantt chart for the garden workshop project.

Project milestones and their implications for project control

There will always be some particular events, tasks or dates in a project schedule that have particular significance and which should be designated as milestones. The plan for a large project might have many milestones. Here are a few examples:

- project start date;
- approval and issue of working drawings for a project or significant part of a project;
- date when work begins on a construction site;
- date when a prototype model passes its testing;
- date when a building is roofed and the internal trades can begin work;
- in the project for compiling this handbook, the date when the publisher receives the finished manuscript;
- project closing date.

Project milestones in relation to controlling progress against the schedule

Milestones are important indicators of progress but they do not contribute significantly to the vital function of controlling progress against the schedule. If a scheduled milestone date is missed, then it might not be possible to accelerate the remaining project tasks and recover the lost time. Always remember that time is a special project resource – once it has been spent it has gone forever.

So how can we exercise control so that every milestone is reached on its scheduled date? The answer to that has several components, the most important of which is to monitor progress on a day-to-day basis, not just by concentrating on milestones. Then the risk of any project task overrunning the schedule is recognized *while that task is still in progress and can be accelerated.* That control has to be exercised over all in-house tasks and also over work put out to subcontractors. Purchasing against schedule is also vital, but that subject is discussed in Chapter 27.

It is important that all project workers are dedicated to project success, and a good project management team can work with the departmental managers so that all workers are suitably motivated and feel that they are part of the project. In project management pride is not a deadly sin! Many projects rely on subcontractors to perform a range of the planned tasks, even including some design work, and it can be a good idea to appoint one or more project liaison engineers, whose responsibilities include visiting relevant subcontractors to monitor progress, answer questions and make the subcontractors feel that they are part the project. The PMO is an ideal home in the organization for liaison engineers.

Project milestones in relation to controlling costs and cash flow

In projects where the client or customer has agreed contractually to make stage or progress payments it often happens that those payment dates are linked to milestones. It can follow that every missed milestone date also means a delay in receiving a stage

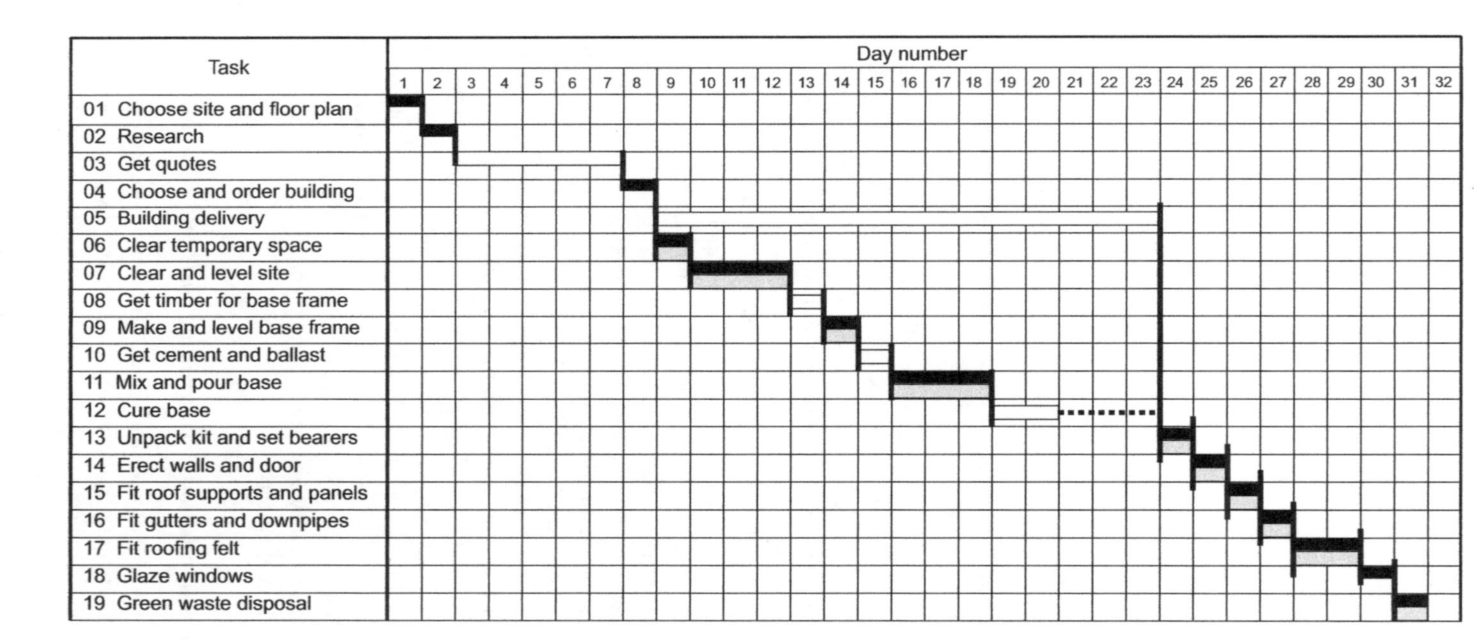

Figure 15.3 Linked Gantt chart for the garden workshop project. Here vertical links have been added to indicate that each task cannot be started before all its preceding tasks have been finished. Although these links are hardly necessary in this small project, they become essential in Gantt charts containing a large number of tasks.

or progress payment. Unfortunately, even though revenues are delayed, the project costs continue to accumulate on a day-by-day basis. If that state of affairs is allowed to continue, the project contractor's bank balance will steadily be eroded. In the worst case that could lead to bankruptcy and failure of the project and its organization.

So the control of progress is closely linked to cost control and cash inflows and I make no excuse for repeating here that any project that is allowed to run late will also overrun its cost budget.

16
CRITICAL PATH PLANNING

Dennis Lock

The previous chapter used a garden workshop project to illustrate the use of Gantt charts. However, although a Gantt chart would be entirely suitable for planning and controlling such a small venture, charts are inflexible to changes and are not really suitable for planning projects with more than a few tasks. Another difficulty with Gantt charts is that they do not readily highlight which tasks should claim priority for the allocation of scarce resources and management attention. The critical path methods described in this chapter can overcome those difficulties.

Introduction to critical path methods for project planning

Critical path networks first came on to the project scene in the early 1950s. Any network diagram comprises a pattern of linking arrows flowing from left to right, intersecting at various points called nodes. In project planning this left-to-right motion represents the flow of project progress against time, although networks do not have to be drawn to a timescale. Two fundamentally different critical path methods emerged from the earliest days, with one system using the arrows to represent project tasks or activities. These arrows joined recognizable achievement events at the nodes. The other systems (including the English Electric link and circle method, and the Roy method from Belgium) placed the project tasks or activities on the network nodes, with the arrows used merely as dependency links between the tasks. That's enough history: I shall now describe the features of both these methods as they are known and used today. Note that 'task' and 'activity' both mean the same thing in the context of project planning.

Activity-on-arrow networks

At one time activity-on-arrow networks (usually abbreviated to arrow networks or arrow diagrams) were the most popular and most widely used critical path planning method. The best way to describe this method is to show the garden workshop plan in arrow network format (Figure 16.1). A fragment of this network is shown in Figure 16.2, which explains the notation used.

Every circle in the diagram represents some recognizable event in the project, and every linking arrow represents the task or activity needed for the project to progress from a preceding event to the succeeding event. Events with special project significance (such as the project start and the project finish) can be designated as milestone events. Everything in all network diagrams always flows from left to right (although in this case that could not be done within the page limits, so the network is shown in two linked parts). The dotted line that links these two parts to create a continuous network is an example of a dummy activity, which is merely a logical link with no work involved and taking up no project time.

Every event is given an ID code, which is especially important when the network calculations are to be processed using a computer. Picking an example at random, the task of making and levelling the base frame is designated activity 7– 8, with event 7 called its preceding event and event 8 the succeeding event.

The name or description of every activity is written below its arrow and the number written above the arrow is that activity's estimated duration (using days as the common time unit in this case). So in the example of Figure 16.1, activity 6–7 'Get base frame timber' is expected to take one day. That estimate is always made for elapsed time (the duration) and might bear no relation to the work content of the task. Suppose that the activity were estimated to need two people for a whole day to get the timber; then although the work content or cost would clearly be two man days the activity duration would still be one day.

The number written above each event circle is the earliest possible time at which that event can be achieved, found by adding the durations of all preceding activities along the relevant path. Where an event has more than one path leading into it, the earliest possible event time must be determined by the longest preceding path. The process of tracing through the network from left to right to find all the earliest possible event times is known as the forward pass. In Figure 16.1 it is seen that the earliest possible project completion (at event 19) is 31 days after the project start. Note that this result agrees with that predicted in the corresponding Gantt charts shown in Figures 15.2 and 15.3.

Now, in the critical path process, we can calculate the latest permissible time for every event to be achieved if project completion is not to be delayed. These latest permissible times, written below their event circles, are found by subtraction beginning at the final task and working backwards through all the trailing network paths (a process called the backward pass). At every event it is the path with the longest duration coming from the right that determines that event's latest permissible time. Knowing the latest permissible time is important for controlling progress, and is also vital information when allocating scarce resources.

This process of forward and backward passes is known as network time analysis. It reveals at least one sequence of events where all the earliest possible and latest permissible times are identical. Achieving these events at their earliest possible times is critical to finishing the project on time. So the path joining the critical events (shown

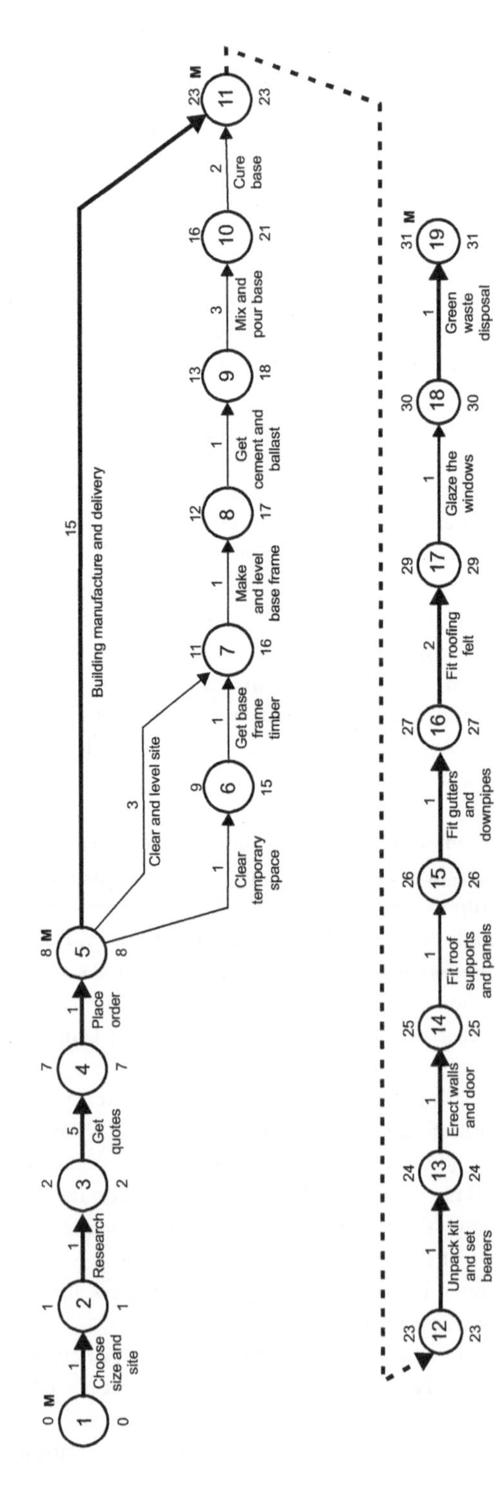

Figure 16.1 Arrow diagram for the garden workshop project.

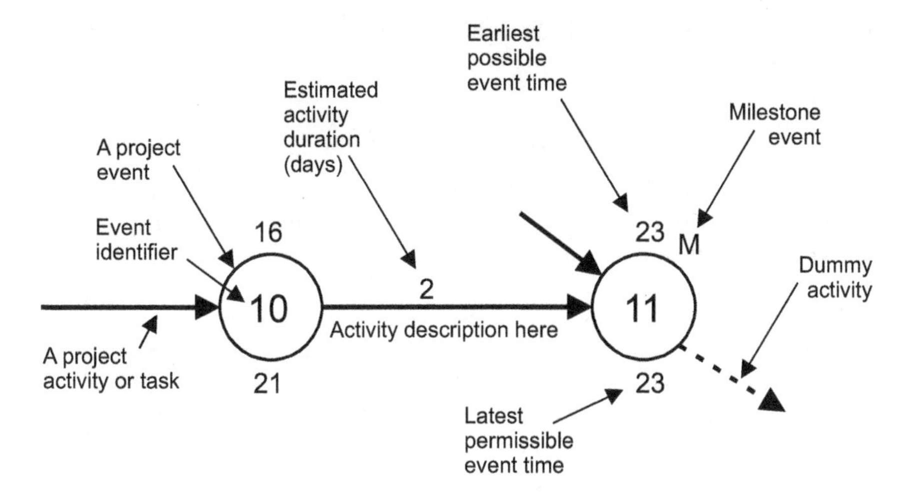

Figure 16.2 Key to the notation used in the arrow diagram in Figure 16.1.

with bold arrows in Figure 16.1) is called the critical path. Delaying any task along the critical path will inevitably delay project completion. That's vital information for project control. Note that in larger networks it is quite possible to find two or more parallel critical paths in the network.

Now look at activity 7–8 in Figure 16.1 (make and level the base frame). The earliest possible finish time for this task at event 8 is day 12. However its latest permissible finish time (at event 8) is day 17, which is five days later. What that means for the project manager is that task 7–8 could be delayed by as much as five days without having any effect on the project completion time at day 31. This five days is called the total float (or total slack) time for task 7–8. However, in this case that total float time is shared by the activities following along the same path. So if (say) the base levelling hit problems and took three days, event 8 would be reached two days late and the float of following activities would be reduced by those two days.

Understanding the significance of float in a project network is important when shared resources are scarce. A project network might contain hundreds of different activities, arranged in complex patterns, with several critical and near-critical paths running in parallel. If, at any time, two or more activities require resources where there are only enough to work on one activity, it is obviously the activity with least float (the most critical) that must claim priority.

Tied activities

Sometimes an activity *must* be scheduled to follow its preceding activity immediately, with no delay allowed. An example is seen in Figure 16.1 where Activity 10–11 *must* be scheduled to follow Activity 9–10 without delay. Clearly the concrete will begin to cure immediately after it has been poured. When a computer is used for processing the network some software will allow the planner to specify any two such activities as *tied activities*.

Different kinds of float

In any project network diagram, whether it is of the activity-on-arrow kind just described or activity-on-node (discussed below), the network patterns mean that different kinds of float will be revealed by time analysis, as follows:

- Total float is the amount by which a given task can be delayed if its preceding tasks take place at their earliest possible times and succeeding tasks can be delayed to their latest permissible times.
- Free float is the amount by which a given task can be delayed when all preceding tasks take place at their earliest possible times and the immediately following tasks can still be carried out at their earliest possible times. Free float is less common than total float.
- Independent float is found when a task or series of tasks lies in parallel with another network task or series of tasks that have longer estimated durations. It is defined as the float available when all preceding tasks take place at their latest permissible times yet succeeding tasks can still take place at their earliest possible times. Independent float is relatively rare.
- Remaining float occurs after one or more preceding tasks have been delayed, either accidentally through late working or deliberately as a result of resource allocation. Once a project has been started it is the remaining float that should interest the project manager most.
- Negative float means that we are in trouble and that the task or tasks under review are late.

Activity-on-node (precedence) network diagrams

The activity-on-node network for the garden workshop project is shown in Figure 16.3. As with the activity-on-arrow version this was again intended to flow continuously from left-to-right but has been fitted to the book page in two linked halves. Activity-on-node networks are more commonly known as precedence diagrams. These networks are far more readily understood by engineers and technicians than their arrow diagram equivalents because they resemble the process flowcharts used in many projects.

The precedence diagram for the garden workshop project really needs very little explanation. The key given on the diagram explains the significance of all the data entered in each activity box. As with arrow diagrams, adding the estimated durations of all activities in a forward pass (from left to right) identifies the earliest possible start and finish for each activity. Where there is a choice of paths leading into an activity it is the path with the longest total duration that determines the earliest possible start for that activity. Subtracting the times in a backward pass (from right to left) along each path reveals the latest permissible times. With the precedence system (unlike arrow

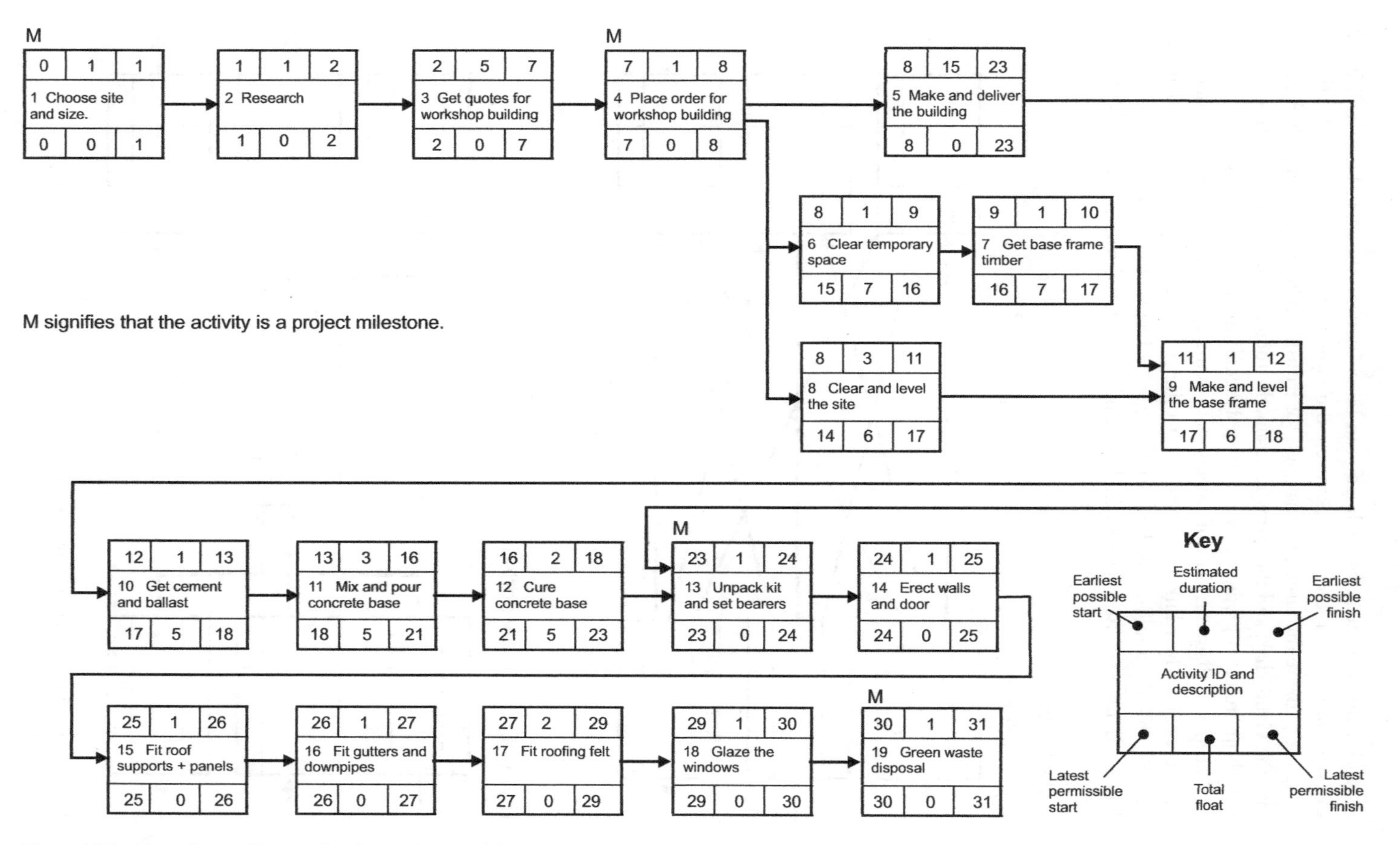

Figure 16.3 Precedence diagram for the garden workshop project.

networks) the total float for each task can be written on the plan. There will be at least one path through the network where the earliest and latest times for all activities along the path are the same, and that path (or paths) will be the critical path. Again, as with the Gantt chart and the arrow diagram, the total project duration for this garden workshop project is predicted by the precedence diagram to be 31 days.

Precedence diagrams do not usually have dummy activities. However, there are rare occasions when link lines cross each other in such complex patterns that the logic is difficult to follow. In those cases it can be helpful to insert a dummy precedence activity box, just to clarify the logic. An example is shown in Figure 16.4.

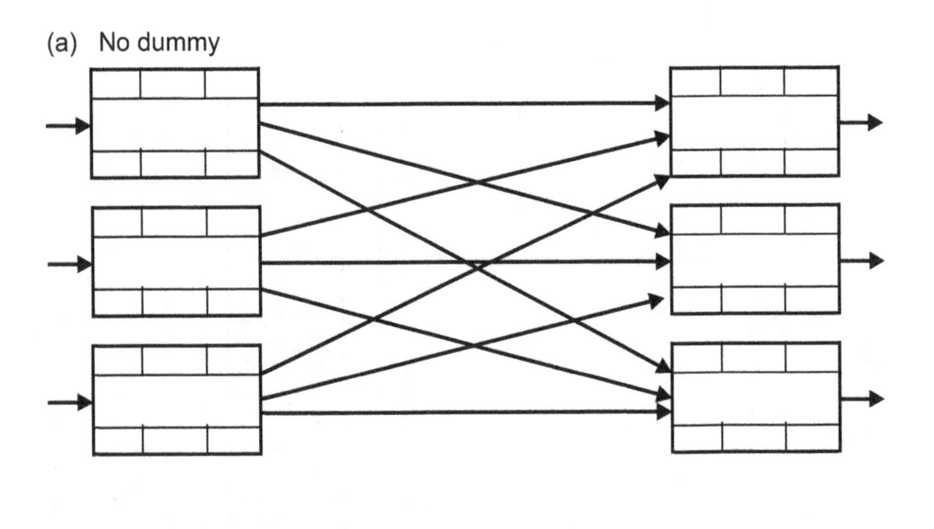

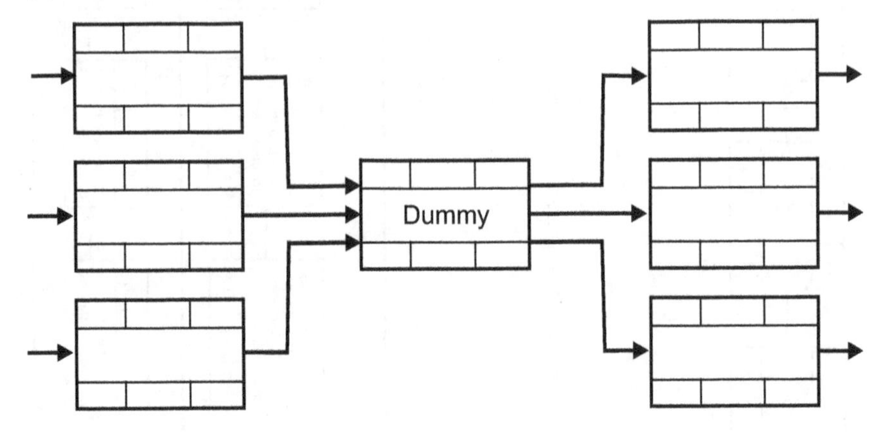

Figure 16.4 A rare occasion when the insertion of a dummy can clarify the logic of a precedence diagram.

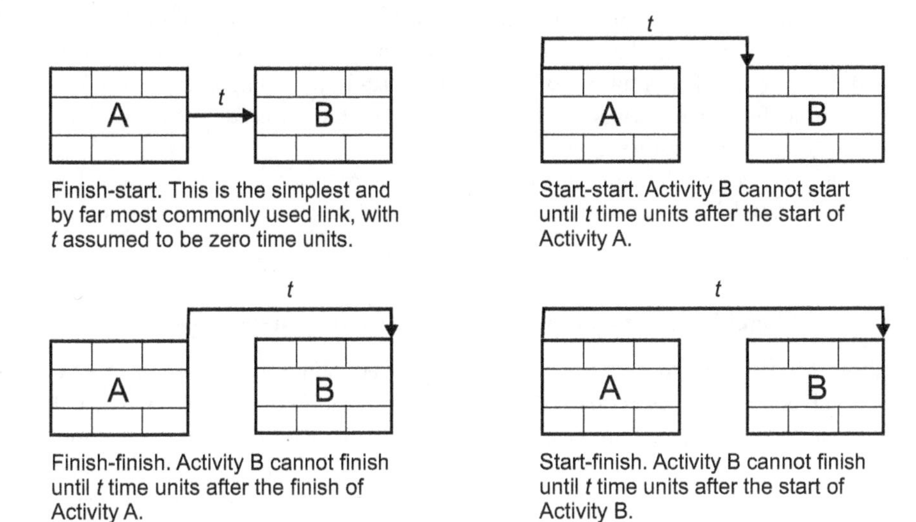

Figure 16.5 Precedence diagrams allow complex relationships between activities.

Precedence network diagrams can depict more complex relationships between tasks than the straightforward and simple finish-start constraints available in activity-on-arrow diagrams. Figure 16.5 shows the possibilities available to the precedence planner. However, these complex constraint patterns can complicate the process of time analysis: also some computer applications might not be able to accept them. Devaux (2015) is a writer with a particularly good grasp of complex precedence relationships.

More detailed descriptions of float and slack

Float and slack are synonymous terms, with slack being more usual in the US and float used more in the UK. Broadly speaking, float is the amount by which a project task could run beyond its earliest possible completion date without delaying the end of the project. But because of complex interdependencies between different tasks in the network that definition needs refining. All categories of float are expressed using the time units used in the network diagram. The definitions given below hold good for both the activity-on-arrow networks described above and for the activity-on-node (precedence) diagrams described later in this chapter.

- *Total float* is the amount of time by which a specified activity can be delayed without affecting the project completion time *provided that* all its immediately preceding task(s) have taken place at their earliest possible times and succeeding tasks can be delayed to their latest permissible completion times.
- *Free float* occurs less frequently than total float and is the amount of time by which an activity can be delayed *provided that* all its preceding tasks take place at their earliest possible times and following tasks can still take place at their earliest possible times.

- *Independent float* is rare and is found when a shorter activity or path of activities is in parallel with another activity or path of activities in the same network. It is defined as the float available when all immediately preceding tasks have taken place at their latest permissible times and immediately succeeding tasks can still take place at their earliest possible times.
- *Remaining float* is the amount of total float left to an activity after its start has been delayed (either by accident or deliberately through the process of resource scheduling).

Sketching an initial network diagram for a new project

When drawing the first network diagram for project control it is necessary to have key people from the project on hand to ensure that no significant task is forgotten and also that the logic depicted in the diagram represents the optimum working method. Those present at the planning meeting should be capable of making sensible estimates for activity durations. In fact, it is always preferable for duration estimates to be made by those who will later manage the tasks and be responsible for getting those tasks done within their estimated times.

Network diagrams, especially when first sketched, can spread over large areas. It will usually be found useful to make the first sketch on a paper roll. It is best if the first sketch is made by someone very familiar with the process (such as a project planning engineer from a PMO), so that the logic as constructed matches the intentions of those present at the meeting and that no errors are introduced.

Arrow diagrams are potentially far easier and faster to sketch freehand than precedence diagrams, and the job will be accelerated if the planner is pre-equipped with a straight edge, some sharpened pencils, an eraser and a roll or sheet of paper that everyone present can see. However, since very few computer applications can now process arrow networks, it is far more likely that the precedence system will be used.

A common and very sensible aid for drafting an initial precedence network during a planning meeting is to use a sticky note (such as a 3M Post-It Note™) for each task. Then the sticky notes can be added to the plan as it develops, and they will be easy to position or reposition as necessary to conform with the network logic determined by the consensus of the meeting. It will be even better and save much time if those sticky notes can be obtained in pads that have been preprinted with the activity box format. That concept is illustrated later in Figure 38.3.

The person leading an initial project planning meeting should know how to ask pertinent questions that can help to prevent logic errors. Such questions can take the form 'We have two tasks coming out of this node, are they really both dependent on the preceding task?'

Calendars and calendar dates

All the project times in this chapter have been given as day numbers. When the calendar start date for any project has been decided, all the day numbers must be converted to calendar dates. Then it has to be decided which days are project work days and which are non-working days. Some organizations have annual holiday shutdowns, and there are always public holidays to be taken into consideration. Some tasks that require no resources (such as paint drying or concrete curing) clearly require no effort and can progress through seven days in each work because paint will continue to dry and concrete will cure whether or not people are around. Other activities will depend on shift patterns and whether or not weekend working takes place. So it is possible (and fairly common) for more than one calendar to apply to a project schedule, with each task being allocated to its intended calendar.

Target dates

Once the start date of the project has been confirmed, that date can be assigned to the first network activity (which is always a milestone). That will enable all the subsequent network activity times to be converted from day numbers to calendar dates.

Sometimes calendar dates are added to milestone activities and events. These imposed dates might coincide with dates for progress meetings or (more commonly) with the dates when progress payments are expected from a project customer. The imposition of target dates on activities within a network clearly can conflict with the times predicted by time analysis. So imposed target dates have to be used with care.

Schedule errors

In network diagrams containing only a few activities it should be easy to avoid errors. However, networks for most projects contain hundreds of activities and then the risk of making mistakes becomes greater. The most risk of errors occurs when the network data are entered into a computer for processing. Then, instead of the expected working schedules the computer report will simply list the logical errors caused by entering incorrect activity identifiers. The following list describes the items that could be listed in an error report.

Dangles

Any network task that has not been specified as a network start or finish milestone will be reported as a dangle if it either has no path leading into it or no path leading from it. Dangles can occur when the planner fails to enter an activity record into the computer or makes an error with an ID code. It is important to designate every start and finish activity as such to avoid these being reported as dangles.

Duplicated records

If two or more activities are fed into the computer with the same ID codes they will be recognized as errors and reported as such.

Loops

If the planner is unfortunate enough to enter the ID codes of one or more activities incorrectly, say by confusing a pair of preceding and succeeding activity ID codes, it is possible to create a continuous cycle of activities or loop. A good computer programme will be able to report all the activities contained round the loop so that the mistake can be found and corrected.

A cautionary tale

Many years ago my department included a tiny PMO. A member of my small team was busy transferring the results of several project planning meetings on to a roll of drawing film which extended beyond the two-metre drawing board width. The activity-on-arrow plan contained over 1000 activities. The unfortunate planner was in the process of allocating ID numbers to all the event nodes from left-to-right when he was called away to answer a telephone. On returning to his task he resumed his event numbering wrongly, so that some 50 ID numbers were duplicated. The resulting computer report from a 132-column line printer comprised a great pile of wasted paper containing multiple warnings of dangles, duplicated tasks and loops. So, great care has to be taken when entering data, to avoid mistakes and the wasting of paper and valuable time.

A case example of planning to rescue a project in distress

This is another story from my distant past when I was a junior manager. Our company was expecting to exhibit a range of products at a medical equipment exhibition in London. We had three product streams, comprising heart-lung simulator pumps, electronic patient monitoring equipment and a range of bulk gas and steam sterilizers. The exhibition stand was booked for about four months ahead but the divisional manager was becoming concerned that all or some of the exhibits would not be ready in time. So he called me in to investigate and try to restore control.

We held a planning meeting attended by the relevant engineers and supervisors. I knew we needed to produce a network plan. But no one at that meeting could tell me how to start sketching the network. The current state of progress could not easily be defined. When drawing any network plan it is important to capture the imagination of the meeting members and motivate them so that they can visualize the project and create a sensible and logical network diagram. So it was necessary to begin planning in this case by asking everyone to visualize the exhibition opening day. My first question was 'What is the last thing you have to do before the visitors arrive?' Answer: Final clean and tidy up. I entered that task at the extreme right hand side of the blackboard.

A practical network diagram emerged through the process of working backwards from that final activity, making certain that everyone's imagination was captured. As if by magic our network branched into the three different product streams as it moved towards the left. The sketch finally ran out of new tasks at the point of actual progress to date. Now we had a plan, which allowed work schedules to be produced. The exhibition stand was ready on time.

References and further reading

There is a vast amount of literature available on project planning, some of which can be highly recommended and one (at least) which I had to review and found to be truly terrible. I have picked just three – two of my own and one from Stephen Devaux (whose work I highly recommend). *Naked Project Management* is a short very basic primer.

Devaux, S.A. (2015), *Total Project Control: A Practitioner's Guide to Managing Projects as Investments*, 2nd edn, New York: CRC Press.

Lock, D. (2013), *Naked Project Management: The Bare Facts*, Farnham: Gower.

Lock, D. (2014), *The Essentials of Project Management*, 4th edn, Farnham: Gower.

17 ACCELERATING THE PROJECT

Tony Marks

Here is a situation that is frequently encountered by project management professionals. Your organization is embarking upon a project for which a completion date has already been agreed. The project might be for an external customer or it could be an internal project within your company. Let's say that a project completion and handover date has been promised for six months ahead. Then, for a variety of possible reasons, when the project starts and you produce a critical path schedule you find that you will apparently not be able to meet that deadline. Or a project that is already in progress has begun to slip behind its schedule. This chapter explains steps that might be considered to accelerate a project and claw back some or all of the excess time.

Reviewing the duration estimates and project milestones

When we talk about estimating accuracy we have to remember that estimates can never be relied upon to be accurate. They are only our best predictions (sometimes even guesses) about what is to come. Project work always contains risk and uncertainty. Indeed, it is really hardly sensible to mention *accuracy* and *estimates* in the same breath.

Consider, for example, the first milestone in the project plan, which will invariably be the project start. That milestone must be given a scheduled (fixed) date, from which the scheduled times of all subsequent tasks can be calculated. However, it very often happens that a project does not start on its scheduled date, or at least is started only in a very feeble way. That can happen for many reasons, particularly when preceding projects have run late and resources needed for the new project cannot be released. Those resources can even include the project manager.

As a project progresses and experience is gained some or all of the task duration estimates can be refined as experience is gained. Trends might become apparent. Figure 17.1 illustrates the concept that estimates can be improved as the project moves through its lifecycle (the 'estimating funnel'). So it is important to keep not

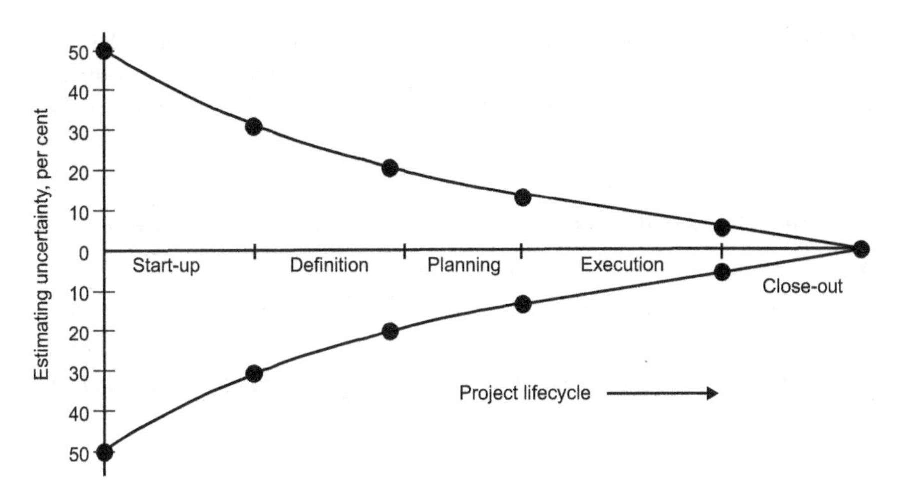

Figure 17.1 The estimating funnel.

only progress, but also estimates under regular review, particularly when there is a pressing need to accelerate the project. Of course, there is a danger that when the estimates are reviewed it might be necessary to increase the estimated durations of one or two tasks.

Accelerating project tasks

We have to assume the existence of a critical path network. So you know which tasks lie on the critical path (or paths) and you also know how much float all the other project tasks possess. Clearly, if you need to accelerate the project, your first attention must be directed to the critical tasks. Note that when a conscious decision is made to speed up a task there is very often a penalty to be paid in the form of increased cost or added risk. Here are a few examples of steps that might be considered to reduce the duration of a single activity or a sequence of activities:

- Allowing weekend or overtime working.
- Hiring equipment that will speed up construction work.
- Subcontracting work to relieve predicted overloads of in-house facilities.
- Putting more people to work on particular tasks, accepting reduced efficiency.
- Allowing a task to start before its immediately preceding task has been fully completed.
- Reducing the elapsed time taken to complete the work of an activity (or activities) by crashing the work (by working longer hours, or by deploying additional resources, or both).
- Reorganizing the work, using 'what-if' scenarios; or
- Overlapping activities that would usually be scheduled sequentially (challenging the logic).

Although what-if? scenarios, challenging the network logic or crashing the work tend to be dealt with separately, in most cases you will need to combine all three of these options in order to sort out a significant problem and achieve a realistic reduction in the overall time for the work.

Considerations to be addressed when attempting to accelerate late work

Taking steps to recover lost time is a process that always comes with risk and uncertainty. Some of the factors to be considered are listed in the following paragraphs.

Diminishing returns associated with increasing the labour on a single task

When considering the use of increased labour to accelerate a project task you have to bear in mind that the law of diminishing returns will apply. To take an extreme case, if it takes one person 100 days to complete a task it is extremely unlikely that 100 people could do the same task in one day. Some actions or subtasks within the main task will probably have to be performed in sequence and people will crowd each other's workspace.

A 'team' of one person has no internal communicating interface. A team of two has one interface over which information must pass. A team of nine people has 36 interfaces. When a task requires many people to work together, some allowance for this effect must be made. Simply applying more resources will rarely reduce the duration of a task proportionately. As a useful rule of thumb, the number of communication channels in a project environment can be calculated as:

$$n(n-1)/2$$

where n represents the number of people requiring communication on a task. For example, if there are four people on a team, the number of communication channels would be six. If we add two more people to this team the number of communication channels would increase to 15. This is clearly important to remember when estimating or reviewing task times in relation to team size because the extra time required in clarification and repeated instructions can have a significant effect on the work estimate and task duration.

Culture

Estimating was considered in depth in earlier chapters, but suppose we are considering an international project where some tasks in the same project network are conducted in different countries. It could be that a company within an international group decides to transfer some project tasks overseas to add resources and speed up the project. The working days and hours, approach to work, rate of working and work expectations of different international cultures vary considerably. Using norms from previous projects in a new culture can be dangerous. Advisers with knowledge of local working customs should be consulted.

Optimism

Human beings tend to be optimistic and people usually think that they will be able to achieve more than they actually can in a given time. People always tend to estimate on the low side and, if not allowed for, that can give rise to activities which run late. So, when looking at ways in which to speed up or recover progress always bear in mind that the promised results might not be achieved.

Variable skill levels

When deciding to use additional or alternative resources to accelerate project tasks, it might be necessary to use people whose skills vary from those originally envisaged. Or, those additional people might be unfamiliar with the project organization and its working practices. So allowances might have to be made, and the intended schedule improvements might not all be achieved.

Using contingency allowances from the initial project estimates and budget

Applying measures to speed project tasks can increase the costs of those tasks above their original estimates (and budgets) for a number of reasons. Funding for these additional costs might be possible by raiding the contingency sums provided in the original project cost estimates. But contingency sums can easily be frittered away and are really only budgeted to cover unexpected emergencies.

Crashing

Crashing is a resource-based methodology that involves throwing additional resources at a project to reduce the overall duration. It is necessary either to work the existing resources harder, or longer, or both. You might also need to find additional resources. This is a compound problem that needs careful analysis, bearing in mind the following:

- Excessive overtime working equates to tired resources and lower productivity.
- Brooks' law (which dates from 1975) tells us that throwing more people at a bad project makes it worse.
- If you accelerate one piece of work by crashing, other bits/resource teams need to be capable of taking advantage of the acceleration. That could well translate to the need for accelerating non-critical areas, so that their resources can be available for the crashed work. You therefore need to consider the *critical resource flows*, which is not dealt with well in the critical path method in conjunction with the inherent logic of the network.

Kelley and Walkers' mission in the late 1950s (which led to the development of the critical path method) was to identify a way of focusing the crashing so that only activities which influenced the overall work duration were crashed (it is obviously a waste of money shortening an activity that does not affect the overall completion).

However, Kelley and Walker clearly understood the complex cost and time relationship in attempting to accelerate work by adding resources. They showed that the costs of a task were lowest at its planned (optimum) duration and could increase not only if the task was accelerated, but also if it took longer than estimated.

If weekend working is to be allowed, at times when the organization would normally be closed for work, then labour costs will increase as a result of overtime payments or special bonuses. Additional hired plant and machinery will attract daily or weekly hire charges. Subcontracting work will probably increase individual task costs above their original estimates.

Beginning a task before its predecessor has been fully completed (such as beginning manufacture or construction before drawings have been fully checked) can sometimes be considered as a means for accelerating progress, but that will introduce obvious risks.

If two different tasks are being considered for expediting and one is critical while the other has 10 days' float, clearly it is the critical task that should be expedited. Accelerating the critical task should shorten the critical path and bring the project completion date forward. Suppose that by spending an additional £1000 on a critical task its duration (and thus the total project duration) could be shortened by five days. That gives a predicted additional cost of £200 per day saved. However, there might be a cheaper option. Suppose that another activity lying on the same critical path could also be crashed by five days but at an additional cost of only £500 (£100 per day saved).

There is another important consideration when crashing critical tasks. Crashing critical tasks can shorten a critical path so much that it is no longer critical, and one or more other paths through the network diagram have become critical. In theory (although not usually in practice) it should be possible to embark on a process of accelerating a project by re-examining all remaining tasks and crashing the critical tasks one by one until all paths through the network have become critical. Then you have a network that could be said to be hypercritical, with zero float everywhere in the system and any task delay would put back the project completion date.

Compensation factors for the costs of accelerating project work

Suppose that a project in danger of overrunning its promised completion date by four weeks can be restored to the original schedule at an estimated increase in direct costs of £10,000. That equates to spending £2,500 per week saved. Perhaps that additional £10,000 can be funded using some of the contingency sum from the original project estimates and budgets, as discussed above. But the notional cost of *not* accelerating the work must also be considered.

A substantial portion of any project organization's indirect costs will be incurred every day, whether or not any work is done at all. So to some extent the additional £10,000 in direct costs used to accelerate the project will be offset by avoiding the increased indirect costs that the organization would otherwise face, simply because the project happened to occupy its facilities and indirect staff for longer than intended.

Another, sometimes more serious, side effect from the late completion of a project is seen when an organization has a full order book and other projects are waiting in the pipeline. Then it is essential that all possible steps are taken to finish the late-running project to ensure facilities and people become available when needed to work on the new projects. This problem can cause serious difficulties when late running is allowed to become the norm and the organization's reputation for delivering its projects on time is damaged.

Conclusion

Effective schedule reduction needs very proficient schedule management and intelligent decision-making based on the best scenario generated by testing various options. It helps to look closely at the management and stakeholder issues that probably triggered the problem in the first place. The alternative is simply to throw money at the problem through fast-tracking and hope for the best.

Looking beyond the financial (fast-tracking) and analytical (schedule compression) dimensions, there are always management issues focused on coordination, communication and motivating the entire project team that must be overcome. Challenges around setting targets to motivate performance against predicting likely completion dates are discussed in Weaver (2011). Unfortunately fast-tracking and schedule compression are facts of life in most projects. The challenge facing schedulers is to keep the schedule realistic and achievable. A schedule is useful when it is used to manage the project but when that schedule loses credibility (that is, the team believes it to be unrealistic) it becomes a complete waste of time and resources.

References and further reading

Brooks, F.P. (1955), *The Mythical Man Month*, MA: Addison-Wesley.

Kelley, J.E. and Walker, M.R. (1959), 'Critical-path planning and scheduling', MA: IRE-ACM 1959 (Eastern) Papers.

Weaver, P. (2011), 'Why critical path scheduling is wildly optimistic and why so many projects finish late!)', 8th Annual PMICOS Scheduling Conference Papers.

18
SCHEDULING PROJECT RESOURCES 1

Dennis Lock and Tony Marks

This is the first of two chapters that deal with the important subject of matching an organization's resources to all of its work commitments.

Introduction

Project controls will clearly fail if the resources available cannot be scheduled to match the organization's requirements. Resource scheduling is the process of balancing the organization's total work commitments with the resources that are available, or which can be made available. In many companies those work commitments include work on multiple projects in addition to 'business-as-usual' operations.

By far the most common requirement for resource scheduling applies to people with various skills, and much of this chapter is concerned with that aspect. However, any project resources that can be specified in simple units of quantity can be scheduled using the methods described here. The essential precondition is a critical path network plan from which the priorities of the various tasks can be quantified in terms of their float, with critical jobs claiming priority for the allocation of the most scarce resources.

Some companies, particularly those concerned with some heavy engineering projects, have complex assembly space needs, where the height and shape of the end products must be taken into consideration (even to the extent of allowing different assemblies to overhang each other). We have to admit that the processes described here will not cope with that problem, but the project manager can use 3D modelling, either using computer applications or using simple scale models.

Liquidity and cash flows are a special (but again neglected) form of resource scheduling and that process comes with its own methods and is discussed later in this chapter. A company might have the best workforce and technical capability in its field, but if it fails to schedule its cash needs and runs into debt, suppliers will stop supplying, subcontractors will stop subcontracting, staff might not even get paid and … well some of us have been there in our careers and would not wish to go there again.

Scheduling people with specific skills for project tasks

The methods described here are illustrated in terms of an organization conducting a single project, but are very easily adaptable for multiproject or programme management.

The first consideration is to decide which resources need to be scheduled. The answer can often be surprisingly simple. Clearly indirect workers (such as managers, supervisors, office staff, project control staff and so on) will not usually have to be scheduled. So we can rule all those people out here. The same argument can also apply to some technical staff. People who check drawings are a vital resource, but if those producing the designs are scheduled to work at a constant, acceptable rate it could be argued that the checkers will receive their work in a smooth flow and will not need to be scheduled separately. A vital rule when scheduling project resources is to keep everything as simple as possible.

A common mistake in resource scheduling is to attempt to schedule everything in precise terms, with every tiny task taken into account. Resource scheduling is not an exact science and the best results are obtained by making realistic decisions and approximations. In a typical department of (say) 50 people, it might be that some of those are absent on their annual leave, some are taking time off for sickness or for dental appointments, and so on. Others might be doing rectification work for earlier projects. Some unscrupulous executives might ask staff to do work privately for them (in one company of my acquaintance a senior draftsman spent continuous months working on the interior design of the chairman's yacht).

Taking all the unpredictable demands on the available time of skilled staff into detailed consideration is an unnecessary complication and a recipe for scheduling failure. So, use common sense and apply averages. A good place from which to start is to assume that in any given department, 15 per cent of the people supposedly available for project work will in fact be otherwise engaged.

Now turn to Figure 18.1. This represents various work conditions for the staff in a department that performs essential project work in an organization. In this case it is a department of electrical fitters, but it could apply equally to any other department where special skills are grouped. In case A, the top chart, the expected departmental strength is shown before any task has been allocated. This shows that five fitters are employed, but that from the third week onwards the department strength will be increased to six by the recruitment of another fitter. If you use a computer (which is essential for anything except the tiniest project) the program should prompt you to enter the total availability of each resource in your organization and should allow you to specify changes in relation to different calendar periods. You will also be expected to allocate an ID code for each resource type and to specify a cost per unit per specified period (often per day).

So it might be thought that in the example given in Figure 18.1 five fitters could be assigned to project tasks for the first two weeks, increasing to six fitters from there

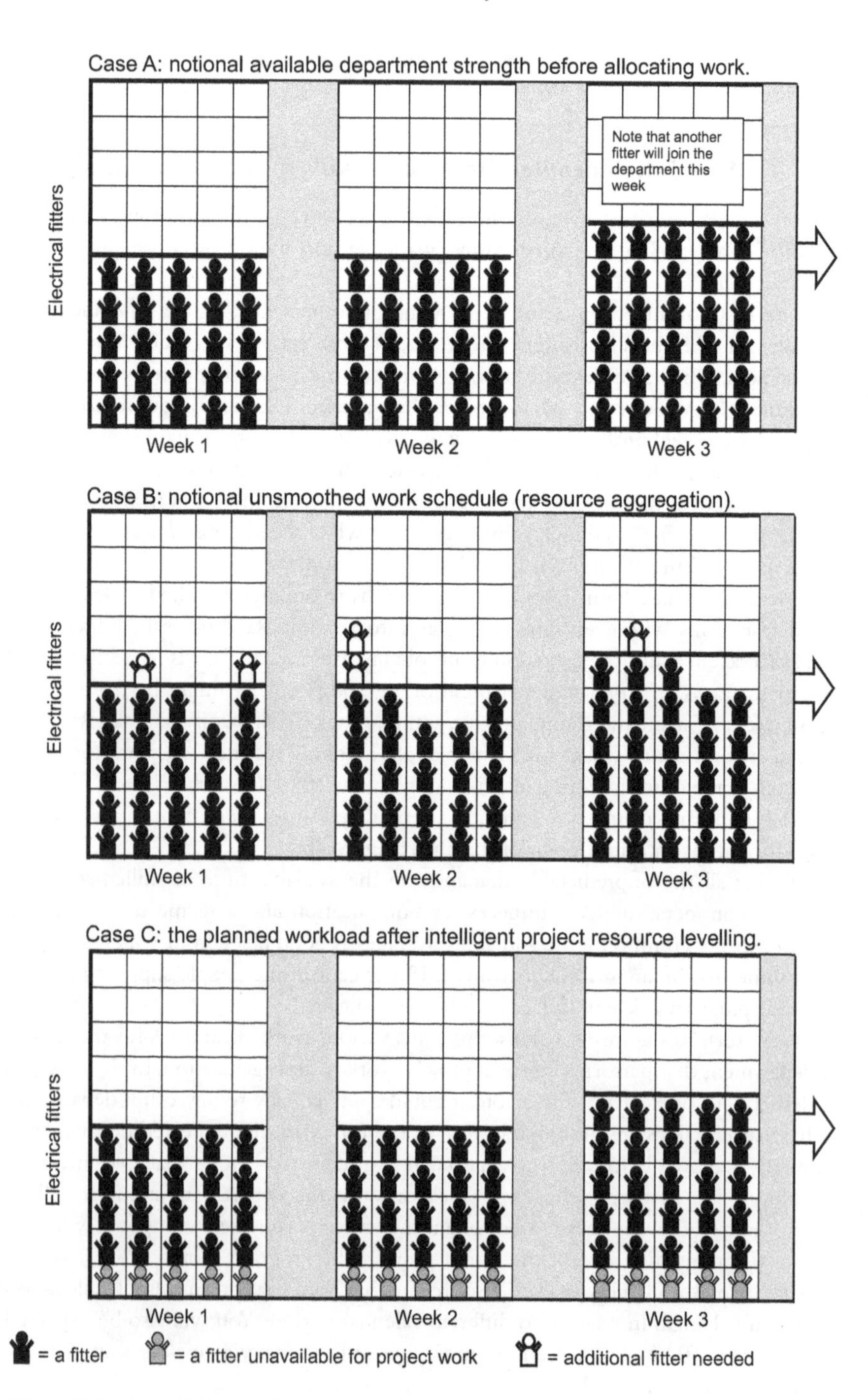

Figure 18.1 Essential principles of project human resource scheduling.

on. Case B shows what the result could be when you attempt to schedule project tasks to this department at their earliest possible starting times calculated from the critical path network. On some days resource overloads are predicted, while on others fitters could be idle. Clearly that is not an economic use of resources. Crude schedules of this type are known as *resource aggregations*.

In Case C in Figure 18.1 the planner (or the computer) has delayed the planned starts of some jobs to ease the department's workload and eliminate the impossible peaks. Here, again, the value of critical path analysis is revealed because only those tasks with sufficient float will be delayed and all critical tasks are still scheduled to take place at their earliest possible times. Planned project completion has not been delayed, yet a smooth resource usage pattern has been achieved. This common sense process is known as resource smoothing or *resource optimization*. A good computer package will be able to do that for you. If you are using a package that cannot, then look for a better package.

Note that in Case C in Figure 18.1 some of the fitter icons are shaded grey. This represents an allowance to compensate for those fitters who will not be available for project work (a sludge factor if you like). In our example this approximates to the allowance of 15 per cent suggested earlier.

Threshold resource levels

Some computer packages will allow you to specify two levels of resource availability. The computer will attempt to use the normal level unless concurrent critical tasks make this impossible without delaying the project (causing negative float). The second, higher resource level will only be used to prevent activities going into negative float. Threshold resources usually come at higher cost.

Sourcing people for threshold levels can be done in a number of ways. For example tasks might be subcontracted, or temporary staff could be engaged. If your company is part of a wider group, you might look to one of the other companies for help in performing the critical tasks. You might be tempted to allow the critical tasks to run into overtime or weekend working, but it is a good general and sensible rule *never to specify any project task for overtime working*. True, out of normal hours work will sometimes be needed, but save those times for real emergencies.

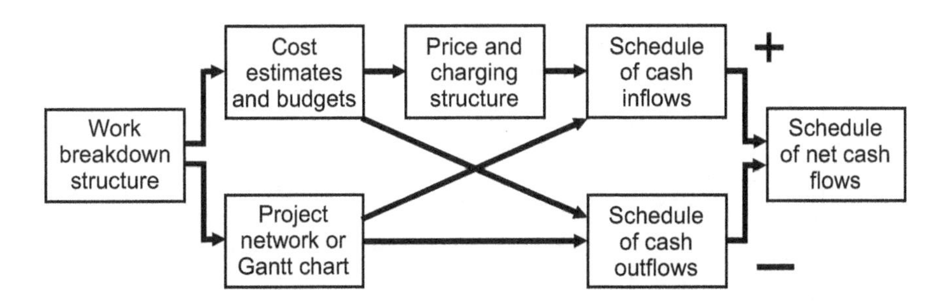

Figure 18.2 Essential elements of a project net cash flow schedule.

PROJECTS UNLIMITED LTD		Loxylene Chemical Plant for Lox Chemical Company														Project number P21900 Issue date March 2012		
Cost item	Cost code	Quarterly periods - all figures £1,000s																Total budget
		2013				2014				2015				2016				
ENGINEERING	A	1	2	3	4	1	2	3	4	1	2	3	4	1	2	3	4	
Design	A105	10	20	50	70	75	50	30	10									315
Support	A110			2	2	2	5	5	5	5	5	3	2	2	2	2	2	44
Commission	A200													2	6	10	10	28
Project management	A500	4	5	7	8	8	8	8	8	7	6	6	4	3	2	2	2	88
PURCHASES	B																	
Main plant	B110					400		500		500				2200			400	4000
Furnaces	B150						60				480						60	600
Ventilation	B175						20					160		10			10	200
Electrical	B200						5	20	25	25	140	5	2	2	1			225
Piping	B300					10	5	5	20	20	40	40	50	20	20			230
Steel	B400			20		80	200	200	200	200								900
Cranes	B500					50							400		50			500
Other	B999		5	25	5	10	20	20	50	5	5	10	5	10	5	2		177
CONSTRUCTION	C																	
Plant hire	C100				2	3	10	10	8	6	4	2	1	1				47
Roads	C200				10	20	40	80	60	5	5							220
External lighting	C250				5				40			5						50
Main building	C300																	
Labour	C310					10	30	100	150	200	400	100	50	40	20	20		1120
Materials	C320					2	10	30	100	200	220	200	100	100	10	5		977
Stores building	C400																	
Labour	C410						5	20	20	20	20	60	60	30	5			240
Materials	C420						2	5	15	5	5	15	30	15	10	5		107
CONTINGENCY SUM	Y999					10	20	25	25	30	30	35	35	45	45	50	50	400
ESCALATION	Z999					35	29	74	59	110	136	70	88	322	30	32	85	1070
TOTALS		14	30	104	102	715	519	1132	795	1338	1496	711	827	2802	206	128	619	11538

Figure 18.3 Cash outflow schedule for a construction project. This shows how the cash outflows are typically presented for a large project (which in this case happens to be for a chemical processing plant). This is the summary sheet for the whole project, showing only the uppermost levels of the WBS.

Scheduling cash flows

Cash flow scheduling might not be important for small inexpensive projects but it is usually an essential control procedure for large capital projects. The principle is simple but the practice can be somewhat complicated. The main project schedule must be available before any accuracy can be guaranteed. Essential data are as follows:

- The project schedule, after resource allocation.
- The expected amounts and dates of all project expenditure (cash outflows).
- The expected amounts and dates of all project revenues (cash inflows).

It is usual to allocate all the individual inflows and outflows to project periods (which might be months or quarters). From those figures both the periodic net cash flows can be forecast and then the ultimate cash flows. The cumulative amount in the final period will forecast the profit or loss for the project.

Cash outflows can be predicted by considering all the project tasks, including purchases, and placing their estimated costs in the appropriate periods. Cash inflows will depend on the contractual arrangements but a useful starting point is to look at all the project milestones, which are often related to incoming progress or stage payments.

Cash flow scheduling can really only be performed with any expectation of accuracy when a considerable amount of project activity has been predicted with some certainty. Cash flow scheduling is a good example of a project control where the project manager provides essential information to the company's finance department. Figures 18.2 and 18.3 (both reproduced from Lock, 2013) summarize and illustrate the project cash flow scheduling process.

References and further reading

Lock, D. (2013), *Project Management*, 10th edn, Farnham: Gower.

19
SCHEDULING PROJECT RESOURCES 2

Tony Marks

This is the second of two chapters that deal with the important subject of matching an organization's resources to all of its work commitments.

Introduction

Earlier chapters discussed network time analysis. Now it is necessary to look more closely at the relevance of that time analysis to the deployment of project resources. In the ideal (but rarely achievable) case all necessary resources will be available so that they can be allocated to activities in the amounts required using the earliest possible start and finish dates of all the relevant tasks. In that case the end date for the project calculated during network analysis should easily be achievable. The critical path (or paths) through the network plan will drive the end date of the project. Thus the project manager can focus on activities that would threaten the end date if they were not achieved on time.

When there are insufficient resources to keep every activity to its earliest possible start and finish dates, suitable computer software can calculate a simple resource aggregation using early and late dates. But that will usually produce a resource usage pattern that contains a mix of impractical peaks and undesirable troughs. Case B in Figure 18.1 of the previous chapter illustrated that case.

Using a far better approach, the computer calculation should enable activities which are not on the critical path to be moved to later times within their float (the different types of float were described earlier in Chapter 16). Some computer software packages will attempt to undertake this process for you. In this chapter I shall take a closer look at the role of computer software in applying algorithms to produce a resource-levelled schedule. However, this is unlikely to resolve all resource conflicts. Much of the available computer software will not always make the best decision for your project. It is therefore important to understand the process being applied and how it will impact on your plan.

Resourcing the plan

There are many project and departmental managers who simply attempt to begin all project tasks at the earliest dates predicted by the critical path network. They assume that all resources can be made available as and when required. But clearly in order to carry out resource planning or scheduling, the availability of each resource must first be determined.

Specifying the availability for each category of resource

The availability of each resource can either be a total (lump) availability or availability spread over a time period. Total lump availability is normally reserved for consumable resources such as concrete and other readily available bulk materials. Specifying availability over a time period is used for other special resources that are not easily or quickly capable of being augmented. So this chapter is principally concerned with scheduling these special resources, which usually means people with specified skills.

One of the first things you will need to determine (and the computer software will need to know) is the level of availability that exists for each resource skill. Resource availability levels in an organization can change with time through planned recruitment (or expected downsizing and redundancies). When declaring availability over one or more time periods the usual approach is to supply the data for each resource skill in the form:

13 engineers will available from 1 Jan 2021 to 1 Apr 2021.
15 engineers will be available from 2 Apr 21 to 31 Jul 21.

The declaration of resource availability must always be realistic. For example, do not assign full-time resource availability for project work for those of a particular skill when you know, realistically, that maximum achievable availability can only be 90 per cent of the total number of people employed (allowing for people to take holidays, work on rectification for previous tasks and so forth).

Another important consideration for resource availability concerns demands made by other projects in the same organization for common resources. Then two different approaches can be considered. Either you can set aside resources separately for each different project or, far better, allow the computer to carry out a full multiproject resource allocation. That might sound difficult but, if the software has the capability to handle a large number of activities, all the calculations will be driven by the criticality (amount of float) possessed by each activity. The criticality of each project as a whole can be determined by fixed dates imposed on its start and finish milestones. Such fixed dates are often also known as scheduled dates.

Assigning resources to individual activities

Resources must be assigned at the lowest level of detail of the schedule, which is to the individual activities. When initially assigning resources, the initial approach to take is to assume that sufficient resources will always be available. You know that those resources might not be available in practice, but deciding how to deal with that problem is a calculation for the computer (usually far too complex for the human brain to solve).

Consideration of variable resource usage rates during the progress of an activity

Suppose that an activity is scheduled to last for three weeks, and that it will need four assembly fitters. That's a total of 12 'assembly fitter weeks'. However, the actual deployment of those fitters might in practice not be at a constant rate. It could, for instance, begin with only two fitters, build to a peak, and then fall off again as the activity nears completion. This condition is illustrated in Figure 19.1. In a very small project that uses few resources such considerations might be relevant. However, most project tasks compete with many other tasks and even other projects for resources

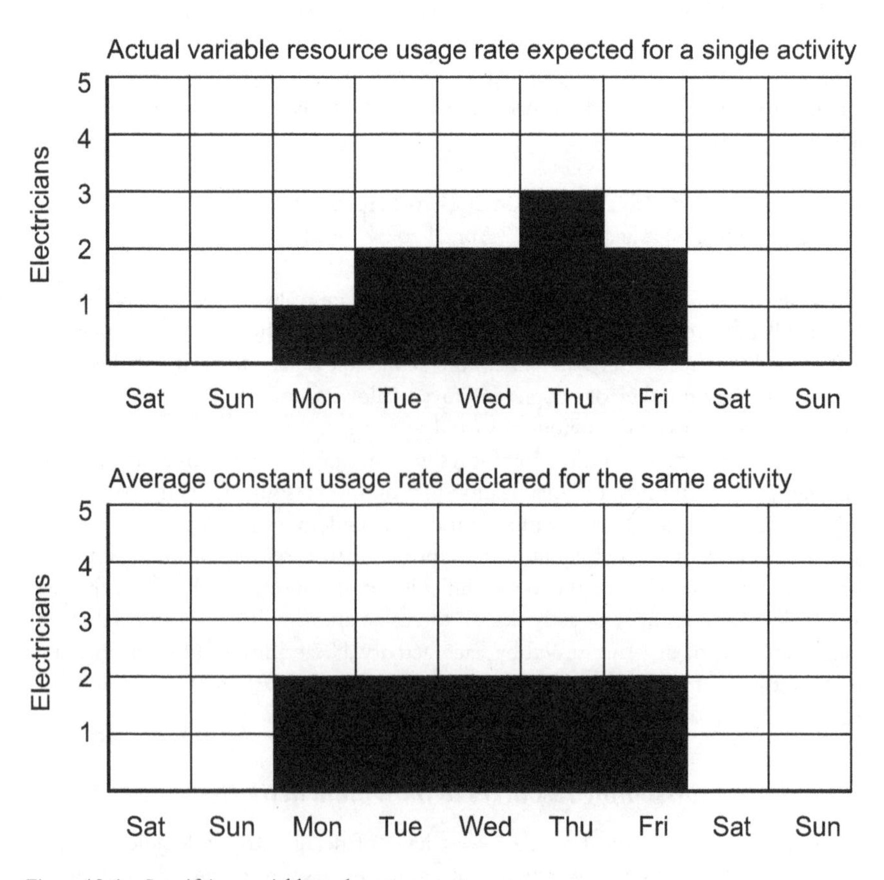

Figure 19.1 Specifying variable and rate constant resources.

from common pools. So these day-to-day variations within a single task can be ignored because, in the total scheme of things, everything gets averaged and smoothed out. *Always remember that resource scheduling is never an exact science.*

Resource planning and aggregation

The objective in planning resources is to optimize the use of those resources. Resource planning is a scheduling process that can be manual or (preferably) automated. If it is automated the resource requirements for the different activities can be added to the network, and the total resources expected to be available for the project can be added to the project data held by the computer. The computer will calculate where there are too many or too few resources available and reschedule the activities accordingly.

To predict the demand for resources against the resources that are available it is necessary to perform a resource aggregation (initially using computer software). This entails looking at the demand for each resource on a day-by-day or week-by-week basis. Once an aggregation has been performed by summing the resource demands, a resource histogram can be plotted. That will highlight any peaks or troughs in the estimated rate of use or, alternatively, show demand versus availability.

The simple chart in Figure 19.2 is a tiny fragment of a project schedule. It depicts the activities that are to be executed over a short period of a project. A daily resource aggregation has been carried out which shows the resource requirement for each day of the schedule. The resource histogram in Figure 19.2 highlights the need for three engineers from the 14th to 18th January where the expected total availability is only two.

It is obvious that it would not be possible to execute the defined tasks within the timescale using only the available resources. The following options can therefore be considered for this case:

1 Redefine the task scope so that fewer resources are needed. That would possibly allow the manning requirement to be reduced but might have a negative effect on quality, risk and safety.

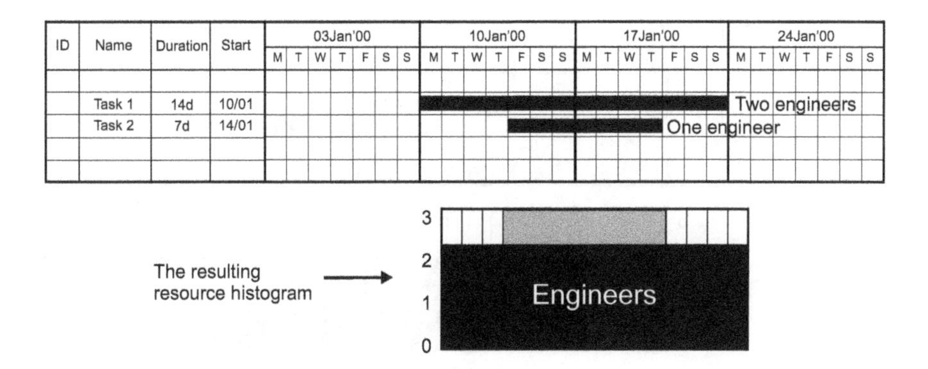

Figure 19.2 Resource aggregation for two project tasks.

2 Change the manning requirement for Task 1 from one person to two, cutting the total peak need for the project from three to two. But that would add time to the project because Task 1 would then take longer. Also it might be impossible because Task 1 might actually need two engineers working together to do the work.
3 Make provision to increase the resources available to the project over this period. That would solve the problem but could cost more money and might not be the best utilization of the organization's resources.
4 Use a different resource (instead of an engineer) for Task 1. That might prove to be a cheaper option but it could also bring negative quality consequences.
5 If the network logic allows it, reschedule the tasks earlier. For example, begin Task 2 before Task 1. No dependencies are shown in Figure 19.2, so this could be a feasible option.
6 Delay the planned task times and try to spread the manpower requirement. For example, Task 2 could be delayed until after Task 1 has been fully completed. However, if Task 2 happened to lie on the critical path this option would delay project completion.

Option 5 (above) would be the only course of action that most software planning packages would automatically consider possible. Should Option 5 be adopted, the effect would be as shown in Figure 19.3. This illustrates the inherent danger of allowing computer software to make all the decisions because one of the other options might be preferable. Different options would have different implications for cost, time, quality or even safety.

If there is a strategic project management plan, that should indicate which options would be preferable to the project stakeholders. For example, if cost cannot be compromised, Option 2 (above) might be preferable. In order to enable Task 2 to be completed using the available resources its start would need to be delayed to 24 January, when the necessary resources become released from work on the other task.

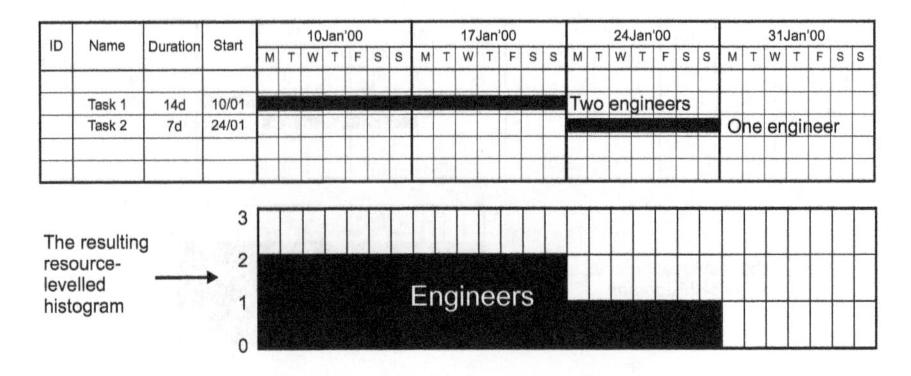

Figure 19.3 Resource-levelled schedule for two project tasks.

Levelling the resource schedule

Resource levelling is used to optimize the distribution of work among resources. Schedule compression is used to find ways of bringing activities that are behind into alignment with the plan. Both of these procedures will now be described.

Developing the schedule is an iterative process. Optimization continues for as many reiterations as necessary until an acceptable schedule is produced. A schedule is deemed acceptable when it satisfies the customer's delivery date requirements, contains a minimum number of staff fluctuation levels and has an achievable build-up and run-down of resources.

Once time analysis has been completed and the schedule has been optimized, a resource constrained (resource-limited or resource-levelled) schedule should be created. Earlier, when network time analysis was performed, only the durations and logic of the network were considered when all the activity start dates and end dates were calculated. Now possible resource limitations have to be taken into account. Conflicts will usually arise, the most common of which is that a combination of time analysis and resource allocation will predict a finish date that lies beyond the project contractual requirements and management objectives. Always bear in mind that projects do not usually stand alone in an organization, and the schedules of subsequent projects will probably be adversely affected if the current project runs late.

If (as often happens) the first scheduling attempt on a project indicates late completion, alternative options must be considered and evaluated. The following approaches can be considered at the single activity level to overcome resource problems:

- Revisit the network logic (for example by scheduling two or more activities in parallel or by relaxing one or more of the logical constraints).
- Review and revise the manpower estimates.
- Provide more resources.
- Re-examine the project scope.
- Suggest actions that would shorten the durations of activities on the critical path.

Sometimes the planned project end date simply cannot be allowed to slip. For example, when planning the various preparation projects leading the opening of the Olympic Games it is clearly essential that everything must be ready on time. When considering resources under such constraints the method used is termed 'resource compression'. This approach will make use of float within the project activities, increasing or decreasing the resources assigned to specific activities accordingly. In this way peaks and troughs in resource usage can be smoothed out. However, if after employing these techniques you are still in danger of missing the end date of the project, your final recourse must be either to:

- employ more resources to protect the project's end date or;
- negotiate a new end date by presenting your evidence to the relevant stakeholders.

Note that in all cases discussed in this chapter the resources under consideration used have been people of various skills. However, machines and materials can usually be treated in the same way.

A simulation example

The bar chart in the upper part of Figure 19.4 shows the plan for a small office relocation project. Numbers written above the bars indicates the resources required for each activity. The histogram in the lower half of Figure 19.4 shows a daily summation of the project resources needed to support the plan (which gives a clearer understanding of the daily resource requirements). The bold horizontal line in the histogram in the lower half of the illustration shows the level of resources normally available. So, in this example the chart indicates that the work is not achievable as it is currently scheduled. So in practice it would be necessary to consider the activities contributing to the overload and decide on the most appropriate action to remove the overload. That could mean delaying some of the work or finding additional resources as a temporary measure.

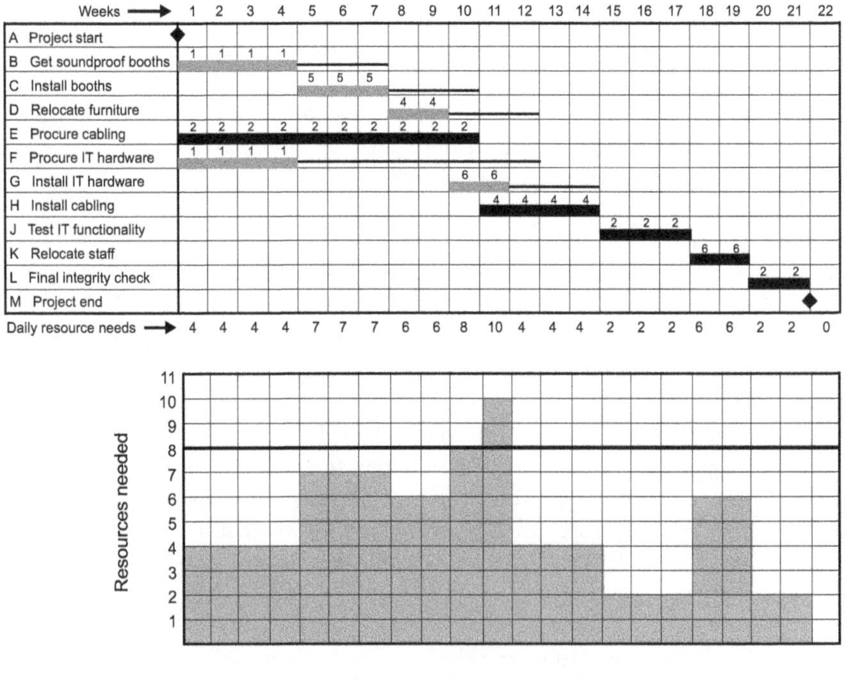

Figure 19.4 Gantt chart and resource histogram for an office relocation project.

The adequacy of any schedule should always be reviewed to ensure that it satisfies the following criteria:

- Clear presentation of the critical path (including all activities which are critical) and areas of risk.
- Identification of any float available in the schedule, along with identified key decision points.
- Key interrelationships between activities and key dates and milestones within the schedule.

Having these criteria in place will enable you to determine the cash forecast and allow you to communicate to the project sponsor the work that must be accomplished. This will in turn provide a sound base for monitoring progress. Importantly you will carry out resource levelling to ensure optimum deployment of staff and equipment.

Predicting expenditure and cash outflows using the project schedule

Treating direct project costs and cash as a resource

Cash is a vital and special form of project resource. Without sufficient funds there can be no project work. So the costs of activities can be taken into account in the resource scheduling process. Here it is not usual to use cash as a limiting resource when attempting to schedule work. The project manager has to assume that sufficient funds will always be available. If that assumption cannot be made the project manager should seriously be considering looking for a job in a more secure organization.

Most project organizations and their financial officers need to be kept informed of the cumulative costs that are committed as a project proceeds. That information is usually compiled by the project manager or PMO in the form of an 'S' curve (so-called because of its shape, beginning with a low rate of rise with time, then steepening and ultimately flattening to become asymptotic with the final total direct project cost). If cost information can be incorporated in the project resource scheduling process, then committed project cost reports can be produced automatically.

In the relocation project just described, suppose that the cost of each resource unit (person) is £1000 per week. Using this cost information and the resource schedule already calculated (shown in Figure 19.4) will allow calculation of the cumulative project costs for each week. So we have been able to tabulate the cumulative costs for this project, and the results are presented both as a table and as its associated S-curve in Figure 19.5.

The schedule baseline

From the work done so far, it is both possible and desirable to create a schedule baseline. This should be regarded as a key component of the overall project management plan. It provides a benchmark from which any variation in performance can be

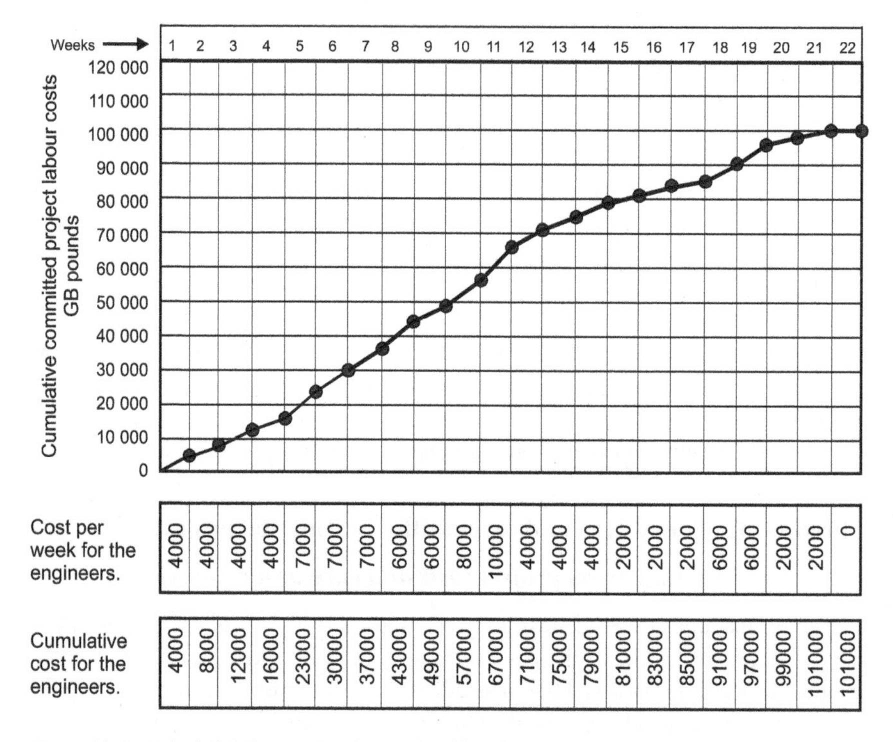

Figure 19.5 Schedule of committed costs for the office relocation project. When the resource schedule and the costs per resource unit are known, an S-curve of committed costs can be drawn. Here the unit daily resource cost in the curve is £1000.

recognized as the project proceeds through its lifecycle. Clearly any deviation from the schedule baseline will have an impact on project performance (in terms of cost, time or quality). The amount of data recorded in the schedule baseline will depend on the particular application but the following will typically be included:

- schedule milestones;
- activity attributes;
- identified assumptions and constraints.

The schedule baseline should be supported by a resource histogram or tabulation, alternative schedules and the scheduling of contingency reserves.

Software options

There are many software solutions available that can be used to support the scheduling of project resources, from inexpensive ubiquitous software tools (such as Microsoft Project) through to more sophisticated Oracle-based solutions (such as Primavera P6

Enterprise Project Portfolio Management). If an organization is regularly involved in projects this choice may already have been made to ensure consistency across an organization's projects and to allow for multiple project or programme management.

With new releases and new products often entering the market, any software recommendations given here would quickly become out of date. Different project management software packages have their own capabilities and weaknesses. These include the following:

- Their ability to schedule multiple projects using a common resource pool.
- The range and presentation quality of graphical reports (such as Gantt charts, S-Curves and plots of the network diagrams).
- The interfaces they support for transferring data between other software.
- The total number of activities they can support.
- Whether or not they are capable of being implemented for a geographically diverse team, or across an enterprise.
- Their ability to support collaboration tools.
- Whether they need to be cloud-based or can be hosted on secure servers in certain cases (an important consideration for commercially-sensitive or defence industry projects).
- Whether or not they support your chosen technology platform(s).
- Whether or not open source tools suit your requirements.

For the simplest projects some project management tools provide free versions, but their limitations (such as the maximum permissible number of activities or projects) are unlikely to satisfy most requirements. There are also an increasingly large number of cheaper 'low-end' software tools, available for desktops or for use in the Cloud, but these have limitations.

Most organizations will be able to provide access to Microsoft Project through their corporate licensing arrangements so this option will often be considered. Microsoft Project does provide detailed options for managing resources and has customizable reports. It can also be scaled at the business enterprise level and has good compatibility with other Microsoft products that are common in the workplace. However, it has limitations, including limited integration with non-Microsoft products and the software does take some training and experience to master fully. There is also an inherent danger, which is that its familiar 'Microsoft look and feel' lulls many into 'jumping in' with inadequate training.

For enterprise-wide applications it is worth reviewing your requirements against higher-end software including (for example) Deltek 'Open Plan' and Primavera P6. In any event, the choice of software is important and the purchase of a 'high-end' package has to be considered carefully. In some cases software that is already used elsewhere in a large group of companies might be favoured by higher management, but they are not likely to have the detailed training and experience at practitioner level that is vital in make the correct choice. So I recommend that the best place for making a final choice is the organization's most senior PMO.

Conclusion

I have described in this chapter how to put together a baseline project plan, resolve scheduling conflicts and create a plan and budget that will be realistic to follow. Creating the baseline plan is critical to everything that follows in the project lifecycle. The planning stage is often too rushed or ill-considered. A good plan can set the project team up for success. A bad plan will assuredly set them up for failure.

The knowledge required to understand project planning and scheduling in detail has become a lost art in some industry sectors since the widespread availability of desktop planning software such as Microsoft Project, and more sophisticated corporate tools such as Primavera for large scale projects. However, these tools cannot think. They cannot apply experience. They rely only on simple mathematical logic. They are essential for planning and managing the data and complexity of large projects but they cannot replace the human ability to make considered choices and decisions. There is a dangerous temptation to let the software tools do all the work. Project managers need to challenge project schedules in all aspects (estimated time, the logic and resourcing), while keeping an ever-watchful eye on risk!

References and further reading

Marks, T. (2012), *20:20 Project Management: How to Deliver on Time, on Budget and on Spec.*, London: Kogan Page.

20

PROJECT SCHEDULE TECHNICAL INTEGRITY

Shane Forth

Project scheduling can involve several people, computers and specialized applications, a considerable amount of decision-making and a great deal of logical thought. From the moment a work schedule is first drafted to the time when the project ends any plan is a good breeding ground for multiple mistakes of different kinds. Maintaining project schedule technical integrity is the process of eliminating mistakes and unwise scheduling choices so that the schedule remains valid and practical throughout the project lifecycle.

Introduction

The expression 'technical integrity' first came to my notice when I was introduced to a technical integrity manager. He worked in a company that was involved in the design and construction of facilities for large international projects. This manager's purpose was to ensure that the engineering design of every project would satisfy all the technical and regulatory requirements and be fit for purpose throughout its operational life. I realized that technical integrity could also be used in the context of network time analysis and project schedules. The *English-language Wiktionary English Dictionary* defines 'technical' as '*Requiring advanced techniques for successful completion*' and 'integrity' as '*The state of being wholesome; unimpaired*'. That made sense to me because network time analysis is an advanced process that must be completed properly for the project schedule to be as flawless as possible to fulfil its purpose in describing the logical sequence of work and acting as a predictive tool for the project timescale and completion date. This is especially the case from my experience of some of the practical issues that arise in the development of network diagrams and project schedules.

The human factor

Brains are incredibly wonderful things. The brains of even small creatures astound us with their capabilities. Birds migrate over incredible distances, returning to former nest sites. Many marvels that we put down to 'instinct' are really underpinned by brain activity. Brainpower reaches its peak in the primates, among which we human beings are thought to be supreme. But even we have our limitations. Chapter 16 of this handbook described project planning using critical path diagrams using a very small garden workshop project as a case example. We, as human beings, can and do make mistakes even when planning and executing such small projects. We are, it is said, 'only human'.

In real life our project plans might contain hundreds or even thousands of tasks, when the risk of making mistakes becomes far greater than that in a small garden workshop project. The data will usually include a great amount of information about task descriptions, estimated times, resources, logical inter-task links and so on. Plenty of opportunities for the introduction of human errors there!

As an example, Figure 20.1 shows what can happen when a planning engineer makes the simple but common error of forgetting to include one or more network links in the initial input data to the computer. Consider first the start and finish milestone activities (Activities 1 and 99 respectively) in this tiny network. They have been declared as start and finish activities, so the computer will (rightly) not regard them as errors. Now look at Activity 30 near the top of this network. There should have been a link from the end of Activity 1 to the start of Activity 30 but the planner has forgotten to key that information into the computer. So Activity 30 is perceived by the computer to be an 'open end' (or a 'start dangle', using the terminology once common among planners). Now look at Activities 4 and 5 in this miniature network. There should have been a link from Activity 4 to Activity 5 (which could either have been declared by 'telling' the computer that Task 4 is succeeded by Task 5 or

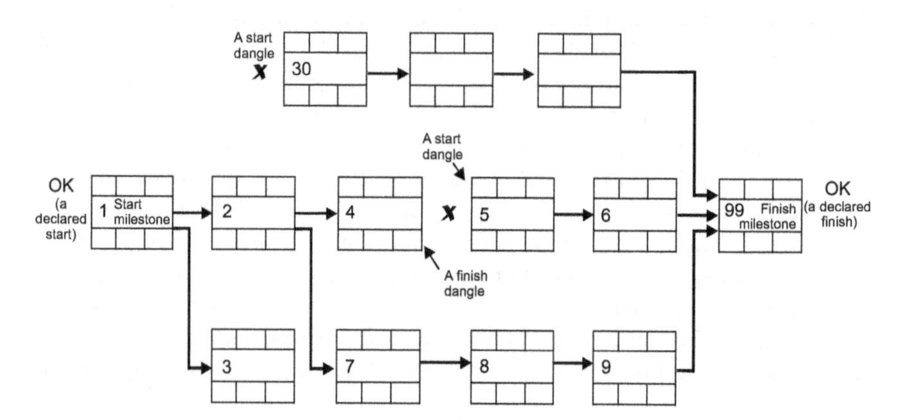

Figure 20.1 Open end errors in a network diagram. This small example of a project network diagram contains three instances of open ends that need correction (two start dangles and one end dangle).

that Task 5 is preceded by Task 4). This human error by the planner means that the computer will perceive and report Activity 4 as an end activity and Activity 5 as a start activity. So the 'sin of omission' here, of just a single link, has created two open end errors.

We are fortunate in having software that can detect and report such mistakes, but some mistakes can go unnoticed. For example, the duration of an activity might be input as one day when it should have been 10 days. One engineer might be specified as the resource needed for an activity when it should have been four engineers, or one mechanical fitter. There is no way that computer software can detect and report errors of that kind. So, in the final event, project schedule technical integrity is at the mercy of human planners and others working in the project.

So, if the human factor is so vital, it must pay to train and treat people in the best possible way to promote their accuracy and efficiency. That means avoiding fatigue by allowing rest breaks, providing a working environment with good, bright, non-glare lighting, adequate ventilation and temperature control, uncramped work conditions and freedom from distraction. Often people are expected to work in open plan offices, with many distractions, interruptions and limited workspace. If possible, ensure that mobile telephones are turned off, and that fixed line telephones are programmed to allow only outgoing calls. Every distraction is a risk that can cause a planning error and jeopardize project schedule technical integrity.

Many difficulties can be encountered in the development and maintenance of project schedules, especially when (in addition to the activity network logic and critical path) resources and progress updates have to be considered. In this chapter I shall describe five common problems and possible sources of mistakes:

1 open ends (referred to as dangles in Chapter 16);
2 imposed dates;
3 negative lags;
4 improper software configuration;
5 too many activities.

Open ends

As shown in Figure 20.1, open ends occur when precedence relationships (links) are missing from the project network. That usually results in activities (other than the first and last) having either no predecessor or no successor relationship. At least one continuous chain of activities (through which the forward and backward passes of time analysis must be conducted) is broken.

Open ends in the original network diagram happen when the practitioner is inexperienced. They are often seen, for example, in work submitted by students. However, even the most experienced planner can make mistakes. A common source of open ends is input errors, when data are being typed into the computer keyboard. It is good practice when entering data for two people to take part, one reading from the network diagram and the other keying the data into the computer and calling out each entry at the same time.

Imposed dates

Project scheduling software will allow start or finish dates to be imposed on activities. The use of these imposed dates invariably overrides the natural results of a forward and backward pass and will usually compromise the calculation of early and late start and finish dates, float and the critical path.

There are several different types of imposed dates, each of which has its own properties. These imposed dates are commonly available for use in most project scheduling software, although the terminology can vary depending on which application is chosen. Figure 20.2 lists commonly available types of imposed dates and their effects, which I have classified as hard, moderate or soft, depending on the severity with which their use might override the project network calculations.

There are a number of risks associated with the use of imposed dates. Lester (1982) urges that they should only be applied to activities when absolutely necessary, Lockyer and Gordon (1996) advise that their use can result in multiple critical paths. Devaux (1999) cautions that they can cause negative float.

Negative lags

The complex relationships between activities used in precedence diagrams described in Chapter 16 and illustrated in Figure 16.5 can include lead times (for start-start and finish-finish relationships) and lag times (for finish-finish and finish-start relationships). Ordinarily, these lead and lag times will be given positive values. However, project scheduling software can allow the assignment of both positive and negative time units

Severity	*Imposed date type*	*Effect on network calculations*
Hard	Mandatory (fixed) start date.	Forces the earliest possible and latest permissible start dates for an activity to the fixed start date. Completely ignores the network logic and does not allow for float calculations.
Moderate	Mandatory (fixed) finish date.	Forces the earliest possible and latest permissible finish dates for an activity to the fixed finish date. Completely ignores the network logic and does not allow for float calculations.
Soft	Must start-on date.	Imposes the same earliest and latest start dates.
	Must finish-on date.	Imposes the same earliest and latest finish dates.
	Start no later than.	Late start pulled forward.
	Finish no later than.	Late finish pulled earlier.
	Start no earlier than.	Early start pulled later.
	Finish no earlier than.	Early finish pulled later.
	As late as possible.	Starts as late as possible without delaying successors and sets early dates to late dates.

Figure 20.2 Typical imposed dates that can be applied to network activities.

in creating the relationships between activities. Thus planning engineers sometimes specify negative lags, as shown in the upper section of Figure 20.3.

Fondahl (1964) contends that specifying a negative lag value means that a following activity might be shown as starting before its preceding activity. That violates a core principle of network scheduling, namely that the lag values on the sequence lines of a precedence diagram must be either positive or zero. I have often seen polarized views between planning engineers regarding the use of negative lags. My experience is confirmed in the first sentence of *The Great Negative Lag Debate* (Douglass III et al., 2006), a paper that considers the arguments for and against negative lags and acknowledges that 'Almost everyone in the planning and scheduling profession has an opinion regarding the use of negative lags in schedule preparation'. The paper makes no clear recommendation and the arguments between planning engineers continue.

I have found that when performing schedule risk analysis (discussed in Chapter 25), the software tends to produce erroneous results when negative lags are present. My recommendation is to avoid them and use start-start and finish-finish relationships as shown in the lower half of Figure 20.3.

Improper software configuration

In recent years, high-end project scheduling software has developed into enterprise-wide solutions with the ability to host all the projects across the business on a single

(1) Negative lag example: backfill needs to start 80 days before excavation finishes.

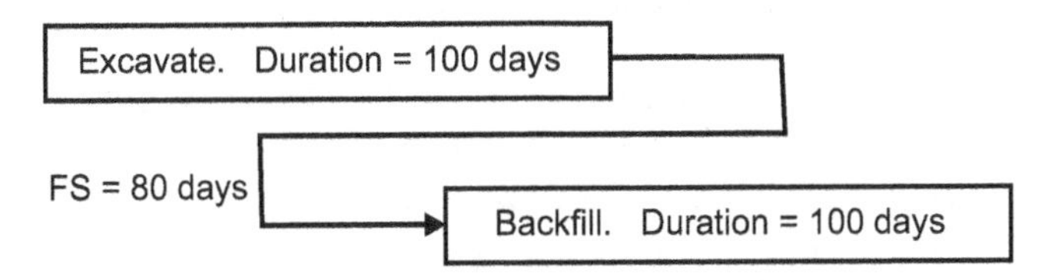

(2) Use of start-start and finish-finish relationships instead of negative lags.

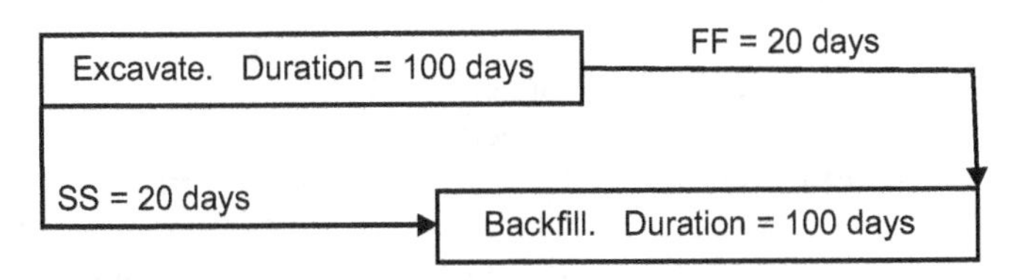

Notes: Network diagrams are not drawn to scale.
FS signifies finish-to-start relationship between the activities.
SS signifies start-to-start relationship between the activities.

Figure 20.3 Two examples of complex relationships.

centralized database. Such enterprise solutions typically include hundreds of database elements such as database structures, dictionaries, resources data, roles, calculation settings, coding structures, user-defined fields, calendars, administrative preferences, templates, methodologies and so on. Competent organizations standardize the settings and rules governing the creation and use many of these data sources (preferably policed from a central PMO).

The traditional approach of an engineer planning one project using a personal computer and software as a single user does not apply when the system is shared across the enterprise. That wider approach means that an organization's planning engineers share access to the system, with some settings and rules already specified and set by someone else (such as a system administrator).

A number of these elements will affect the forward and backward pass calculations in different ways, which can lead to errors or inaccuracies in calculating early and late start and finish dates, critical path and float. A few typical examples of such errors are described below.

Activity duration types

Activity duration types take effect once resources have been added to the activities. They can take one of the following forms and are something of a minefield:

- fixed units/time;
- fixed duration and units;
- fixed units.
- fixed duration and units/time;

Choosing the most appropriate option can be a challenge. Depending on which choice is made the associated software calculations can change activity durations and resource assignments without this being initiated directly by the planning engineer. That effectively takes the control away from the engineer. That can be a very serious problem, especially when a project network has thousands of activities. Suppose that an activity and its details are entered as follows:

- duration = 10 days;
- calendar = 8 hours per day, 7 working days per week;
- activity duration type = fixed duration and units/time;
- a single resource ID is added to the activity;
- the resource ID is assigned 80 resource units (man-hours).

This activity would have the equivalent of one person working full time (10 days × 8-hour day = 80 man-hours). However, owing to the activity duration type being 'Fixed duration and units/time', if the planning engineer were to change the activity duration, the software would automatically change the resource units. Here are two examples:

1 Suppose the activity duration originally entered were to be reduced from 10 days to 5 days. That would automatically reduce the originally entered resource units from 80 to 40 man-hours (5 x 8-hour days = 40 hours = 1 person for 1 week).
2 Doubling the duration originally entered from 10 days to 20 days would have the opposite effect. The originally entered resource units would increase proportionately and be doubled from 80 man-hours to 160 man-hours (20 x 8-hour days = 160 hours = 1 person for 4 weeks).

In most organizations, a relevant functional engineer or manager estimates the initial activity durations and the resource units. Although the planning engineer may adjust activity durations to finalize the project schedule, the man-hours (work content) should only be changed by agreement (or at the very least, consultation) with the responsible function lead. 'Fixed duration and units' is the only activity duration type that satisfies this requirement and is therefore the recommended setting in most circumstances.

Choosing the scheduling option 'Make open-ended activities critical'

If the 'Make open-ended activities critical' option is selected; the late finish date will be the same as the early finish date for activities without successors (they will have zero float). Furthermore, if the network logic is traced back from the open-ended activity, the float calculations for some other earlier activities in preceding logic paths might be affected. Now suppose that the 'Make open-ended activities critical' option is not selected:

- Open-ended activities with no predecessor will default to an early start at the beginning of the project network. Then they will invariably be out of their logical sequence and have excessive total float.
- Open-ended activities with no successor will default to their total float calculated from the early finish of the last activity in the network. That will invariably result in excessive total float.

Choosing the scheduling option 'Use retained logic or progress override'

When updating progress at the end of a period, suppose that one or more activities are started before all of their logical predecessors have been completed (known as out of sequence progress). Then the planner has a choice between two scheduling options (which will affect project network calculations):

- Retained logic - the remaining duration of the activities will not begin until all their logical predecessors are complete.
- Progress override – having started, the remaining duration of the activity will continue from the status date, ignoring the logic of its incomplete predecessors.

In a similar manner to the negative lags discussed earlier, the retained logic versus progress override debate can result in intense discussion between planning engineers. A word of caution here is that with the retained logic scheduling option, having a large number of out of sequence progress activities tends to result in a longer schedule.

Precedence networks are not an exact science and, given their imprecise nature, human factors come into play. Planning engineers determine the times for leads and lags in the precedence relationships between activities. These estimated times can differ according to the individual planner, who might also have specified some preferential (soft) logic in addition to the hard logic that must be adhered to. For this reason 'progress override' is often a more realistic setting to reflect what can actually happen. In any case, provided that when the schedule is updated to reflect progress the network logic for activities in progress and remaining is reviewed (and modified if necessary) then the argument between retained logic and progress override becomes academic.

Too many activities

Examples of networks that were too large to be handled and managed without difficulty date from the earliest cases, such as those developed at duPont for a $10m chemical plant. Kelley and Walker (1959) describe how engineers produced the project network diagram in much more detail than had been expected. They saw this trend continue on a new plant construction project, where the project network diagram had to be reduced from 1,800 events and over 2,200 activities to 920 events and 1,600 activities because of limitations in the computer processing power that was then available.

Faulkner and Hillstead (1968) describe how Atomic Power Construction Ltd. implemented network analysis techniques during the construction of the Dungeness B Nuclear Power Station with project networks that were considered unwieldy and cumbersome to maintain, while at the same time lacking the necessary detail on critical and complex activities. Clearly project managers and planning engineers must be realistic and include only the amount of activity detail needed to estimate durations and allow the gathering of progress.

Burke (1992) highlights the common client requirement for contractors to provide a detailed project schedule within three to six weeks of contract award. However, a number of writers have pointed out the pitfalls of attempting detailed planning before the necessary information is available. Andersen (1996) does not believe it is possible to identify all the activities at an early stage and recommends focusing on milestones using a goal-orientated approach. Senior (2003), seeing an imperfect world in which anything and everything can go wrong at any moment, postulated that detailed planning is pointless and that planning and control systems are set to fail because they cannot normally identify these occurrences in advance. He adds that the detail in the project network diagram should be in accordance with the currently available information.

Review of good practice

A wealth of advice and practical guidance is available from professional organizations (including those listed at the end of Chapter 1). Project schedule technical integrity advice exists on issues such as open ends, imposed dates, negative lags and too many activities.

Open ends, imposed dates or negative lags in the network logic

Recent guidance from the professional organizations (summarized in Figure 20.4) either prohibits or at the very least discourages the use of open ends, which if used at all should have the reasons documented formally. These professional associations also discourage or prohibit the use of imposed dates (which they refer to as constraints) but accept that they cannot be avoided altogether. If used at all they should be sparingly, again with the reasons formally documented and justified (see Figure 20.5). The same associations discourage or prohibit the use of negative lags for several very important reasons (Figure 20.6). Wherever possible, the logic in project network diagrams should be driven by simple finish-start relationships between activities and, where this is not done, the reasons should be explained in the schedule documentation.

Networks containing too many activities

How big is too big? Cases of networks containing too many activities have already been mentioned in this chapter, citing an early DuPont project as just one example. Networks of unmanageable size can still be a problem and demonstrate a fairly common failing in developing network diagrams. Various writers have recommended some solutions, but the problem of oversized networks persists and tells us that essential lessons are often not being learned.

A long-established approach used in the engineering and construction industry to limit the size of individual networks, recommended by Harrison (1992), is to break the overall project plan down in hierarchical fashion. The top layer of the tree need contain only summary information. The breakdown into greater detail that will eventually be needed for project control can be arranged in a hierarchy of subnetworks, with increasing levels of activity detail at the lower levels. Woodgate (1977) also advocates this approach. Pinto (2010) uses the term 'Meta-Network' as the equivalent of Level 1 or higher-level detail. Hackney (1992) urges caution not to subdivide the project too much.

Practices, descriptions and the number of breakdown levels will vary from one organization to another, but the four examples tabled in Figure 20.7 are typical of the general approach. The example from Riggs (1986) is particularly helpful because it give recommendations for the maximum number of activities at Levels 1, 2 and 3).

Guidance from the American Association of Cost Engineers (AACE International, 2010a) is that hierarchical schedules must be applied appropriately to be effective and that they should be closely aligned with or identical to the WBS. For large engineering, procurement and construction (EPC) projects AACE International specify

AACE International (2010b).	**Prohibited!** Aside from the one activity starting the network and another finishing the network, open-ended activities 'break' the logical network sequence and may not exhibit correct float calculations. Some CPM software such as those published by Primavera can automatically cause open-ended activities to be critical path activities even though they are not on the longest path.
Association for Project Management (2015).	**Prohibited!** A network should have one start milestone and one finish milestone. All other activities are connected, directly or via preceding or successor activities, respectively, to the start and finish activity by logic.
Project Management Institute (2014).	**Discouraged.** The best practice for a project schedule is to link every activity with at least one predecessor and one successor activity except the project start and completion milestones.

Figure 20.4 What the professional organizations have to say about open ends.

AACE International (2010b).	**Discouraged.** Strongly recommends against the use of mandatory constraints as they lead to illogical results where activities are scheduled to occur even if preceding work is incomplete. (Other constraints are also discouraged).
Association for Project Management (2015).	**Discouraged.** All activities should ideally be linked by logic; but in some instances this may not be possible - or desirable, e.g. the start of something where nothing precedes it. 'Start on site' or 'Appoint contractor' may be examples. Similarly, finish dates may need to be fixed without anything following them, e.g. sectional completion dates.
Project Management Institute (2014).	**Discouraged.** The best practice regarding the use of constraints is to keep the number of constraints in a schedule to a minimum using normal predecessor and successor logic links and lags for each activity, avoiding open ends ... It is difficult not to use some constraints in a schedule, so use them sparingly and correctly.

Figure 20.5 What the professional bodies have to say about imposed dates (constraints).

AACE International (2010b).	**Discouraged.** During schedule calculations, overlapping activity relationships, such as SS with lags, FF with lags, and FS with negative lags, that do not provide a relationship tie to/ or from the balance of that activity, may cause a portion of the activity to become open ended.
Association for Project Management (2015).	**Prohibited.** A lead is a negative lag, and illogical, as time only runs forward (so far as we know).
Project Management Institute (2014).	**Discouraged.** Negative lags can hamper risk analysis in some software. Negative lags may prevent the project team from being aware of the underlying 'logic' being used to represent the activities, as it is counter intuitive.

Figure 20.6 Some negative views about negative lags in guidance from professional organizations.

that detailed work schedules at Level 4 (which are associated with the Level 3 critical path schedule) should be prepared as 'stand alone', so as not to overload the Level 3 schedule with too much detail and thus compromise schedule maintenance.

Another solution to the problem of too many activities in the project network is a methodology known as 'rolling wave planning', which is outlined in the next chapter.

Documenting the schedule basis

Something that is often forgotten (but should have become clear from this chapter) is the importance of recording the schedule basis in a written description or specification. Failure to do that could lead to subsequent misunderstandings and arguments between project stakeholders, both within and outside the organization (such as clients and contractors). The size of a 'basis of schedule' document will depend on the complexity of the project and its schedule. Often three or four pages will be more than sufficient. Typical contents should include the following items:

- milestones and key dates;
- organization – WBS and Structure;
- detail;
- use of imposed dates;
- critical path;
- key issues (for example, reference to long-lead items);
- risks and assumptions;
- work patterns (calendars);
- seasonal implications;
- resource availability and levelling;
- schedule risks.

From the 1960s (Walton, 1968):
Level 1: Geographic, technical or other.
Level 2: Functional or technical activities from Level 1.
Level 3: Level 2 activities divided into subnetworks.
Level 4: Daily assignment.

From the 1980s (Riggs, 1986):
Level 1: Project approval (25 to 75 activities).
Level 2: Project management (100 to 400 activities).
Level 3: Planning analysis (3000 to 4000 activities).
Level 4: Daily assignment.

From the 1990s (Bent and Humphreys, 1996):
Level 1: Master milestone schedule.
Level 2: Project bar chart schedule.
Level 3: Project master schedule.
Level 4: Daily assignment.

From the 2000s (AACE International 2010a):
Level 1: Engineering, procurement and construction (EPC) Level 1.
Level 2: EPC Level 2.
Level 3: EPC Level 3.
Level 4: EPC Level 4.

Figure 20.7 Examples of recommended hierarchical planning levels.

Useful guidance for documenting the schedule basis is provided in AACE (2009).

Maintaining project schedule technical integrity throughout the project lifecycle

Suppose that you are very experienced and adept at producing an effective project schedule. You are eminently capable of creating a network diagram that is free from open ends, negative lags, illogical imposed dates and all the other traps that lie in wait for the newcomer to our profession. Now you have just finished supervising the creation of a well-nigh perfect network of a thousand activities for a brand new project. And so the project is launched on its lifecycle and work begins.

Your project schedule will reside digitally in a computer that might also house all kinds of other data. Other people will have access to the computer at any time for their own purposes. Your project will begin to progress through its planned schedule as time passes and new data from timesheets, progress reports and possible changes (some authorized and others not) will add to schedule complexity. The integrity of your perfect schedule is now at the mercy of many risks - logic and data errors created unwittingly by people with insufficient network planning experience. Mistakes, malfunctions and misfortunes can include:

- fire or failure in the computer;
- timesheet data entered incorrectly;

- activities reported as finished, out of logical sequence;
- schedule data (such as remaining durations, resource quantities, progress status) entered incorrectly;
- one or more schedules for different projects with different managers entered into the same database making claims on common resources;
- and so forth.

So, you need to consider all the risks, and take steps to preserve your project schedule technical integrity. Some access to the database by all and sundry might be permissible – even desirable – provided that only persons with the essential skills and authority are allowed to change the logic and system parameters of your project schedule.

Competent project software should provide access to the system at a range of levels, specified so that only project planning professionals working on the particular project are allowed to change the network logic or declared quantities (such as resource information). On the other hand, swift and accurate communications are essential for modern project control, which means that many people working on the project should have some kind of access to report tasks as completed and so on.

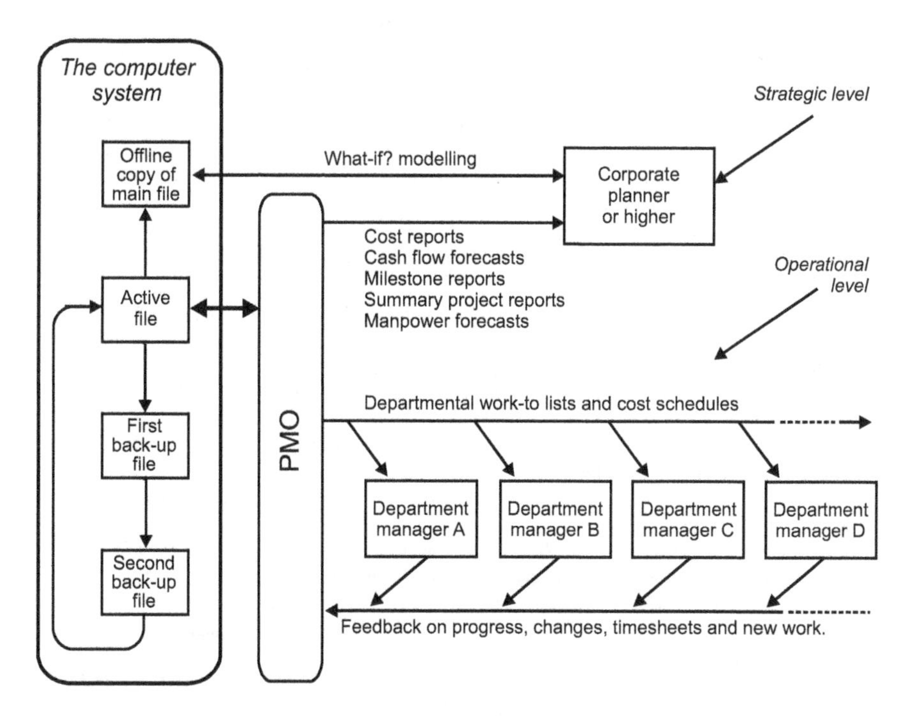

Figure 20.8 Organizing schedule communications to preserve integrity. The integrity of the schedule must be preserved. A communications framework such as this can help to prevent errors and preserve schedule integrity.
Source: Adapted from Lock (2013).

Your organization might contain a long-range (strategic) planning group or PMO, whose duty is to ask questions in the 'What if?' category about the possible effects of introducing one or more new projects (sales opportunities or management change projects) into the organization's committed workload.

Figure 20.8 illustrates a possible solution to this dilemma. Like so many other project management solutions this depends on having a competent PMO. People can be encouraged to enter progress and timesheet data, filtered through supervisors to ensure accuracy and honesty. Those in the secretive, esoteric, strategic, forward planning group or PMO might be given access to a copy of the principal database so that they can test the effects on resources and running projects should new project opportunities become reality.

Here, then, is a situation when a PMO staffed with experts (not too many of them) acts as a security buffer between sensitive or easily corruptible data and the wider organization. The PMO ensures the distribution of schedule information to those who need it, encourages and supports enquiries from those authorized to make them but prevents unauthorized changes and inept interference. Depending on the location of the computer, the PMO might rely on an IT department to safeguard the database itself, by maintaining a complete and reliable back up source.

References and further reading

AACE International (2009), *Recommended Practice No. 38R-06 Documenting the Schedule Basis.* Morgantown, WV: AACE International.

AACE International (2010a), *Recommended Practice No. 37R-06 Schedule Level of Detail - As Applied in Engineering, Procurement and Construction.* Morgantown, WV: AACE International.

AACE International (2010b), 4–6. Recommended Practice No. 49R-06 Identifying the Critical Path. Morgantown, WV: AACE International.

Ahuja, H.N. (1976), *Construction Performance Control by Networks.* New York: Wiley.

Andersen, E.S. (1996), 'Warning, activity planning is hazardous to your project's health'. *International Journal of Project Management*, 14(2), pp. 89–94.

Antill, J.M. and Woodhead, R.W. (1970), *Critical Path Methods in Construction Practice*, 2nd edn. New York: Wiley.

APM (2015), Planning, *Scheduling, Monitoring and Control; The Practical Project Management of Time, Cost and Risk.* Princes Risborough: Association for Project Management.

Battersby, A. (1964), *Network Analysis for Planning and Scheduling.* London: MacMillan.

Bent, J. and Humphreys, K. (1996), *Effective Project Management Through Applied Cost and Schedule Control.* New York: Marcel Dekker.

Bowers, J. (1985), 'Network Analysis on a Micro'. *Journal of the Operational Research Society*, 36(7), pp. 609–611.

British Standards Institution (1981), *BS6046-2 Use of Network Techniques in Project Management Part 2. Guide to the use of Graphical and Estimating Techniques.* London: BSI.

Burke, R. (1997), *Project Planning and Control*, 2nd edn. Chichester: Wiley.

Chartered Institute of Builders (2011), *Guide to Good Practice in the Management of Time in Complex Projects.* Chichester: Wiley.

Chartered Institute of Builders (2018), *Guide to Good Practice in the Management of Time in Major Projects: Dynamic Time Modelling.* Chichester: Wiley.

Devaux, S. (1999), *Total Project Control: A Managers Guide to Integrated Project Planning, Measuring and Tracking.* New York: Wiley.

Douglas III, E.E., Calvey, T.T., McDonald, Jr., D.F. and Winter, R.M. (2006), 'The Great Negative Lag Debate', *in AACE International Transactions*, pp. 1–7.

Faulkner, H.J. and Hillstead, J. (1968), 'Application of critical path planning techniques, Dungeness 'B' nuclear power station', in Brennan, J.A.S. (ed.), *Applications of Critical Path Techniques*. London: English Universities Press.

Fondahl, J.W. (1964), *Methods for Extending the Range of Non-Computer Critical Path Applications*. www.dtic.mil/dtic/tr/fulltext/u2/454007.pdf. [Accessed 7 June 2012].

Government Accountability Office (2015), GAO Schedule Assessment Guide Best Practices for project schedules GAO-16-89G.

Hackney, J. (1992), *Control and Management of Capital Projects*, 2nd edn. New York. McGraw Hill.

Harrison, F.L. (1992), *Advanced Project Management*, 3rd edn. Aldershot: Gower.

Kelley, Jr., J.E. and Walker, M. (1959), 'Critical-path planning and scheduling' in *Papers presented at the December 1–3, 1959, eastern joint IRE-AIEE-ACM computer conference*. Boston, MA: ACM Press, pp. 160–173.

Lester, A. (1982), *Project Planning and Control*. London: Butterworth.

Lock, D. (2013), *Project Management*, 10th edn. Farnham: Gower.

Lockyer, K. and Gordon, J. (1996), *Project Management and Project Network Techniques*, 6th edn. London: Financial Times Management.

McLaren, K. and Bluesnel, E. (1969), *Network Analysis in Project Management*. London: Cassell.

Moder, J., Phillips, C. and Davis, E. (1983), *Project Management with CPM, PERT and Precedence Diagramming*, 3rd edn. Wisconsin. Blitz Publishing Company, 1995.

Pinto, J. (2010), *Project Management Achieving Competitive Advantage*, 2nd edn. N.J: Prentice Hall.

PMI (2014), *CPM Scheduling for Construction Best Practice and Guidelines*. Newtown Square, PA: Project Management Institute.

Riggs, L.S. (1986), *Cost and Schedule Control in Industrial Construction*. Bureau of Engineering Research, Austin, TX: University of Texas.

Senior, B.A. (2003), *Critical Path Method Implementation Drawbacks : A Discussion Using Action Theory*, Colorado, BO: Colorado State University.

Walton, H. (1968), 'Optimum Level of Detail', in Thornley, G. (ed.), *Critical Path Analysis in Practice; Collected Papers on Project Control*. London: Tavistock Publications, pp. 31–33.

Williams, D. (1968), 'Introduction to the Basic Method', in Thornley, G. (ed.), *Critical Path Analysis in Practice; Collected Papers on Project Control*. London: Tavistock Publications, pp. 7–16.

Woodgate, H. (1977), *Planning by Network*, 3rd edn. London: Business Books.

21
CONTROLLING PROJECT MANUFACTURING

Aydin Nassehi

The purpose of manufacturing planning and control is to ensure that components, subassemblies and main assemblies, including those specially manufactured for a project, will be available on time, in the right quantities and of the right quality for installation in the project when they are needed. Production variables such as changes in customers' decisions, suppliers not delivering on time, machine breakdowns or staff shortages can cause things to happen differently from the plan. Control is the process of coping with such variables. Effective control will allow the planned objectives to be met, even when the assumptions on which the plan was based are no longer applicable.

The P:D ratio

The P:D ratio is defined as the ratio of total throughput time (P) to the time it takes for the product or service to be delivered to the customer after the order is placed (D). The lower the P:D ratio, the lower is the uncertainty and risk in production planning and control. Figure 21.1 shows how various approaches to meeting customer demand can affect the P:D ratio.

Sequencing

Sequencing is the process of determining the order and time in which various jobs should be carried out, taking into consideration the limited resources available. Acceptable sequencing methods will take one or more of the factors listed in Figure 21.2 into account.

For project practitioners priorities are usually derived from the project network plan, and quantified by the amount of float remaining. However, priorities sometimes also depend on management decisions, such as giving the highest priority to jobs for customers who pay their bills on time or are otherwise highly valued and (at the other

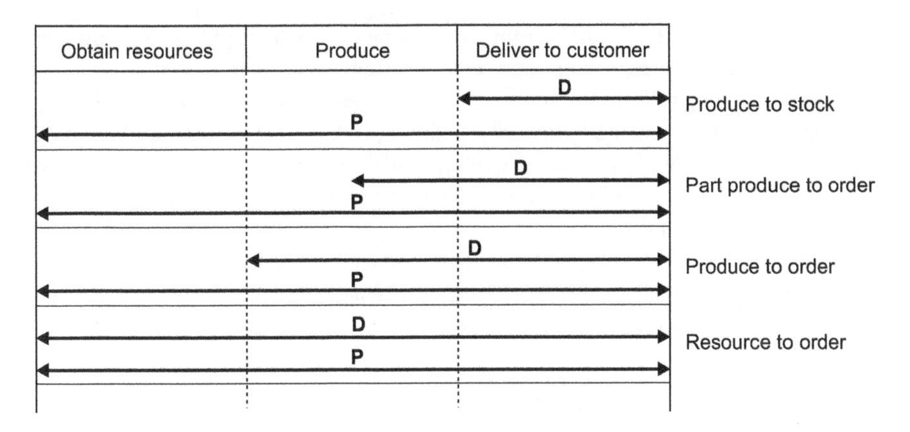

Figure 21.1 Explanation of the manufacturing P:D ratio.

Sequencing rule	*Explanation*
Physical constraint.	Sometimes the mix of work arriving at part of an operation can determine the priority given to jobs. For example, when fabric is cut to a required size and shape in garment manufacture, the surplus fabric would be wasted if not used for another product. Thus jobs that physically fit together might be scheduled together to minimize waste.
Customer priority.	Customer priority sequencing can be used to allow work for an important or aggrieved customer to be processed before other work, irrespective of the order of arrival of the customer or item.
Due date (DD).	Prioritizing by due date means that work is sequenced according to when it is needed for use on the project or for delivery, irrespective of the size of the job or importance of the customer.
Last in: first out (LIFO).	Last in: first out is a sequencing method usually chosen for practical reasons. For example, unloading an elevator which has only one set of doors is far more convenient because there is only once entrance and exit.
First in, first out (FIFO).	Some operations serve the project or customers in exactly the sequence in which they arrive. This is also sometimes called first come, first served (FCFS).
Longest operation time (LOT).	Operations managers might feel obliged to sequence the longest duration jobs first. This has the advantage of occupying work centres for long uninterrupted periods.
Shortest operation time (SOT).	Operations might become cash constrained at some stages. In these cases the sequencing rules can be adjusted to tackle the short duration jobs first. Those jobs can then be invoiced as early as possible, so that the resulting payments will ease any cash flow problems.

Figure 21.2 Sequencing rules that can be applied to manufacturing tasks.

extreme) holding back jobs for customers who owe money from outstanding invoices or are considered at high risk of going out of business.

The sequencing rules aim to improve the performance objectives of dependability, speed and cost. Forward and backward scheduling is then used to develop the timing of individual activities. In forward scheduling, tasks are completed as early as possible. This offers high labour utilization and flexibility. In backward scheduling tasks are completed as late as possible, which offers lower material costs and less exposure to risks in case of order changes. Minimizing the 'makespan' (the time from the start of the first job to the finishing of the last job), and the lateness (how far tasks go beyond their deadlines, averaged and summed) are among the important goals in sequencing.

Johnson's sequencing method

Johnson's algorithm for sequencing applies to the progression of jobs through two consecutive workstations (in other words, determining the order of jobs when there are two serial machines). The algorithm is as follows:

1 Look for the smallest processing time in the remaining jobs.
2 If that time is associated with the first workstation, then schedule that job first or as near to first as possible. If that time is associated with the second workstation, sequence the job last or as near to last as possible.
3 Delete the scheduled job from the task list.
4 If there are any tasks remaining, go to 1.

Production management strategies

Production management strategies are concerned with deciding the manner in which signals are given to workstations in a production system for starting and stopping work. These strategies are mainly divided into push and pull strategies (Figure 21.3).

Push strategies are those in which the production signals are generated in a top-down manner based on demand forecasts. The signals are then given to workstations, which 'push' processed parts to the subsequent workstation. Material requirements planning (MRP), master production schedule (MPS), manufacturing resource planning (MRPII) and enterprise resource planning (ERP) are examples of push strategies.

Pull strategies are those in which the production signals are generated downstream at the point of delivery to the customer and are transferred upstream to workstations in the production system. Starting with the final workstation, each station 'pulls' the processed parts from the previous station. Just-in-time and lean manufacturing are concepts associated with pull strategies.

Material requirements planning

MRP is a push system that utilizes the bill of materials (BoM). The BoM is a hierarchical listing of all the raw materials, parts, subassemblies and assemblies needed to

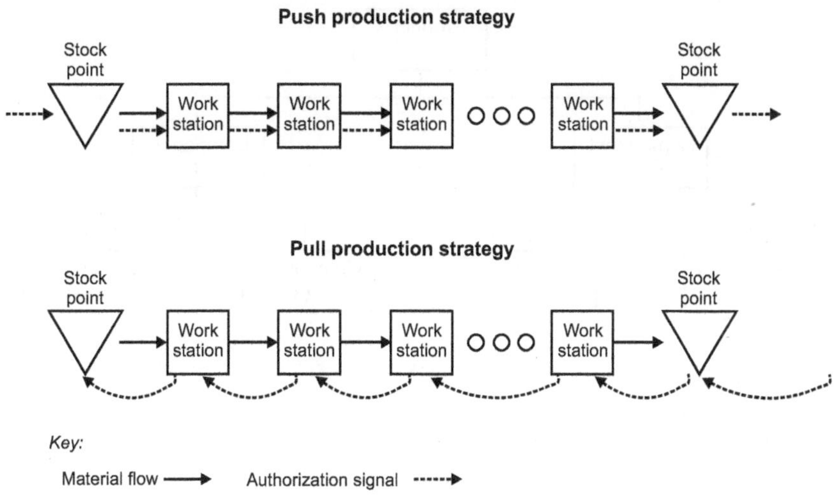

Push production strategy

Stock point → Work station → Work station → Work station ○○○ Work station → Stock point

Pull production strategy

Stock point → Work station → Work station → Work station ○○○ Work station → Stock point

Key:

Material flow ⟶ Authorization signal ------▸

Figure 21.3 Push and pull production scheduling strategies.

produce one unit of a product. It is similar in some respects to the WBS familiar to project practitioners. The BoM concept is illustrated in Figure 21.4, where the BoM shows all the components needed to make one unit of Product X. MRP also requires a master production schedule to make timing and volume calculations to meet the forecast demand.

To illustrate how to calculate the total quantities of each component needed for one complete product X, take the example of component E in Figure 21.4. The requirements are as follows:

- Component C requires two component Es.
- Each component B requires a total of 5Es, but because two Bs are needed for the product that's a further 10 Es altogether.
- Thus each unit of product X requires 12 component Es.

In MRP the BoM is traversed from top to bottom. The following quantities of items pertaining to each subassembly are calculated for every time period:

1 The gross requirements: the total quantity of the item needed in the time period
2 The scheduled receipts: the quantity that will be received from orders placed before the start of the planning period.
3 The projected on-hand inventory, which is the quantity of the item that is expected to exist at the end of the given time period.
4 Planned receipts, which are the total quantities of items to be received during the time period as the result of orders in the planning period.
5 Planned order releases: the total quantity ordered in the time period.

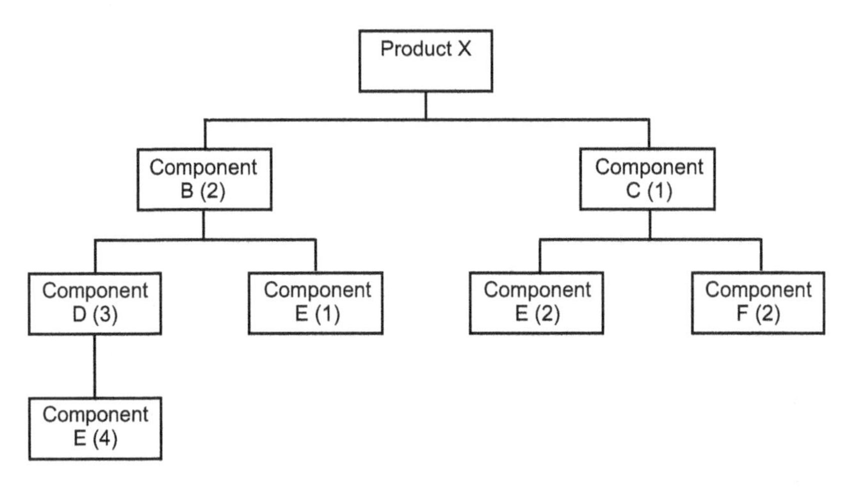

Figure 21.4 Example of a bill of materials (BoM). This BoM shows all the components needed to make one unit of Product X. Figures in brackets are the quantities required.

Working in a top-to-bottom manner, would allow the orders needed for all parts of the product to be specified.

An MRP example

Consider the BoM for product A, which is shown in Figure 21.5. Suppose that a master schedule calls for the following quantities of product A:

- 250 units in week 6;
- 140 in week 8;
- 200 in week 9.

On-hand levels are: A = 5, B = 5, C = 0, D = 0 and E = 5. Another five units of D are scheduled to be received in week 1 and 5 units of E in week 2. Order quantities for A and C are low-for-lot. B and E have to be ordered in multiples of 10. A minimum order size of 150 applies for D. Question: find the planned order releases for all items using MRP. The solution is shown in Figure 21.6.

Manufacturing resource planning

MRPII is a development of MRP in which MRP and its related information systems are integrated in a single framework. A single database is thus held for systems such as inventory management, capacity management and production planning.

Enterprise resource planning

ERP is the extension of MRPII outside the company. It allows the resources within the entire enterprise to be managed within a single framework. In ERP, systems such

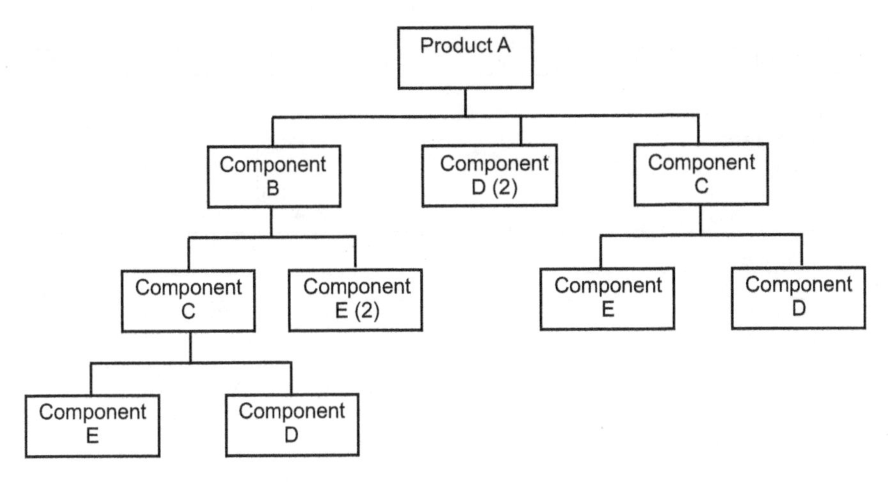

Figure 21.5 Simplified bill of materials for 'product A'. This is the BoM for the MRP worked example in the text.

as human resources, finance and so forth are integrated with production management strategies. The most important advantage of ERP is perhaps that an enterprise's ERP can communicate with the ERP in other enterprises throughout the supply chain. This clearly introduces unprecedented levels of flexibility in supply chain management.

Lean production

Lean production started as Just-in-Time (JIT) in Japan. Also called Lean Synchronization, the aim of this approach is to meet demand instantaneously with perfect quality and zero waste. This involves supplying products in perfect synchronization with the demand.

In the traditional push planning approach, if a problem occurs in one part of the production system, it would not be immediately visible in the other parts of the system because these are isolated from each other by buffer inventories. In the Lean approach, if something happens in one part the effects will be felt immediately in subsequent stages and, soon after, in preceding stages. This would mean that solving the problem at a specific stage is no longer the sole responsibility of people working in that stage - instead it becomes a job for the entire production line. This increased exposure is one of the defining properties of Lean synchronization because buffers (which are considered as blankets of obscurity by Lean practitioners) no longer protect the other sections from disruptions.

The analogy of a river with rocks is often used to illustrate the two approaches. The volume of the water represents the inventory which hides the rocks (problems in the production system). In the Lean approach the volume of inventory is lower, so that the problems become apparent more quickly. It is noteworthy that the benefits of Lean come at the cost of capacity utilization. If you consider the traditional approach, when stoppages occur the buffers allow other parts of the system to continue functioning

Component A

LT = 1Q = L4L	Period	1	2	3	4	5	6	7	8	9	10
Gross requirement							250		140	200	
Scheduled receipts											
Quantity on hand		5	5	5	5	5	5	0	0	0	0
Net requirements							245		140	200	
Planned order receipts							245		140	200	
Planned order releases						245		140	200		

Component B

LT = 1Q = multiples of 10	Period	1	2	3	4	5	6	7	8	9	10
Gross requirement						245		140	200		
Scheduled receipts											
Quantity on hand		0	0	0	0	0	5	5	5	5	5
Net requirements						245		135	195		
Planned order receipts						250		140	200		
Planned order releases					250		140	200			

Component C

LT = 2Q = L4L	Period	1	2	3	4	5	6	7	8	9	10
Gross requirement					250	245	140	340	200		
Scheduled receipts											
Quantity on hand		0	0	0	0	0	0	0	0	0	0
Net requirements					250	245	140	340	200		
Planned order receipts					250	245	140	340	200		
Planned order releases			250	245	140	340	200				

Component D

LT = 1Q = minimum 150	Period	1	2	3	4	5	6	7	8	9	10
Gross requirement			250	245	140	830	200	280	400		
Scheduled receipts		5									
Quantity on hand		0	5	0	0	10	0	0	0		
Net requirements			245	245	140	820	200	280	400		
Planned order receipts			245	245	150	820	200	280	400		
Planned order releases		245	245	150	820	200	280	400			

Component E

LT = 1Q = batches of 10	Period	1	2	3	4	5	6	7	8	9	10
Gross requirement			250	245	640	340	480	400			
Scheduled receipts			5								
Quantity on hand		5	5	5	5	5	5	5	5	5	5
Net requirements			240	245	635	335	475	395			
Planned order receipts			240	250	640	340	480	400			
Planned order releases		240	250	640	340	480	400				

Key: LT = lead time. Q = order quantity rules.

Figure 21.6 Solution to the MRP exercise.

and thus utilize their capacity. However, higher utilization does not usually translate into more production as a whole.

The Lean philosophy is about getting the entire enterprise to take action to eliminate problems in the production system. The following basic working practices have thus been developed to 'involve everyone':

- Discipline in following work standards, critical for safety and quality.
- Flexibility to increase people's responsibility to their abilities.
- Equality - because one person's problem is really everyone's. Some Lean companies mandate the same uniform for everyone regardless of rank in the organization.
- Autonomy - delegating responsibility to people so that they can make direct decisions.
- Development - a more capable workforce will increase productivity.
- Quality of working life, which includes involving people in decision-making, increasing job security and the enjoyment of facilities.
- Creativity, which is an indispensable element of motivation, and;
- Total people involvement - staff will take on more responsibility in dealing with the supply chain, recruitment, quality issues and spending on improvement budgets.

Continuous improvement

An important principle in Lean is that improvement towards the ideal state of 'instantly meeting demand with perfect quality and no waste' is a never-ending process. The Japanese term 'kaizen' is often used to express this principle of continuous improvement.

Waste is defined as any activity that does not add value. There are studies showing that only a small subset of the activities in an organization directly add value. A good analogy is when someone applies for a driving licence. It takes mere minutes for the request to be processed but it can take weeks for the documents to be returned. The Japanese words 'muda', 'mura' and 'muri' are used to convey the three main causes for waste, as follows:

- *Muda*. These are activities where no value is added to the production or the customer. Poorly communicated objectives are often the main cause. Not understanding customer requirements is an example cause.
- *Mura*. This is the lack of consistency. For example, if tasks are not properly documented, when they are performed by different people the results are different.
- *Muri*. This means making unreasonable demands from the available resources. Often resulting from poor planning, these difficulties can be eliminated by sequencing, scheduling and considering resource loadings in production planning.

Seven types of waste can be identified that are often present in production, as follows:

1. Over-production - producing more than what is immediately needed.
2. Waiting time - efficiency measures can be used to find out the idle time of resources. The waiting times of products in the pipeline are somewhat more difficult to calculate.
3. Transport - moving items during the operation adds no value but can add cost and should be minimized.
4. Process - some processes might only exist owing to poor design of the product or the process chain.

5 Inventory - keeping raw materials, work in progress or finished products is a waste of valuable space. It also ties up working capital.
6 Motion - an operator who is moving around a lot does not necessarily add a lot of value.
7 Defectives - products that do not meet quality requirements are a major waste in production systems.

Eliminating waste

Various approaches can be employed to identify and eliminate waste. These include the following:

Streamlined flow is the smooth flow of people, products and information is integral to the Lean philosophy. Long process routes are often those where delays and inconsistencies can creep in.

In order to ensure a smooth flow, all elements of the throughput time should be closely investigated.

Value stream mapping is an approach used to create a visualization of the production path from start to finish; this enables identification of value-adding and non-value-adding operations. First, the value stream is identified and then the 'current state' map of the process is drawn. The problems are then diagnosed and changes are suggested in a 'future state' map.

Another useful suite of tools for streamlining the flow are visual management techniques that employ visual signs, computer screens or simply lights to convey critical information about the production flow to everyone.

Yet another approach for streamlining the flow is to employ several small-scale simple process technologies in parallel instead of relying on larger and more complex ones. Then, if one of the machines performing a simple process breaks down, the line will still continue operating with reduced effectiveness rather than stopping altogether. Also, machines that are simpler and thus smaller are easier to move and service.

Matching supply and demand

By using pull control, it is possible to ensure that no part is produced unless there is a need for it. Richard Hall, an authority on Lean operations says 'Don't send nothing nowhere, make 'em come and get it'.

The use of kanbans (Japanese for cards) or signals is one method of creating pull control. In its most basic form, a kanban is a card used by a customer to instruct its supplier to send more items. Usually each kanban corresponds to a single item; so three kanbans signal the need for producing three items. Kanbans serve three purposes:

1 They instruct the preceding stage to send more.
2 They serve as visual controls to show up areas of over-production.
3 Reducing the number of kanbans becomes a tool for kaizen.

Instead of cards, spaces indicated on the floor or containers may also be used.

Flexible processes

Responding instantaneously to exact customer demand requires flexible processes that can change what they do and how often they do it without incurring great costs or requiring a long time. Reducing changeover activities is an example of making processes more flexible.

Minimizing variability

Variability often disrupts flow and prevents lean synchronization. There are a number of techniques to minimize variability. These include:

- Levelling flow between stages. Keeping the mix and volume of flow between stages at an even flow over time reduces variability. This can require batch sizes to be decreased and could have significant effects on planning and control of the production system.
- Levelling delivery schedules. This is similar to the above but applies to delivery stages instead of production stages. Delivering in smaller (more frequent) batches keeps the inventory low and also allows production to respond to changes in demand much more quickly.

Hybrid strategies

MRP and JIT can coexist and they can be combined in several ways to form a hybrid system. The way in which they should be combined depends on the complexity of product structures, the complexity of product routing, the volume-variety characteristics and the level of control required. In general, MRP is better for planning and JIT is better for control. Various ways in which these two systems can be combined include using separate systems for different products and using MRP for overall control and JIT for internal control.

Interfacing production management and project management

Many industrial projects require the use of manufacturing facilities, which is why this chapter has been included in this handbook. The main difference between manufacturing for stock or general sale and manufacturing for a project is that the project (or the project manager) becomes an internal 'customer'. Manufacturing priorities should be determined by the project schedule, which in all well-run projects will be based on a critical path plan. That project plan will generate all the priorities and 'required by' dates needed for input into the manufacturing plant schedules, so that information must be communicated to the manufacturing management.

As in all project work, good communications between all the managers and departments are essential. For most project tasks the project manager has some authority, even in a balanced matrix organization. The situation when a project requires manufactured components can sometimes present difficulties, because the organization's manufacturing plant is usually a common service that has to satisfy not only other company departments but also external customers who, because they are (rightly) valued and are paying cash, will usually be given priority.

References and further reading

Heizer, J., Render, B. and Munson, C. (2017), *Operations Management, Sustainability and Supply Chain Management*, 12th edn, Harlow: Pearson.

Slack, N. and Brandon-Jones, A. (2019), *Operations Management*, 9th edn, Harlow: Pearson Higher Education.

22
MORE SPECIALIZED SCHEDULING

Dennis Lock

This chapter describes some less common project scheduling methods. These techniques are not usually described in the project management literature but they can be useful to project practitioners. Modular networks and templates require some ingenuity, but the rewards can be very great, saving time and money.

Standard project start-up plans

Suppose that you are about to manage a large capital project. The only project definition available comes from the project specification and a business proposal. Your project will have a budget and a schedule, but that schedule can only be an outline Gantt chart. So when you get the go-ahead to begin work you have to produce a more detailed work schedule, and produce it quickly. You have experience and can therefore predict with reasonable accuracy what needs to be done for the first few weeks surrounding the kick-off meeting. But, beyond that time, you can have no precise idea of all the detailed tasks that lie ahead.

Now here is your dilemma. You must prepare detailed work schedules for at least a few months ahead so that resources can be marshalled and tasks can be identified, described and prioritized. You need a network diagram right now so that work can begin without delay. Some companies will have checklists of project start-up tasks, compiled for the purpose of ensuring that no significant task is overlooked. So, why not adapt your checklist of start-up tasks to form a standard start-up network diagram? This was done successfully, for example, in a heavy engineering company that manufactured special-purpose heavy machine tools for the automotive and other industries. Figure 22.1 illustrates the principle. Every company that regularly carries out projects of any kind should be able to devise a standard start-up network schedule to suit its own projects and purposes.

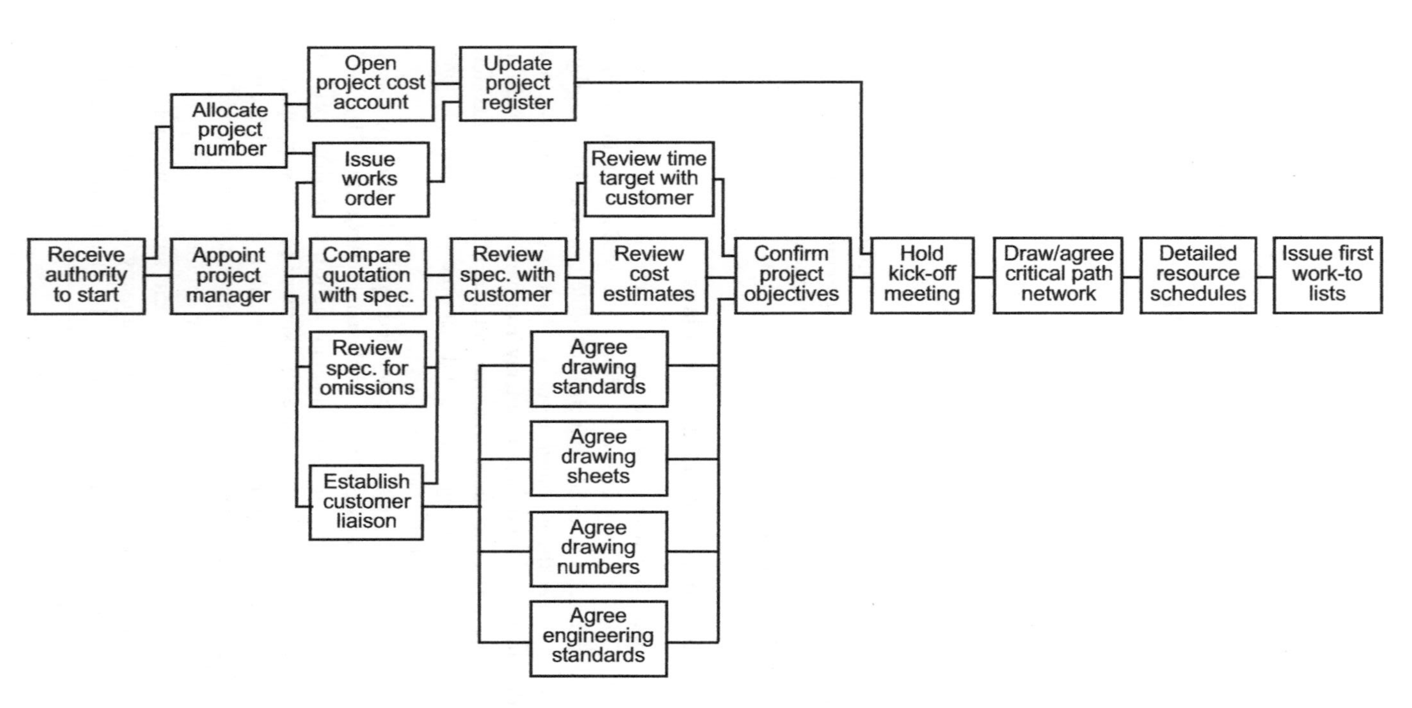

Figure 22.1 A standard project start-up network.Most companies should be able to develop a project start-up chart in the form of a standard network diagram. This is simply a checklist for early project tasks and there is no need to estimate task times or carry out critical path analysis. However, this will ensure that tasks are carried out in their logical sequence and that no important task is overlooked.
Source: Lock (2013).

Rolling wave planning

Imagine that you have been appointed project manager and that your project has a target or promised completion date that is (say) five or even more years ahead. If you have a standard project start-up checklist or, much better, a standard project start-up network diagram such as that described above and shown in Figure 22.1, you will have a week or two in which to begin drawing up the working schedule for the entire project. So, you ask your PMO manager for help and you organize the first project planning meeting. Such meetings often resemble brainstorming sessions, with questions such as 'OK, when this task is done, what comes next?' or 'Can we really start this task here – doesn't anything have to come before it?' However, as your planning session considers tasks farther into the future the picture will become much less clear until, when looking a year or so ahead you can only make unhelpful guesses about all the small but essential tasks that will have to be scheduled and performed.

Critical path analysis (and the determination of critical activities and other priorities) is an essential part of planning any modern project. You need that information in order to prioritize tasks, allocate scarce resources and make sure that the schedule is practicable. But usually you cannot plan in the required level of detail for tasks that lie several years into the future. The only known fact in that distant future time is the promised project completion date, which is your organization's commitment and will be the final project milestone.

Any complete project planning method depends on having at least one continuous path through from project start to the finish milestone. But all you have to begin with is a standard start-up network. The solution to this apparently very difficult problem lies in a method known as rolling wave planning. Like many management techniques this is really based on common sense. This is how it works.

You begin your project schedule with the standard start-up network as the basis and, as more information becomes known, you can extend that schedule by adding all the newly defined tasks, say covering six months or a year ahead. A single link connection to the final project milestone from the last of these known tasks will provide the continuous path needed for the forward and backward passes to establish time analysis. So all your scheduled times for current and known future tasks will be consistent with achieving the final project milestone date. As the project progresses and its engineering design is developed, you will be able to extend the detailed front end of network diagram and work schedules more into the future. Thus your scheduling can be developed month-by-month with greater detail and accuracy, flowing into the future like a rolling wave. The principle is illustrated in Figure 22.2.

Line of balance techniques for construction projects

Line of balance is a name given to a group of long-established methods that aim to present graphically a quantified assessment of how much of a project should have been completed at a particular review date. One line of balance method is particularly applicable to manufacturing projects that require the manufacture and procurement of components and their subsequent assembly in two or more repeating batches.

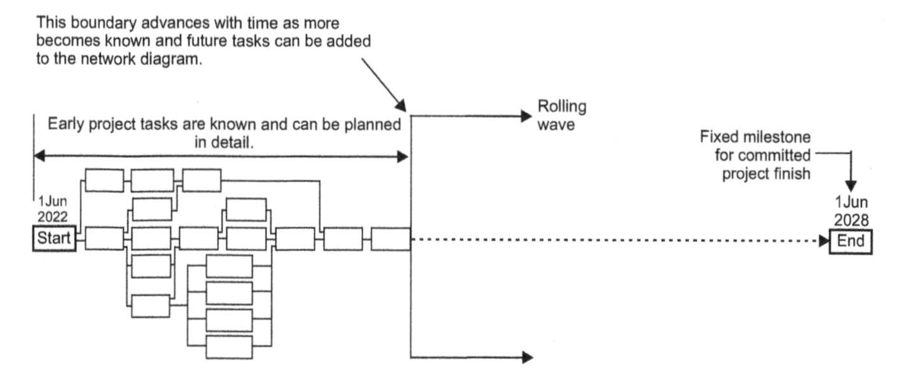

Figure 22.2 The concept of rolling wave planning.

Other completely different line of balance techniques exist for construction projects where there is a strong repetitive element (such as when building a number of very similar dwellings on a new housing estate). Line of balance methods allow the project manager to compare the actual progress on a calendar date with the quantities that should have been completed on that date if the project promises are to be fulfilled. There are two different line of balance approaches for construction projects, and the choice depends on the number of units to be built. Both of the following examples are derived from Lock (2007). Note that later editions of that book do not contain line of balance examples.

Line of balance charts for a small number of units in a construction project

A good example of a relatively simple line of balance chart for a construction project is for a project where a number of houses have to be erected for a new housing estate. Figure 22.3 is a simple Gantt chart (or bar chart) for such a project. The appearance of these houses will vary slightly from one to the next so as to give a sense of individuality for the purchasers. However, these differences will only be cosmetic and all the construction designs are essentially the same. So the same pattern of tasks must be repeated for each house. Work on houses 2, 3, 4 and 5 has been progressively delayed to allow for the various trade groups to move in logical sequence from one house to the next. A vertical cursor line placed on the chart at the current date would allow a rough and ready idea of the state of progress, because everything to the left of the cursor should have been finished. Thus, it could be argued with some justification that even doing that would provide a simple line of balance method.

Line of balance is a refinement of the date cursor method that gives a more accurate picture of planned and actual progress on the day of measurement. A line

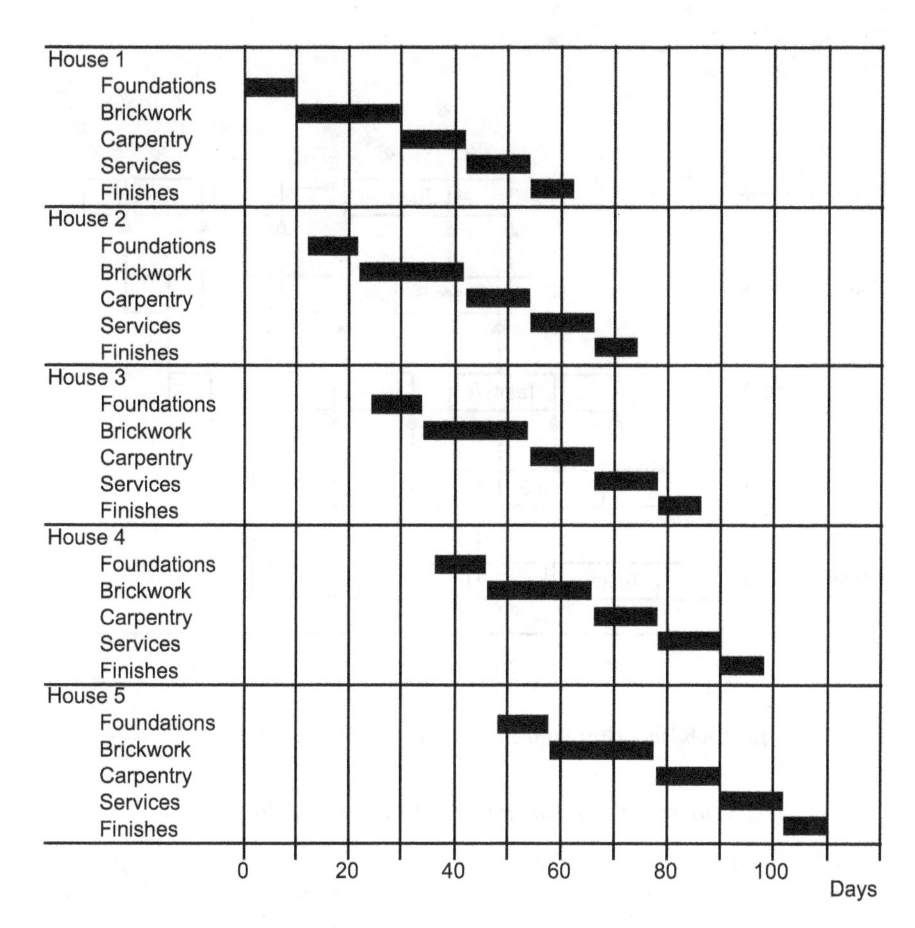

Figure 22.3 Gantt chart for a five-house construction project. This provides the data for the line of balance chart in Figure 22.4.

of balance chart can be drawn for this small construction project by rearranging the Gantt chart from Figure 22.3 to the pattern shown in Figure 22.4. Notice that there are vertical arrows showing the logical links between consecutive tasks. These are practical constraints arising from the use of shared resources. They indicate that each task requiring a particular trade should be finished before the corresponding task can be started on the next house. Brickwork is the longest task, for which two separate teams (Teams A and B) have been assigned to speed up this small project.

Anyone attempting to draw such a chart will soon notice a slight scheduling problem, caused because not all tasks have the same duration. The trades engaged on the shortest tasks might have to stand by and wait while their busier colleagues catch up. As shown in Figure 22.4, one or more intentional delays (buffers) might have to be inserted in the plan to accommodate that difficulty.

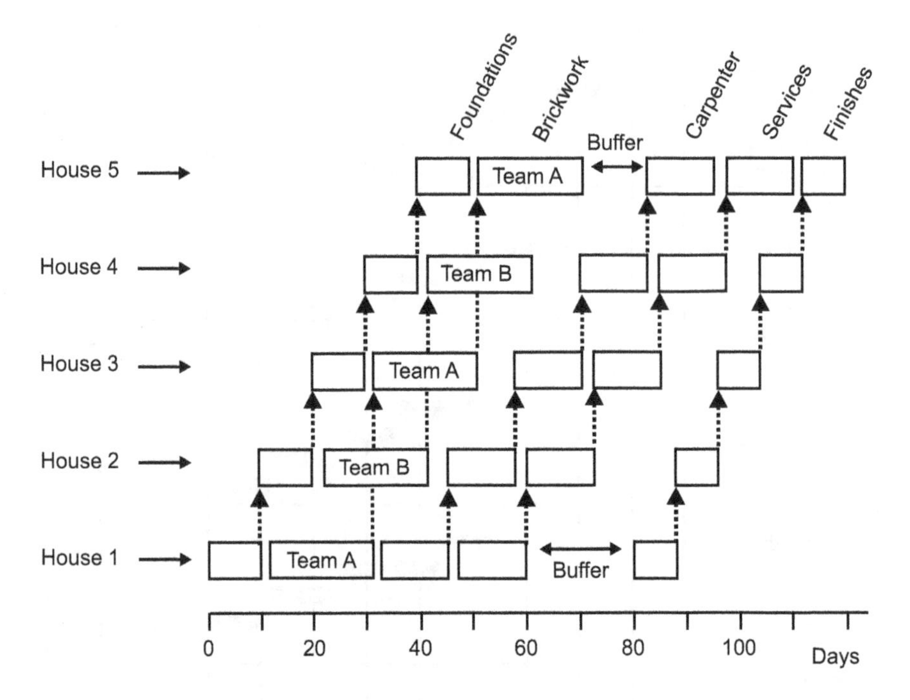

Figure 22.4 Line of balance chart for the five-house construction project.

Line of balance for a project to build 80 houses

Figure 22.5 is different form of a line of balance chart for a construction project, but this project is for 80 houses. Here, not only are there many more houses than in the previous example, but the tasks have been designated in slightly more detail so that there are 15 separate tasks in the plan for each house. Although this chart looks very different from the charts used in the five-house project, it uses a similar method. The principal difference is that, rather than endure the tedium of drawing a huge Gantt chart showing all the 1,200 separate tasks needed to construct all 80 houses, the drawing has been greatly simplified.

Every set of 80 identical stepped and linked tasks is represented in Figure 22.5 by a single sloping line. The thickness of each line is proportional to the duration of the task at each house. The slope of each line depends on the time allowed by the planner for completion of all 80 identical tasks. The bent line for the glazing tasks has been caused deliberately because the planning engineer has introduced the time buffer that is necessary in most line of balance charts for construction projects (mentioned above in the simple five-house example).

When a vertical 'today's date' cursor is placed on the chart, it will intersect each bar to show the number of houses for which the various tasks should have been completed on that day. The scale should be chosen so that the bars slope at the maximum possible angle from the vertical. Then the angle between the cursor line and the bars will be greater, allowing greater accuracy when reading off the results.

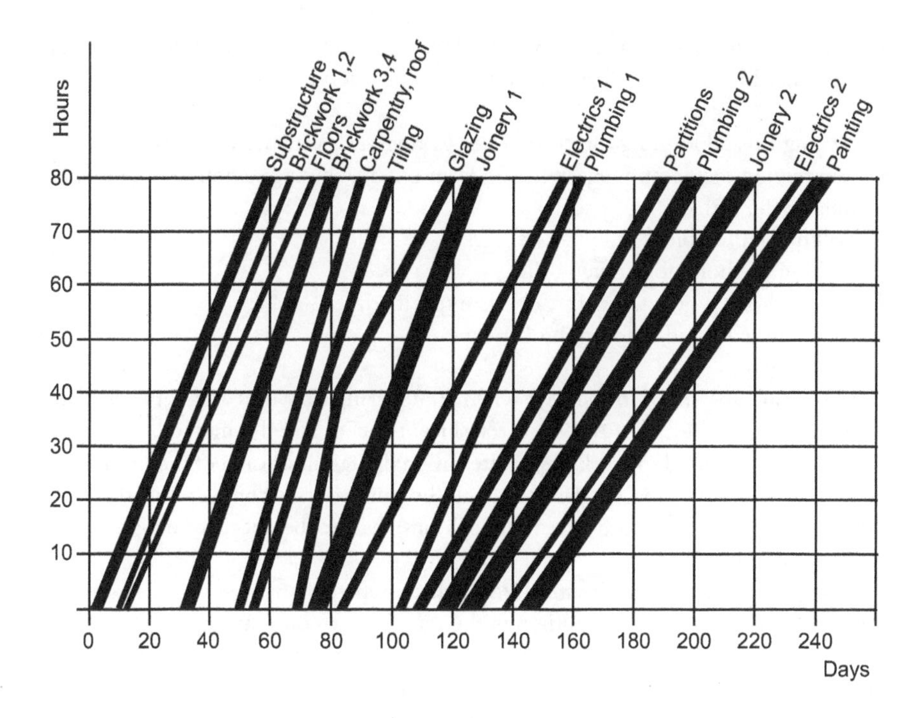

Figure 22.5 80-house project: line of balance chart.

Critical path network modules and templates

In organizations that regularly carry our similar projects it should soon become apparent to the project managers and planners that some groups of tasks tend to repeat from one project to the next. The first example of this of which I am aware dates back to the 1970s, when a construction company in the English Midlands had set up a tiny PMO. Critical path planning was the primary function of this PMO and they had perfected the art of repeating groups of activities from one project to the next for tasks such as concreting and bricklaying. So, suppose they wanted to schedule the tasks for laying a concrete slab foundation. They could simply specify one task named 'lay concrete slab' and the tasks for ordering the materials and carrying out the work would be attached automatically from information stored in the computer. That practice has now become more common and is an example of a process sometimes described as using network modules or standard templates. The following two cases are from my own experience.

Modular network diagrams in heavy machine tool engineering projects

The products range of a British heavy engineering company included the design, supply and commissioning of transfer line machinery for the precision machining of automotive components such as cylinder blocks, cylinder heads and gearbox cases. The transfer line might perform the following operations:

- picking up and locating the component casting very accurately in a jig;
- transferring the component in the jig from one machining station to the next;
- milling the surfaces;
- drilling all the holes;
- probing the holes for broken drills;
- tapping holes where screw threads were specified;
- probing the tapped holes for broken taps.

This example goes back to the days when engineering drawings and network plans were stored in the form of AO sized translucent sheets, and the methods described here have been proved to work even better using digital files. Some of the machining heads (such as those used for drilling) could be purchased as complete off-the-shelf units but others (particularly the milling heads) had to be specially designed and manufactured in house.

It soon became clear to the small PMO that a repetitive pattern was emerging in the project network diagrams. Although the transfer line composition and layout was specific for each project, component design, the portions of the networks machining heads for the machining heads were similar. So it was possible to produce templates (network elements) that could be stored ready for use as small self-adhesive translucent sheets for the machining heads that could simply be stuck on to an AO sized translucent drawing sheet or a roll to build up the network pattern. The part of the network for designing and manufacturing the lining transfer line was also very similar from one project to the next. So the PMO could paste up the network diagram for each new project and all the project manager had to do was to check the network logic (removing any duplicated design work) and add duration estimates to the tasks.

These were activity-on-arrow networks, but the method works in exactly the same way for precedence networks. All the event numbers were preprinted on the forms as three digit numbers. But when the planner prefixed all of these with a three-digit project identifier, that prevented duplicate numbers appearing in the total schedule model when all the projects were scheduled together as one large multiproject.

The template-derived critical path networks were processed using a mainframe computer. The resulting departmental work-to lists (produced after resource levelling), needed only the following resource types:

1 Design engineers.
2 Drawing office staff, including tracers.
3 Heavy machining.
4 Light machining.

A daily cost rate was specified for each of the above resources. The time units used were whole days, with weekends and public holidays taken out of the schedule. Note the very simple choice of resource grades, which proved entirely adequate. Further breakdown of the machining tasks was left to the production control department. This method proved so successful that:

- non-technical clerical staff could 'construct' a draft network using standard templates, using only the machine schematic diagram that was drawn for the original sales engineering purposes;
- the time taken to draw the network logic was reduced from several days to hours;
- logic errors were reduced or eliminated;
- schedule reliability was greatly improved;
- total project cost estimates emerging from this automated scheduling process were within ± 5 per cent of those produced independently by the company's cost estimating department.

Of course the project managers had to be involved in the process, but all they had to do was check the overall network before processing, review the estimated times and (in some cases) remove design tasks where existing designs could be used again. Figure 22.6 shows one of the standard modules.

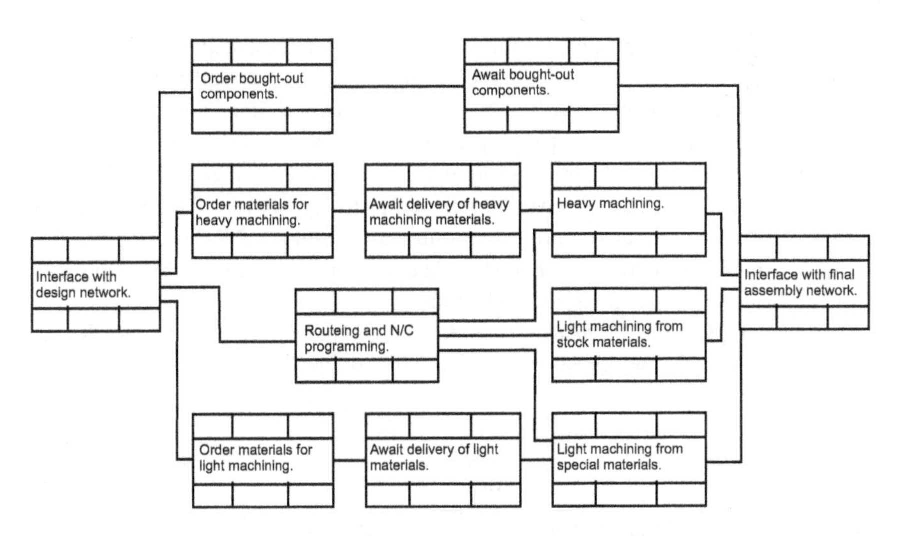

Figure 22.6 Procurement and machining network template for a project to produce automotive transfer line machinery. The very simple template shown here was sufficient for scheduling the machining of all the components needed for each specially manufactured assembly (such as a milling head) in a transfer line machine project. The project ID number prefixed every task ID number (not shown here to save space). The left and right hand interfaces connect the logic automatically with the design and assembly networks, respectively.

Case example 2: Modular network templates for automotive turbocharger projects

This case example describes briefly the use of modular networks that arose from a consultancy assignment which I undertook some 20 years ago. The client company specialized in the design and manufacture of turbochargers for engines fitted in heavy goods vehicles. This company's problem was that they usually had over 100 small projects in various stages of progress, ranging from active projects to those put on hold awaiting customer decisions. The company's activities were heavily dependent on the changing priorities of its customers, so the plans had to be versatile. All these projects demanded the use of common engineering design, pilot manufacturing and test-bay resources.

The turbochargers for all these projects broke down into the following five main components:

1 A central shaft and bearings unit.
2 Compressor wheel (mounted on one end of the central shaft).
3 Compressor wheel housing.
4 Turbine wheel (mounted on the other end of the shaft).
5 Turbine wheel housing.

Figure 22.7 is a very greatly simplified and stylized representation of the above components. Each new project could require anything from a complete turbocharger design or (more frequently) a redesign of either the compressor components or the turbine components. This company's problem was solved using template modules, stored in their computer. There was one standard network template for each of the above components plus one standard linking template that interfaced with whichever modules were needed for the project, as outlined in Figure 22.8. The software was chosen with great care to ensure that it could cope with the large number of projects and carry out full cost and resource scheduling.

So, by using modular templates, stored in the computer, the network logic for any new project could be produced very quickly. Then the project manager had to check

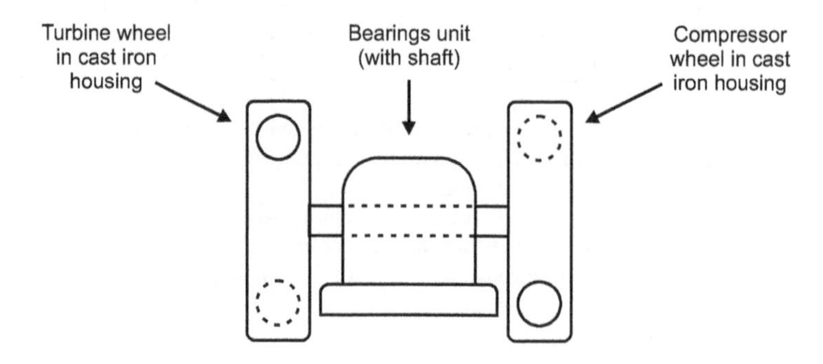

Figure 22.7 Principal components of a turbocharger.

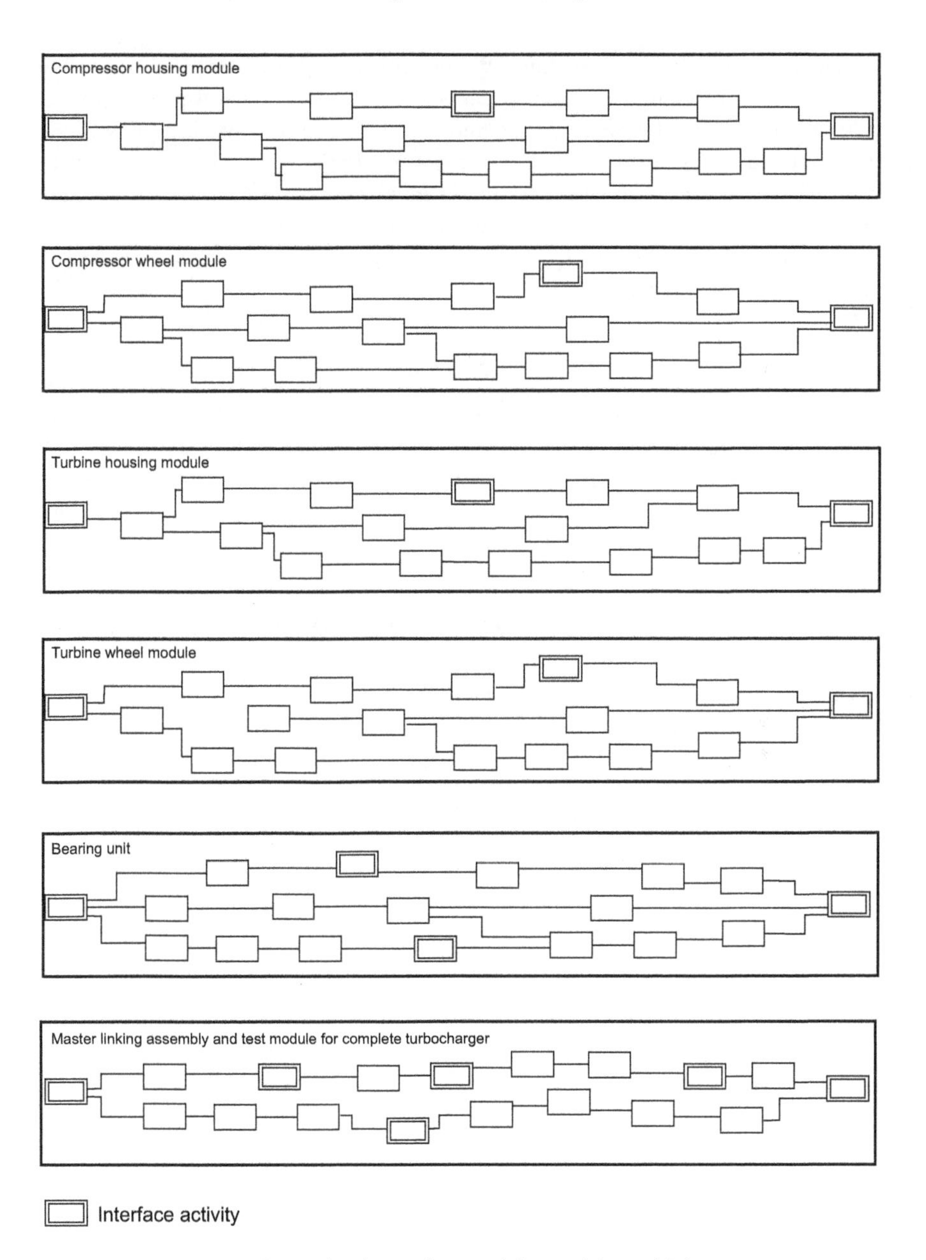

Figure 22.8 Concept of critical path template modules used for multiple automotive turbocharger engineering projects.

the standard estimates, in case there was something special about the new project. The complete multiproject model could thus be made available within a day or so of receiving each new project order.

There was another positive spin-off from this consultancy assignment. During the course of designing the standard network modules the critical path logic was subjected to scrutiny by the senior engineers in a couple of brainstorming sessions. That resulted in a decision to remove one unnecessary cycle of prototype manufacture and testing, which cut about six weeks from every project schedule, saving the company hundreds of thousands of pounds in staff time and materials. The chosen proprietary software also incorporated a time-sheet recording and analysis facility, which was another benefit.

Conclusion

The examples given in this chapter owed their success to the skills and imagination of dedicated people working in PMOs and engineering design. I recommend that, if it has not already done so, every company that engages in a series of similar projects should seek to save time and money by exploring the possibility of using standard network modules or templates. The benefits could be very significant.

References and further reading

Lock, D. (2007), *Project Management*, 9th edn, Aldershot: Gower.
Lock, D. (2013), *Project Management*, 10th edn, Farnham: Gower.

PART V

Risk management

23

INTRODUCTION TO RISK MANAGEMENT

Tony Marks

Project risk management is a proactive approach for identifying, assessing and (where possible) mitigating or averting the effects of risks that could prevent successful project delivery.

Introduction

The identification and management of project risks used to be ignored but are now recognized as a cornerstones of successful project delivery. When risks become actual events they can affect any or all areas of the cost, time and performance objectives set within the project specification or business case. Identification and management of risks should be a continuous activity, beginning with a risk analysis very early in the project lifecycle, with monitoring until project delivery. For internal capital projects that means carrying out an analysis of potential risks during the feasibility study stage, with the results recorded as part of the business case before considering project authorization.

Project changes, which can happen through a variety of reasons and at any time, pose risks such as scope creep, overspent budgets and late completion. However, the management and control of project changes is covered in Chapter 32, so is ignored here.

One definition declares that a risk event is 'an uncertain event that, should it occur, will have an effect on achievement of one or more of the project's objectives'. Risks of this type can be identified, assessed and managed through the risk management processes described in this handbook. The following are examples of risk check questions that should be asked - and answered:

- Will the new product launch date be achieved?
- Will our delivery of components be here on time?
- Can our IT systems cope with the new project demands?

- Can our staff deliver the objectives we have set them?
- Will our design satisfy the client's needs?

Project risk is the cumulative effect of risk events and other sources of uncertainty. Although, for practitioner controls it is important to focus on the risk events themselves, the overall project risk should be the principal concern and that must be managed accordingly.

Risks are not always negative events. Occasionally they can provide opportunities. For example, the delay of a major project could (depending on the contract and reasons for the delay) result in additional revenues for the contractor. Whether or not a risk is a threat or opportunity will often depend on which side of the contract you happen to be.

Risk is very much 'in the eye of the beholder'. For example, if the project budget cannot be altered, even the smallest apparent risk could become a major risk. If the project delivery date cannot be altered, then any risk to project progress and completion could be serious. How a risk is perceived and categorized will therefore be dependent on the needs and expectations of the various project stakeholders.

All of us, either consciously or inadvertently, practise risk management in our lives on a day-to-day basis. We buy insurance for our cars, use pedestrian crossings when crossing roads, put our money in secure bank accounts, store food when there is risk of future shortages, have regular health checks and so forth. All of these are all simple day-to-day actions representing risk management. For project stakeholders risks could affect one or more of the following project aspects:

- schedule and delivery;
- budget and costs;
- quality, reliability or performance objectives;
- client satisfaction;
- project staff morale and motivation;
- the environment;
- health and safety.
- contractor's reputation and market competitiveness.

So, clearly, project risk analysis and risk management have to be taken very seriously.

Formal project risk analysis is a fairly recent development when compared with other aspects of project management. It has arisen partly in response to the recognition of the magnitude of risks faced on many projects, the need to reduce those risks, and of the need to put measures in place so that we can be prepared should risks materialize into actual events. Many project contracts now mandate the use of risk management for the parties to the contract, including consultants, managing contractors, main contractors and subcontractors. Very often, risk management is spread across the different organizational structures within a project.

Insuring against risk

A large commercial industry exists for insuring against risks of all conceivable kinds. This industry employs many capable underwriters and their agents. Yet project management texts and project risk managers seem to be completely unaware of the protection that insurance can offer. However legislation provides that some risks, such as those concerning personal injury and professional liability *must* be insured against and failure to do so would render the organization liable to prosecution, with serious consequences if found guilty.

Examples of optional insurable risks include such things as pecuniary loss, thefts, accidental damage and fire. Key person insurance can provide cover for the unforeseen illness, injury or death of the people such as the project manager. At least one professional insurance agent should be included in the address book of every project risk manager.

Introduction to project risk management methods

All current project risk management methods require assessment of several risk factors, which include:

1. Identity – recognizing the existence of a possible risk and giving that risk a name.
2. Probability – how likely it is that the risk will actually become reality and happen.
3. Impact – the effect on the project should the risk actually happen.
4. Mitigation – the steps that could be taken in advance to lessen the negative impact should the risk event happen.

Identifying possible risks means giving careful thought to lessons learned from past projects (internal and external) and also using imagination. Brainstorming is one process for identifying possible risks.

Lack of certainty implies that there is a probability (but not a certainty) associated with the risk. If a perceived risk is certain to happen, then it is not a risk but is instead a problem (or, in popular current management jargon, an issue) to be dealt with using traditional project management and planning methods.

The impact of a risk is a measure, estimate or prediction of its seriousness in terms of the effect that its occurrence would have on the project. Account must be taken here not only of the project's primary objectives, but also of the possible effect on issues such as safety, reliability, the organization's overall business and value for money. If there is no expected impact then the risk is irrelevant and can be ignored.

Identifying and listing possible risk mitigation measures is an important stage in risk analysis because it allows project management to be prepared in advance and have measures in place should the worst happen and a risk actually materialize.

Qualitative and quantitative approaches to risk

There are two distinctly different approaches to risk management. These are *qualitative* risk management (discussed in Chapter 24) and *quantitative* risk management (Chapter 25). Qualitative and quantitative are somewhat similar words and it is easy (but quite wrong) to confuse their meanings.

The qualitative approach to risk involves first listing and analysing the types of risk that could occur in a project. Then their sources, the nature and severity of their primary impacts and possible secondary (knock-on) impacts are registered and considered. This kind of analysis depends to a great extent on the expertise, experience, knowledge and judgement of the individual team members who take part in the risk analysis.

The quantitative approach takes the process a few stages beyond the qualitative method. It attempts to predict not only the likely occurrence of project risks, but attribute some value to the risk's predicted impact on the project and to rank each risk in order of its overall severity. This method is best performed by processing vast quantities of information, using a computer and custom-made software. But quantitative risk management is also commonly undertaken on the basis of human judgement and experience.

Risk cultures

There is no textbook answer on how to handle risk other than to prepare a risk mitigation strategy. The project manager or the organization must rely on sound judgement and the use of appropriate tools for identifying possible risks, forecasting the probability and impact of their occurrence and devising measures to minimize negative impacts. The ultimate decision on how to deal with risk is most often based upon the project manager's tolerance of risk. Three categories used to describe an individual person's tolerance to risk are:

- Risk averter (or avoider).
- Risk neutral.
- Risk taker or risk seeker.

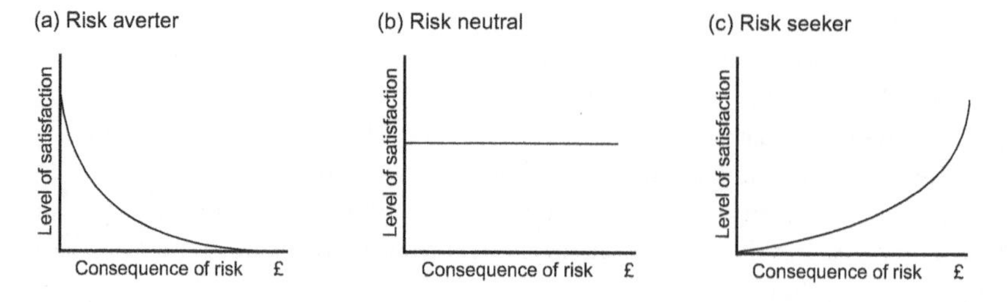

Figure 23.1 Three different attitudes to risk.

With the risk averter, depicted graphically in Figure 23.1(a), personal satisfaction is reduced when greater risk is at stake. The risk averter is afraid or wary of risk. A risk neutral person is less influenced by the possibility of risk and personal satisfaction is not affected by the level of risk present or accepted (Figure 23.1(b). A risk seeker is willing to live dangerously, prefers the more uncertain outcomes and will demand a premium to accept a risk (Figure 23.1(c). Clearly it is useful for senior managers to have some idea of which of these categories applies to their senior staff (including the person appointed to manage the project).

Risk management and opportunity

Although we generally think of risks as threats, some risks can also be looked on as opportunities. A risk opportunity can be realized in several ways.

If a risk management process is properly implemented and managed it can provide a business opportunity that might not have existed previously. For example, it could be prudent to bring a piece of subcontracted work back in-house and so gain greater control over enhancements and response to change. That, in turn, would offer a greater opportunity to the business in terms of flexibility in changing and responding to customers' needs.

A risk that has been defined as a threat to the project could also be turned around and exploited as an opportunity. For example, suppose that someone were to be injured as a result of project work. That clearly would be a matter for regret and in-depth investigation. However, that risk event might provide a reason and an opportunity for developing some new safety mechanism or procedure that could give the company a future competitive advantage. The resulting invention of some safety mechanism or equipment could even result in a new product that could be marketed for profit.

One problem of risk management is identifying how well it is being carried out. If a risk does not happen it does not mean that the analysis was wrong or an unnecessary waste of effort, time and money. The effectiveness of planning risk mitigation measures only becomes apparent when a risk event actually happens. Risk assessment is something like gambling because decisions must usually be assessed on the basis of uncertain or incomplete data. All risk forecasts can only be best assessments. But those assessments and preparedness are far better than relying on blind faith and unsupported optimism, where just hoping that risks will never materialize would result in your doing nothing.

Costs and benefits of risk management

All the risk management procedures introduced in this chapter will incur a cost in some way or another. That cost might be just the cost of time incurred in the risk management process or it could also include the cost of insurance, additional or alternative procurement, or even the cost of totally redefining the project. In all cases a cost-benefit analysis should be conducted, looking at:

- the cost of managing a risk;
- any resulting cost saving;
- the cost of ignoring the risk.

In some cases, the cost of a proposed mitigation strategy for managing a risk (or a series of risks) might be considered as excessive, even possibly exceeding the overall project budget. Then a decision might have to be made on whether or not the project should be allowed to proceed. Conducting high-level risk management early in the project lifecycle (at the inception stage) will allow risk management costs to be identified and provided for by inclusion in the bidding or procurement strategy.

Figure 23.2 illustrates some aspects of the relationship between time and costs associated with project risk. As any project proceeds the sunk costs will accumulate, so that the cost of an actual risk event that causes work to be scrapped and repeated will become greater as the project proceeds. Conversely, the potential of a positive risk event to add value to the project will reduce from maximum to zero as the project progresses through its lifecycle phases.

Some risk management benefits could be tangible, saving both time and costs. Other benefits might be intangible, such as a strengthened team, improved morale and so forth. To conclude this chapter, here are some of the benefits that can be delivered by effective risk management:

- More realistic plans and budgets.
- Identification of alternative approaches or methods.
- Improved team spirit and morale.
- Increased probability of the project delivering what it was supposed to deliver.

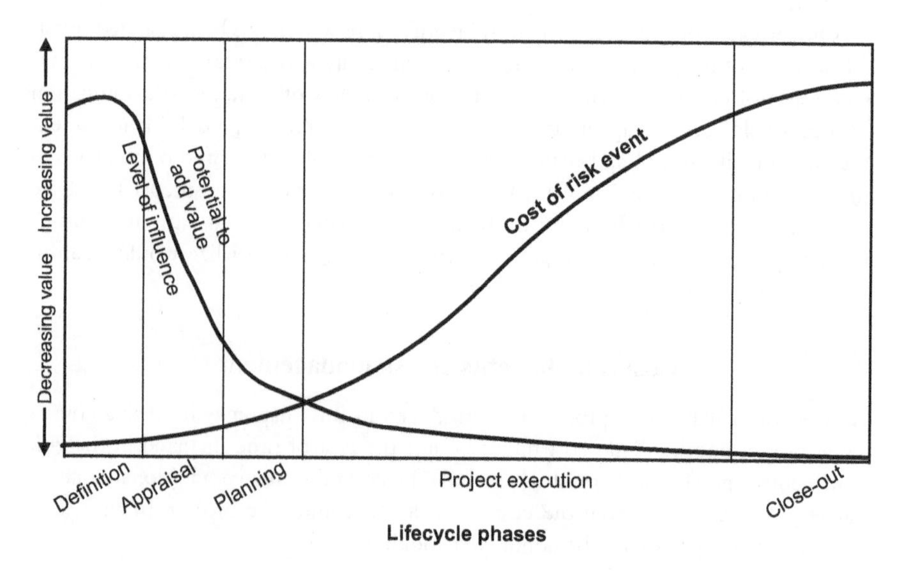

Figure 23.2 Typical time, cost and benefits relationships of project risk events.

- Increased customer and team confidence in the project data.
- Assistance for future projects.
- Assistance with the justification of contingencies and additional spending.
- Possible justification for not proceeding with a project (pulling the plug).
- Help with focusing the team on the project objectives and key project issues.
- Helping to eliminate the luck factor.
- Helping to ensure that sound contracts are awarded.
- Giving more credibility with customers and clients.
- Improving some aspects of communications within the project.
- Increasing knowledge and experience within the project team.
- Possible assistance in gaining future business.

The important subject of project risk management is continued in the following chapter (Chapter 24) in which the qualitative approach is explained.

24
QUALITATIVE RISK MANAGEMENT

Tony Marks

Qualitative risk management is a structured process that seeks to identify all the possible risk events that could adversely affect project execution and planned outcomes and, where possible, specify the actions to be taken to prevent each risk from occurring or otherwise mitigate its possible negative consequences.

Introduction to risk management processes

The aim of qualitative risk management is to foresee all the potential project risks and where possible prevent their occurrence or otherwise reduce the undesirable consequences (impact) if a risk should materialize. This can be a difficult, sometimes impossible, process, liberally furnished with unknown quantities and lack of data. There will always be some hidden risks lurking in the background that could suddenly materialize and cause damage to the project. However, if possible risks are identified in advance, mitigation measures can often be devised and held in readiness so that, should the risk happen, its impact can be negated or reduced to an insignificant level.

The essential document for administering the risk management process is a risk register, which, for qualitative risk management, should use a page design such as that shown in the example of Figure 24.1. The qualitative risk management process should be conducted in the following logical sequence:

1. Nominate a member of the project staff who can administer the process as 'risk manager'.
2. Design and prepare pages for the project risk register. An example of a risk register page for qualitative risk management is given in Figure 24.1.
3. Identify and list all the events that could occur and have a significant, undesirable impact on the project. In other words, prepare a risk list and enter the risk names

Risk register								
Project ID: SP 4521				*Project name:*	*Office relocation*	*Project manager: Harry McFarlane*		
Risk ID	*Risk description*	*Responsible (risk owner)*	*Risk category*	*Impact (I)*	*Probability (P)*	*Exposure (I x P)*	*Planned risk response*	*Date recorded*
1.1.1	Materials late on site	Project manager	Client	2	3	6	Expedite and monitor with client.	17 Jan 21
1.1.2	IT hardware required obsolete	Procurement	Technical	3	1	3	Buy as soon as possible defining minimum specification.	14 Feb 21

Figure 24.1 Example of a risk register page for pre-emptive qualitative risk management. This is a simple format that allows risks to be listed and considered with little or no attempt at quantifying their significance.

in the risk register. Figure 24.2 is a checklist of keywords that should jog memories and help to identify possible risks. There is more on risk identification later in this chapter.

4 Establish or estimate the negative impact that each risk would have on the project, should it occur. A common way of doing this is to use a scale of 1 to 5, where 1 represents the least and 5 the most serious factor or impact value.
5 Estimate the probability of each risk event occurring and, again, allocate a score. This could also be on a scale of 1 to 5, where 1 represents the lowest probability.
6 Calculate the exposure of the project to each identified risk, which is found by multiplying the relevant impact and probability factors.
7 Where appropriate and possible, identify and record actions that should be taken to prevent risk from occurring or, if prevention is not possible, to reduce the risk impact.
8 Name the risk owner, who is the person responsible for taking mitigating action for each risk.
9 Subsequently, review the effectiveness of proposed mitigation actions and look for any risk exposure changes.

This process will help to ensure that time, money and effort are allocated to the features that pose the greatest threat to successful completion of the project.

Three key steps or stages can be identified in the risk management process. These are:

1 Identify the possible risks.
2 Quantify the risk effects (the risk impact).
3 Manage to mitigate or avoid the risk.

These steps are illustrated and amplified in Figure 24.3 and described in the following three main sections.

Stage 1: risk identification

The thoroughness with which risk management is accomplished will depend to a great extent on being able to identify (foresee) all the possible risks. Several methods can be used for identifying risks. The approach outlined here is described with reference to the risk identification framework shown in Figure 24.4. Risk management must be seen in its broadest sense. Many risks are outside the project manager's control, such as interest rate changes, fluctuations in foreign exchange rates, commodity prices and so on. The arrow in Figure 24.4 indicates that the nearer we go into the project environment, the more manageable risk should become. However, it is the project manager's job to explore all the areas of risk and ensure that, wherever possible, effective mitigation strategies are identified, documented and implemented.

Assets	Health and safety	Quality
Behaviour	Industrial relations	Recruitment
Capability	IT	Reliability
Changes	Infrastructure	Reporting
Client	Innovation	Reputation
Climate/weather	Language	Resources
Commercial/contracts	Legal and regulatory	Schedule
Communications	Logistics	Security
Competition	Management	Shutdown
Construction	Manufacturing	Social/ethical
Currency	Operations	Subcontractors
Delivery	Partners	Suppliers
Design	People	Technical
Economic	Planning	Terrorism
Engineering	Plant and equipment	Testing
Environment	Political/national	Trades unions
Finance	Project management	Vendors

Figure 24.2 Keywords that could help to forecast possible risk elements.

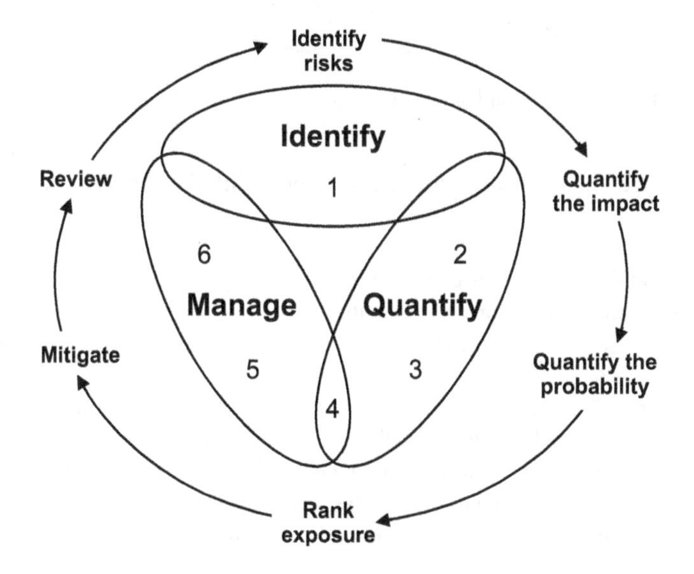

Figure 24.3 Summary of the risk management process.

Risk within the project environment

Risk within the project environment can be associated with factors such as the following:

- difficult or impossible project objectives;
- difficulties caused by the client;

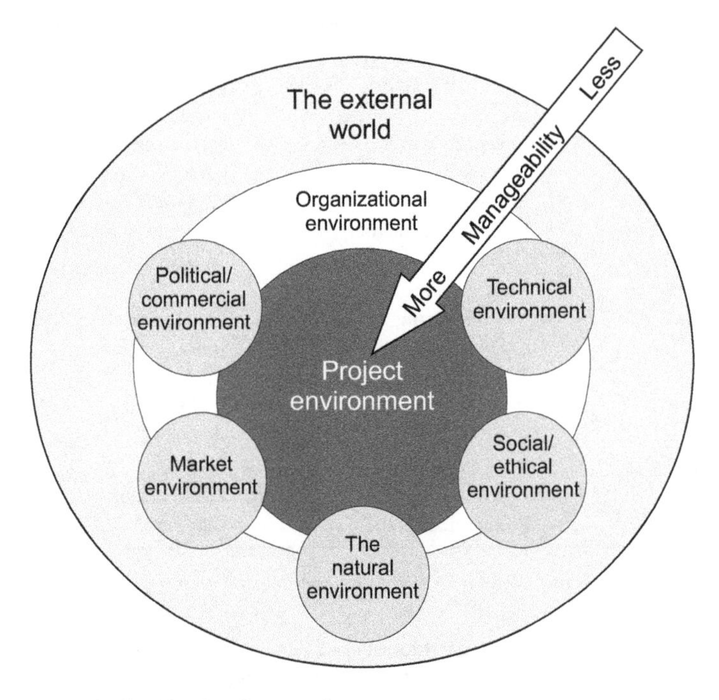

Figure 24.4 Risk identification framework.

- human resources (particularly shortage of people with the necessary skills);
- disputes;
- design or build errors;
- problems with vendors or subcontractors;
- problems with specialists and consultants;
- failure of key equipment or machinery.

Much of the risk in this area should be manageable and within the control of a competent project manager. However, not all project managers are competent. Even those who are can be subject to illness or other personal risks themselves. If the project manager includes severe rock climbing in her external activities, or loves playing golf during electric storms, you need to worry and ensure that you have a reliable understudy waiting in the wings.

Risk in the organizational environment

This category includes risks associated with strategic problems, finance and wider resources, senior management decisions, shareholders and so on. Many risks in this area may be under the project manager's control, but others will not (such as if the project organization runs into cash flow difficulties and cannot pay its suppliers and subcontractors).

Risk from the external world

Risks from the world outside the project and its organization can be associated with social, political and economic factors, legal and regulatory bodies, environmental pressure groups and so on. For projects that include working in the open air or at sea, bad weather can sometimes have devastating results. Clearly many of these risks will be outside the control of project management, but the project manager might be able to devise and deploy appropriate mitigation strategies in some cases.

Risk ownership

Identifying and documenting risk is essential to ensure that every project risk is monitored and controlled, with the responses monitored and measured. Every perceived risk must be allocated to a risk owner. Then the owner becomes accountable for that risk and has to ensure that, should the risk materialize, the relevant mitigating actions are carried out and documented. It will be of no benefit to the project or the business if the risk identification, evaluation and registration processes are carried out without allocating risk ownership to guarantee that the relevant mitigating actions will be taken.

It is possible that a risk could have more than one owner. For instance, there might be someone involved in devising a risk response but, should the risk happen, a different person will be asked to carry out the necessary actions.

Stage 2: risk impact analysis

Once all the possible risks have been identified the risk management process moves into the next stage, which is to assess the probable impact of each risk. The aim is here to identify those risks in the risk register that would have a significant detrimental impact on the project.

If the perceived impact from an identified risk is considered to be insignificant then that risk does not need to be managed. It can be removed from the risk register. Similarly, if the impact of a risk would not be detrimental to the project then, again, effort need not be spent in monitoring it.

Impact types

It is often easier to analyse the effects of a risk occurrence by concentrating on the primary impact types. These primary types are:

- time;
- cost;
- quality (including reliability and performance);
- health and safety;
- environment;
- reputation.

A risk in any of these categories could have an impact on the project alone or at a more strategic level. The purpose of this stage in risk management is to define the impact type of each risk so that the impact can be assessed in the context of the project and its business constraints.

Estimating the impact of each risk

Now it is necessary to estimate the relative impact that each risk could have on the project. There are several simple, yet effective approaches to quantifying the risk impact. One method, suitable for a large project, uses a high/medium/low matrix such as that shown in Figure 24.5.

In many cases it is more reasonable to apply a percentage figure to represent the level of impact. This method can be more useful than the high/medium/low matrix because it allows for a greater level of detail in definition. Also, the same categorization can be used for different projects, because it does not allocate actual figures. However, some people find percentages very abstract and also tend to confuse them with the probabilities of risk occurrence.

It is also often possible to relate all risks to an ultimate financial cost, should they occur. This method is useful in considering project cost contingencies but it undermines the importance of safety and environmental risks.

Some of the softer categories (such as quality, safety and environment) can be difficult to define. When thinking in terms of these categories many people will use their knowledge and experience and go with their instinct.

Stage 3: risk probability analysis

This stage is concerned with assessing the probability of each risk occurring. This will need to be revisited at each risk review to confirm or revise the probability recorded in the risk register.

Every risk carries with it an uncertainty or, in other words, a probability that it might occur. Probabilities are usually expressed as percentages. If a risk has a 100 per cent probability of occurring then it is not a risk, but a certainty. It might still be an undesirable event but we know it will happen, so it should be accommodated in the project plans. Similarly, if a perceived risk has a zero probability of occurring, then it

Level	*Cost*	*Time delay*	*Quality*	*Safety*
High	> £1m	> 4 months	> 50 defects	> 10 LTIs*
Medium	£100K-£1m	1-4 months	20-50 defects	5-10 LTIs
Low	< £100K	< 1 month	< 20 defects	< 5 LTIs

Figure 24.5 A useful risk assessment matrix.

Note: * LTI = lost time incident

will not happen and can be ignored. Most risks that have a probability close to zero can be ignored. However a risk with low probability but a high predicted impact cannot be ignored.

Estimating the probability of a risk

There are tools that can help in determining and ranking the probability of a risk occurrence. One method uses a very simple high/medium/low probability matrix in which each risk is ranked as follows:

- high (likely to happen on projects of this type);
- medium (might happen - has been known to occur on similar projects);
- low (unlikely to happen).

Allocating a percentage chance of occurrence is another way of ranking risk probability. This method can be more useful than the high/medium/low matrix because, even though the percentage figures are only estimates and might have no factual basis, it allows for a greater level of detail in ranking probabilities. This method can be adapted to take account of the distribution of people's views regarding the level of probability. The idea is to obtain a set of estimated probability values from people who have some knowledge and experience of the particular type of risk. Once the numbers have all been submitted, a weighted average can be calculated as follows:

$$P_E = (E_O + 4E_A + E_P)/6$$

where E_O is the most optimistic estimate, E_A is the average estimate and E_P is the most pessimistic estimate, giving P_E as the estimated probability.

Stage 4: calculating and using risk exposure values

When all the steps described above have been performed it is possible to produce an exposure catalogue. The threat of (or exposure to) any risk is a combination of the impact it would have and the probability of its occurrence. The question to be asked is always: 'Is a high impact, low probability risk more or less desirable than a low impact, high probability risk?'

Risk exposure calculations (explained below) are intended to resolve this question. When the relative exposure that the project has against each of the identified risks has been assessed, we can choose those risks which present greatest threat to the project and concentrate our efforts on reducing and controlling those. An exposure catalogue lists the risks in a ranked order, with the high exposure risks at the top and those with the lower exposure values at the bottom. When the risks have been selected and ordered in that way we can decide on the extent to which time and money should be spent in managing each of them.

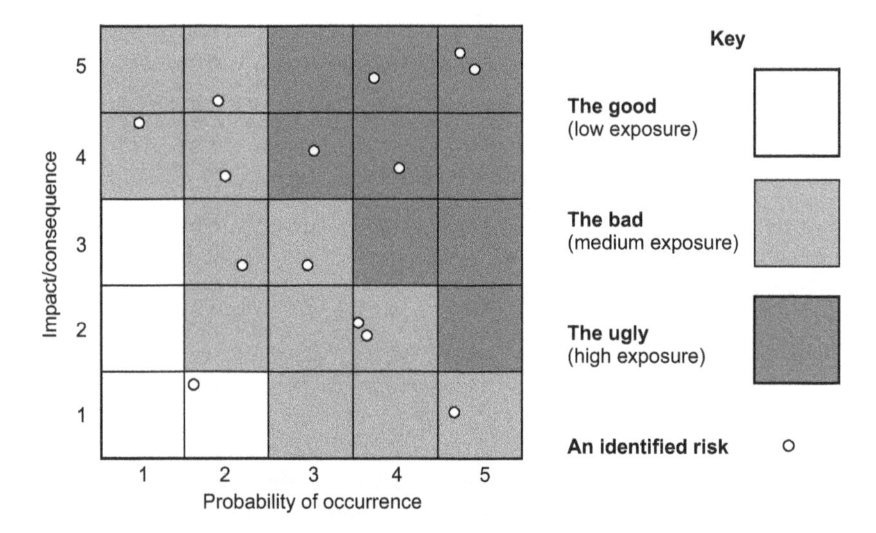

Figure 24.6 A risk exposure matrix.

Risk exposure calculations

There is a very simple method for determining a project's exposure to each risk. This involves multiplying the forecast risk impact by its probability of occurrence. Thus:

exposure = impact × probability

If the risk impact was expressed as a cost value, then this exposure calculation can also provide an input to estimating a financial contingency for the project.

If either the impact or the probability has been defined in non-numeric values, then a risk exposure matrix is the best way to represent the information. By plotting all risks in the matrix, each cell in the matrix will define the relative level of exposure to each risk. In its simplest form the exposure matrix can be used as shown here in Figure 24.6. However, this method does not separate quality, cost or time. Also, it ranks high probability, medium impact risks in the same category as high impact, high probability risks. This may be suitable for some projects, but generally a greater degree of quantification and differentiation is required.

Stage 5: risk mitigation

It is often not feasible to attempt to manage all risks remaining on the project risk register at the same time. Thus it is important to select a manageable subset of the top grouping from the exposure catalogue for active monitoring and risk mitigation.

The first part of the risk mitigation stage is to identify all risks that are considered a threat to the next stage of the project. Then the most appropriate mitigation plans for minimizing probability and impact of these risks have to be devised and documented.

In essence this process attempts to shift all the risks identified in the exposure catalogue from the top right hand corner (high risk, high impact) of the risk exposure matrix to the bottom left corner (low risk, low impact). However, this will not always be possible.

Mitigation strategy

There are several ways to reduce the exposure of a risk. The purpose of this step is to define the most appropriate and cost-effective mitigation strategy for each risk remaining in the exposure catalogue.

The activities at this stage should be ordered so that the most desirable mitigation strategies are considered first, before working down to the unavoidable options. In general (subject to cost considerations) mitigation should be conducted in the following sequence of steps.

- **Risk removal:** is it possible to remove the causes of risk. For example, could the project plan be changed? Would it be possible to change a relevant specification, such as specifying a pump driven by an electric motor for one driven by a diesel engine? It is important to remember that by changing a specification in this way you might create a new set of risks related to the revised specification.
- **Risk transference:** is it possible to transfer the risk away from the project? For example, could the risk be insured against? Could it be transferred via contractual terms to the client or a subcontractor? The problem with trying to transfer a risk by subcontracting is that if the risk is of a high severity the subcontractor might not be willing to take the job on.
- **Risk reduction:** is it possible to reduce the exposure of a risk (reduce its probability, or its impact, of both of these)?
- **Manage:** where none of the above options is available we must ensure that the remaining risks are carefully managed. That means ensuring early detection should they occur so that contingency plans can be enacted.
- **Accept the risk:** the reason for accepting a risk might be that the cost to mitigate the risk could be greater than putting a risk mitigation strategy in place.

Some mitigation strategies will introduce secondary risks. These will need to be fully evaluated. Thus there could several iterations back through stages 1 to 3 of the risk management procedure for some of the more complex risks.

Stage 6: risk monitoring and review

Risk monitoring is carried out by ensuring that adequate reporting mechanisms are put in place.

The task is to ensure that all the risks are adequately monitored so that timely action can take place. It is equally important to recognize both if a risk occurs and if it does not. If a risk did not occur, it does not indicate that the initial analysis was wrong. Often the opposite is the case: it could mean that the mitigation strategy was effective. Risk monitoring takes place on two levels, as follows:

1 Proactive monitoring, where we continuously assess the effectiveness of our mitigation strategy for each risk. In other words is the mitigation strategy we have in place effective in containing the risk?
2 Reactive monitoring, which happens after a risk has materialized and positive (but hopefully anticipated) mitigating action is needed.

This monitoring stage will ensure that effective and frequent reviews are carried out and, where necessary, appropriate action is triggered.

Risk reviews

Risk reviews should be carried out at regular intervals. The purpose of each risk review is to:

- identify which risks have occurred during the period under review, and whether or not any contingency action taken was adequate;
- identify those risks that could have occurred during the period but did not;
- monitor the effectiveness of mitigation measures on risks that are still active;
- check risks that might occur during the next review period and confirm that the mitigation strategy for them is still appropriate;
- identify any new risks that could occur and instigate the necessary analysis (by going through stages 1 to 4 described above);
- keep the risk records updated.

Risk evaluation review

Risk management, as with any other management discipline, relies heavily on experience. Some elements can be automated or reduced to mathematical modelling, but prediction based on past performance is still the most reliable method and source of data. Throughout the project lifecycle, the risk analysis and management methods outlined in this chapter should capture and record the decision-making processes and the conditions that existed in bringing about the elimination or realization of each risk.

The review stage, carried out after the project has been completed, is designed to review the effectiveness of the risk analysis techniques and risk management process employed. If we successfully avoided a risk occurring, then the procedures put in place on this project need to be made available to others. If our analysis was inaccurate or our mitigation strategy was ineffective, then we need to analyse why, and ensure that mistakes will not be repeated in the future. Here are some check questions that need to be asked in a post-project review:

- Did risks occur that were not identified? If so, what mechanism could be put in place to increase the chances of identification next time?
- Were risks identified but inaccurately analysed? If so, what mechanisms can be put in place to improve both the impact analysis and probability estimates?
- Was any risk identified too late? In other words, did we fail to recognize in time that a risk had materialized? If so, how can we improve our detection procedures?
- Were our mitigation strategies effective? If not, why not?
- Was the level of risk analysis and monitoring appropriate?
- Did we spend too much/too little time on risk management?
- How might we improve the effectiveness of the analysis?
- How can we improve our techniques for exposure catalogue cut-offs?

The important role of risk registers

To conclude this chapter it is appropriate to stress the important role that risk registers play in the risk management process. The project risk register is a central control document, a kind of road map, for risk management during the project lifecycle. Here is a recapitulation of the entries that need to be made on each risk register page.

Reference ID:	A unique number or code that identifies each risk entered in the register;
Description:	A concise but complete description of the risk;
Responsibility:	Name of the person assigned to oversee the risk and ensure appropriate action to minimize exposure to the project;
Risk category:	This could either be by reference to the WBS or to a generic function such as commercial, financial, people, management, processes and so on;
Impact:	This entry can be a simple high, medium or low rating (from 1 to 5 for example). It could also be more explanatory, such as relating each risk impact to quality or to probable delays or costs;
Probability:	This is usually expressed by rating the probability, perhaps using a scale of 1 to 5;
Phase:	Not seen in all risk registers, this entry could have use in projects with long lifecycles, giving an indication of when and for how long mitigation strategies need to be in force. It allows risk entries to be removed after their particular phase has passed.
Risk response:	A key entry defining mitigation actions and the person responsible for carrying them out.

Corporate risk register

Ideally, but by no means commonly, every company that routinely conducts projects should maintain a corporate risk register. This should be a digest of information from the company's past and present projects. Then the corporate risk register can be used as a reference to help in managing and registering risks on new and future projects. As time progresses, the corporate risk register becomes a valuable resource for the company.

25

QUANTITATIVE RISK MANAGEMENT

Tony Marks

This chapter continues the discussion of project risk management.

Introduction

Quantitative risk management can take more than one form but, as described in this chapter it is a statistically-based approach that uses 'What if?' scenarios. The method requires historical data that are representative, accurate and of significant volume.

Estimates and assumptions for each project activity can change during the project lifecycle, affecting both costs and durations. During a what-if scenario analysis, assumptions are made that review both the costs (the budget) and the estimated durations (the schedule). The most popular 'What-if?' scenario technique is the Monte Carlo method.

The Monte Carlo method (or probabilistic analysis) is an incredibly useful tool when there are good historical data (or 'norms') and an experienced practitioner is involved. In these cases the technique can be highly sophisticated and give meaningful cost and schedule predictions. However, a lack of sound historical data or experience can lead to meaningless outputs, so the method must be used with care. This chapter illustrates the Monte Carlo method using very small amounts of sample data. To use this technique on a significant project would need the use of specialist software.

Monte Carlo simulation

The objective of producing a risk model is to inject a degree of realism into plans. This is done by applying ranges of possible costs and durations to the deterministic (single point) estimates. These ranges are called *input distributions* and are covered in more detail later in this chapter.

First, consider the simplest input distribution, which is a triangular distribution. This distribution comprises three values, as follows:

- minimum possible value;
- the most likely value;
- maximum possible value.

A triangular distribution is shown in Figure 25.1. Here, the probability of any value occurring increases at a linear rate from the minimum value to the most likely value. It then decreases at a (possibly different) linear rate from the most likely value to the maximum value. When drawn, the graph is triangular, hence the name triangular distribution. The best way to illustrate a Monte Carlo simulation is by means of a project example. Within the limits of these pages this has to be a small project, so I have chosen to build a small garden shed. The estimated costs for this tiny project are as follows:

One shed kit	£100
Concrete slabs	£42
Sand	£6
Cement	£6
Van hire	£15
Total estimated cost	£169

This estimate is admittedly unreliable because it was undertaken without reference to any primary data such as shop catalogues or advertisements. However, for this project a rough estimate was needed quickly because I am buying from a small trader who deals only in cash, and I need to withdraw the cash before the weekend. So how much money do I need to take out? Many estimators would use a rule of thumb margin of error, such as one of the following:

- Project estimate based on actual prices: ± 10 per cent.
- Project estimate with no supporting data: ± 20 per cent.

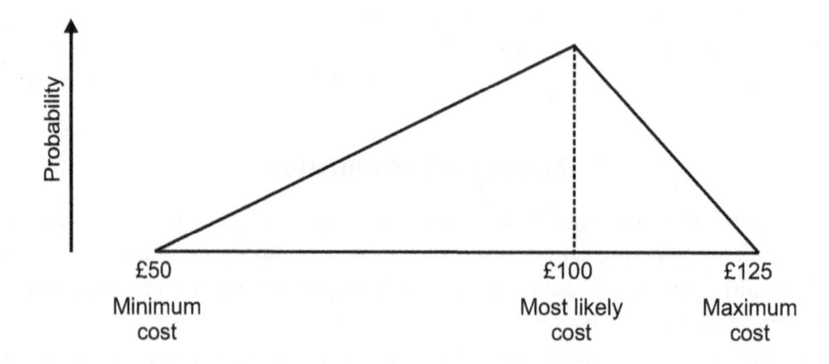

Figure 25.1 A triangular distribution curve.

However, using a Monte Carlo simulation can give a better idea of what the likely cost could be, and what I could expect to pay. In this case example I am only going to consider the risk of uncertain estimates and I shall exclude other specific risks, such as the following:

- My wife might decide that the shed needs to be painted a different colour. So the cost of paint would have to be added to the estimate.
- The shop might have sold out of 7x4ft sheds and I would have to buy a larger shed, which would be more expensive and also require additional concrete slabs, sand and cement.

Minimum, most likely and maximum possible estimates

My next step in this garden shed project is to consider each item in the estimate and apply a range of possible costs. As I am only considering triangular distributions, this estimate can be in the form of a minimum, most likely and maximum cost for each item.

It has to be borne in mind that these minimum and maximum values should be the extremes. Suppose that the range of possible shed costs is considered. It might be possible to find a store offering sheds at discounted prices. Assume that we can find a shed with a 50 per cent price discount in a closing down sale. So the minimum possible cost for this shed becomes £50. However, on the other hand, we could be unlucky and find that there is a timber shortage and shed prices have risen, putting the maximum possible cost at £125. But if we are confident in the original estimate, the most likely cost estimate for the shed will remain at £100. That is the case shown in the triangle of Figure 25.1.

Carrying on down the list of materials, we can enter cost estimate distributions for all the items, as tabulated in Figure 25.2. So now we have a very simple risk model that only considers the risk of uncertainty in the cost estimates (ignoring other risks such as thefts, fire and so on).

Now we need a computer armed with a suitable Monte Carlo simulation tool. The computer will run through the project as many times (iterations) as we instruct it to, randomly selecting values between the minimum and maximum for each item in

Item	*Estimated cost*	*Minimum possible cost*	*Most likely cost*	*Maximum possible cost*
Shed	£100	£50	£100	£125
Slabs	£42	£30	£42	£63
Sand	£6	£5	£6	£8
Cement	£6	£5	£6	£8
Van hire	£15	£10	£15	£20
Total	£169	£100	£169	£224

Figure 25.2 Cost estimate distribution for the garden shed project.

a game of chance that is rather like using a roulette wheel – hence the name Monte Carlo analysis. Because of the chosen shape of the distribution, values close to the most likely value are more likely to be selected than those near the minimum or maximum. On the first iteration the computer might select £60 for the price of the shed, £30 for the slabs, £5 for the sand and so on. On the second run the shed cost chosen could be £100 and so on. Note that the computer will make these choices completely at random, and will not necessarily choose values in whole numbers of pounds. However, here I have rounded all the values to the nearest pound for simplicity. The computer programme will allow the operator to choose the number of iterations. The usual number chosen will probably be around 3,000. The more iterations run, the more the distribution of values selected will approach the defined triangle shape. The histogram in Figure 25.3 shows the results for this shed project.

Even from this simple example there is quite a range of possible costs. By running thousands of iterations, the computer has built a picture of all, or at least most, of the possible outcomes and the likelihood of each outcome being achieved. The model has followed our instructions and the results sit within an envelope that resembles a triangle with a minimum shed price of £50, a maximum price of £120 and a most likely price of £100. The most likely shed price is the price corresponding to the greatest number of iterations (the highest bar on the chart). The *expected cost*, which is the arithmetical mean calculated from all 3,000 results, would be about £90. Combining all the line items (the shed, slabs and cement, and so on) will give the computer far more data to work with and will produce a much smoother histogram, like that shown in Figure 25.4. It is sometimes more beneficial to show the distributions as a cumulative percentage, which in this case would produce the S curve that is superimposed on the histogram bars in Figure 25.4.

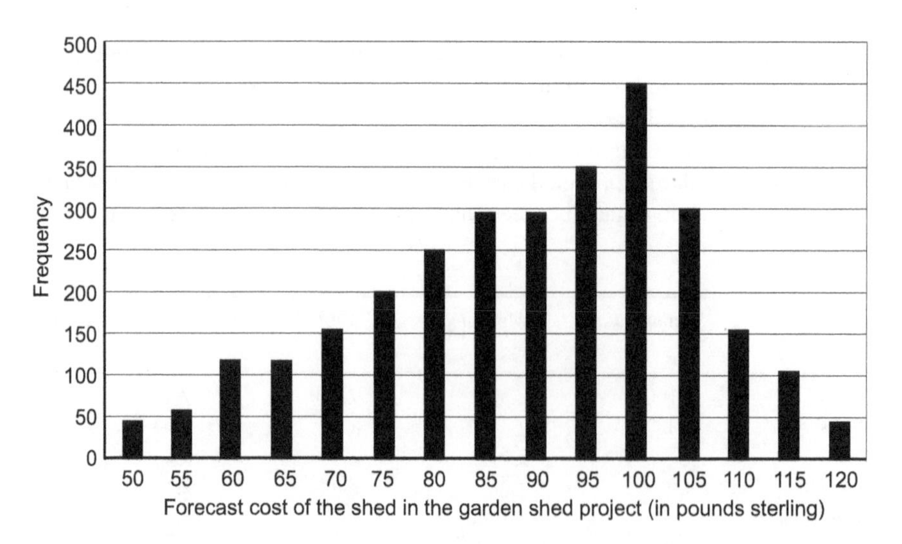

Figure 25.3 Distribution of results from a Monte Carlo simulation for the garden shed cost estimates. Three thousand iterations were run to produce this result.

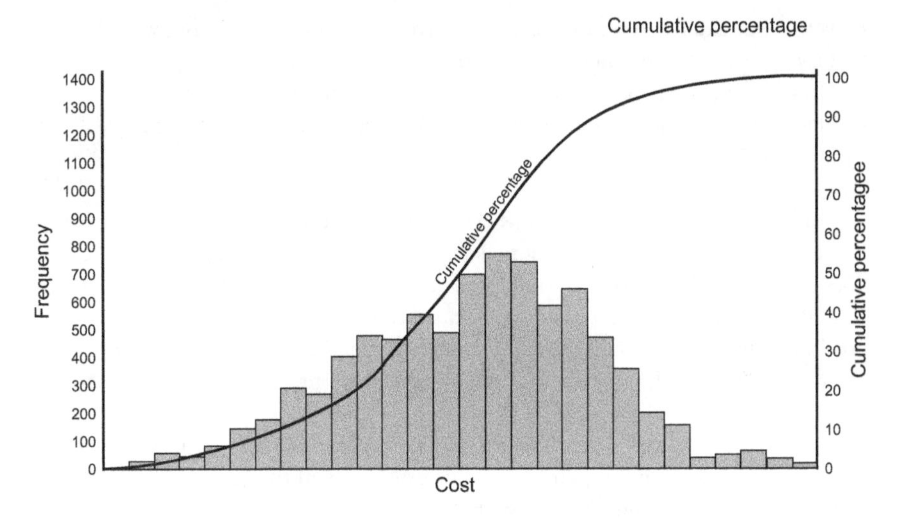

Figure 25.4 Distribution of results from a Monte Carlo simulation for the complete garden shed project using all the probability cost estimate data. With more data to work with the computer can produce a smoother curve.

Mapping the risks

Two distinct types of uncertainty can be included in the risk model. These are known as:

1 General uncertainty in the estimates and;
2 Specific risk impacts.

General uncertainty in the estimates

When producing a cost or time estimate the estimator has had to make many assumptions as to the state of the future. For example, if you noticed your car's fuel tank is nearly empty you would try to find a garage to fill it up. You know roughly that it will cost about £40. If you had to estimate the cost to within 10 per cent, you would find it quite difficult because of several uncertainties. For example:

- You don't know exactly how much fuel will be needed to fill the tank.
- You don't know exactly how much the garage will charge for the fuel.
- You don't how far away the next garage is, so don't know how much more fuel will be used to get there.

Imagine, therefore, how much more difficult it would be to estimate the cost of a large project such as Crossrail or HS2. One way in which to improve the accuracy of estimates is to compile a list check of questions that you can attempt to answer each time you make an estimate. Here are some examples:

- Is the estimate based on firm prices received from suppliers or contractor?
- Have you built a contingency allowance into your estimates?
- Do the estimates make allowances for cost inflation?
- Has our organization undertaken similar tasks before that can be used as a reference?

Another question you could ask is 'How important is this cost to the project?' Returning to the case of the car fuel tank, if you intend to pay by credit card it doesn't really matter if you can only estimate the amount to an accuracy of ±10 per cent. But when estimating project costs, the estimator should spend the highest proportion of time pricing the key 'major' items of expenditure. The costs of minor items can be estimated with less care and can afford to have greater uncertainty associated with them.

So, how do you apply this uncertainty? Generally, cost uncertainty is expressed in terms of a plus or minus percentage. For example, you thought that it would cost approximately £30 to fill your car with fuel. Taking into account the factors already mentioned you could comfortably say that the actual cost would be within ± 20 per cent. Thus the maximum cost you could expect to pay would be £36 and the minimum would be £24. When it comes to producing the risk model for a project, it is vitally important to be sure that you have captured the full range of possibilities.

Schedule risk uncertainty is applied in a similar way. Take the same example of filling a car with petrol. A Formula One team could fill a tank in about eight seconds. It will probably take you five minutes. However, there could be queues at the pumps and it could take you up to 15 minutes. So, your accuracy range on the estimated duration for filling your car tank could be seven minutes minus two minutes or plus eight minutes.

It is always good practice to consider the error ranges of estimates in actual times or costs, as well as by percentages. It is easy to write down an estimate credibly quoting a range of uncertainty of ± 10 per cent, but if the estimate happened to be for the cost price of a helicopter at around £5 million, then ± 10 per cent hides the proportionate reality.

Specific risk impacts

Specific risks are more closely related to the register. In fact, it is a good idea to relate each specific risk directly to the risk register, should anyone in the team start to ask difficult questions along the lines of 'Where did that cost come from?' One way of doing this is to include a 'How modelled' field in the risk register, which provides an audit trail from the register to the model.

Specific risks normally have two aspects: the likelihood of occurrence and the severity of the impact. A common example of this is found in the oil and gas industry. When installing an oil rig at sea, many important items of equipment must be lowered into position by a crane mounted on a barge. There are many documented cases of cranes dropping key items of equipment. The resulting repair or replacement costs amount to millions of pounds.

The oil rig example could be modelled as, say, a 10 per cent likelihood of occurrence with a possible impact of at (minimum) £200,000, most likely £1,000,000, and a worst case cost of £2,000,000. The least expensive risk impact would be the crane dropping the item when it was only a short distance from the deck. The worst case would relate to a drop over the sea, where the item could be lost. Then the associated costs of delay while awaiting a new part to be delivered would be added to the impact.

So far, I have only discussed three-point estimates (worst case, most likely case and best case). There are numerous other probability distributions available to the professional statistician but here I need only cover distributions that are likely to be applicable to project risk analysis.

Uniform distributions

Uniform distributions are used when the cost or duration could be anywhere between two points and there is no possible indication of what the most likely value is. So, in this case, the computer simulation tool will, on each iteration, randomly select a value between the two known extremes.

Discrete distributions

Discrete distributions are used when there is a set of discrete values that could be selected, such as in an 'either or' situation. This can apply when there is a risk that might or might not occur. Suppose you have an identified risk that would cost nothing if it did not materialize but would cost £100,000 if it did. Percentage likelihoods can be attached to each value, for example there might be an 80 per cent chance of the risk costing zero and a 20 per cent chance it would cost £100,000. Note that the percentages chosen (20 and 80 in this case) when added should be equal to 100.

Custom distributions

Custom distributions can be used when there are no specific data on which to base your model and/or the data cannot be easily fitted to one of the standard statistical distributions.

For example: a pipeline must be laid in a particularly hazardous stretch of water. The pipe-laying vessel can only work if the wave heights are less than 1.5 metres high. Weather statistics are available and give the probability of wave heights for each month. An example for June is tabulated in Figure 25.5.

Wave height in metres	<0.5	0.5–1	1–1.5	1.5–2	>2
Probability	20	25	40	10	5
Cumulative probability	40	45	85	95	100

Figure 25.5 Estimated wave heights for a marine pipe-laying project.

These data can be added straight into the model as a custom distribution. The values are entered as shown in the table. When sampling, the simulation will treat individual elements in the table as a mini-uniform distribution. That means 40 per cent of the sample should lie between 1 and 1.5 metres, spread evenly between these two extremes.

Normal distributions

Normal distributions are bell shaped curves that tend to represent many natural events. The spread of results from a student's examination might fit a normal distribution. Normal distributions are not used very frequently but are defined as a function of the mean value, and the standard deviation. One problem with using the normal distribution is that the data we need to enter are skewed towards the pessimistic values. The normal distribution is, by definition, symmetrical. We also must take account of the possibility that negative values might be sampled, which is usually not sensible when talking about estimated costs or durations.

References and further reading

Hillson, D. (2009), *Managing Risk in Projects*, Abingdon: Routledge.
Hillson, D. (2016), *The Risk Management Handbook*, London: Kogan Page.
Kerzner, H. (2009), *Project Management: A Systems Approach to Planning, Scheduling and Controlling*, 10th edn, Hoboken, NJ: Wiley.
Morris, P.W.G. and Pinto, J.K. (2004), *The Wiley Guide to Managing Projects*, Hoboken, NJ: Wiley.

26
USEFUL RISK MANAGEMENT TOOLS

Tony Marks

This chapter continues the important subject of project risk management by describing some more tools that can be used to predict, identify and investigate risk events.

Introduction

A wide range of risk management tools is available to project managers. Some can be used alone for specific purposes but these tools can also be used in different combinations. Some of the more complex tools might be applicable only to complex, high risk or high value projects where investigating in considerable detail is worthwhile. In other cases, some of the simpler tools described here might be more appropriate.

Brainstorming

Brainstorming can identify risks, opportunities and possible solutions to problems by concentrating the thoughts of a group of people in a short time. It is particularly applicable to identifying possible risks and for suggesting alternative methods that are less risk prone. Brainstorming encourages the free association of ideas to open new avenues of thought and invites participants to contribute individually and build on the ideas suggested by others. Early project scheduling meetings can often be similar to brainstorming meetings. A well-conducted brainstorming session can enhance teamwork and participation and encourage participants to identify and evaluate ideas or problems (often in unconventional or innovative ways). Brainstorming can be used whenever there is doubt about how to proceed and solve a problem. It can be used to suggest opportunities for improvement (in value engineering for example), identify possible root causes of problems, identify alternative approaches and compile action plans for corrective action. The procedures for a brainstorming session are well-established, as follows:

- A competent person must be chosen to lead and encourage the session.
- Everyone participates, either randomly or in turn.
- As many thoughts and ideas as possible are written on a flip chart or board.
- Absolutely every idea or suggestion is recorded, with nothing ruled out. No time is considered to be right, wrong or ridiculous at this stage.
- Nothing is discussed in detail.
- Ideas and suggestions will build as the session proceeds.
- Any violation of these brainstorming rules has to be pointed out.

When no more suggestions or ideas are forthcoming, the responses can be evaluated and categorized. Then priorities and a plan of action can be identified.

Evaluating risk and reliability using cause and effect diagrams

Cause and effect diagrams (also known as fishbone or Ishikawa diagrams) are a powerful investigative graphical method for predicting and analysing possible causes of a defect or problem. They combine analytical and creative thought with team effort and provide a way for breaking problems into smaller pieces that are easier to manage and understand. Cause and effect diagrams allow searches to identify the root causes of problems (actual or possible) by asking who? what? where? when? how? and why? questions and adding answers in a visible way. Fishbone diagrams can:

- Help individuals and groups to generate ideas.
- Record those ideas.
- Provide a visual display of the relationships between conditions, causes and effects.
- Reveal previously undetected relationships.
- Contain details that indicate how thoroughly a problem has been investigated.
- Provide a historical document of the thought processes used to identify risk effects, causes and possible solutions.

The Ishikawa fishbone diagram

An investigation using fishbone diagrams uses the following steps:

1. Identify and state the problem or effect. Be clear, concise and specific. That constitutes 'the statement'.
2. Write the statement in a box on the right-hand side of a sheet of paper.
3. Decide the major items to be included and write these categories in boxes above and below a main line leading to the statement box.
4. Connect these major item boxes with lines that slant toward the main line.
5. Organize or brainstorm all possible causes of the problem.
6. Continue to break down the possible causes by asking questions (Who? What? Where? When? How? and Why?) until there are no new ideas.
7. Evaluate and analyse all the possible causes. Look for common causes.

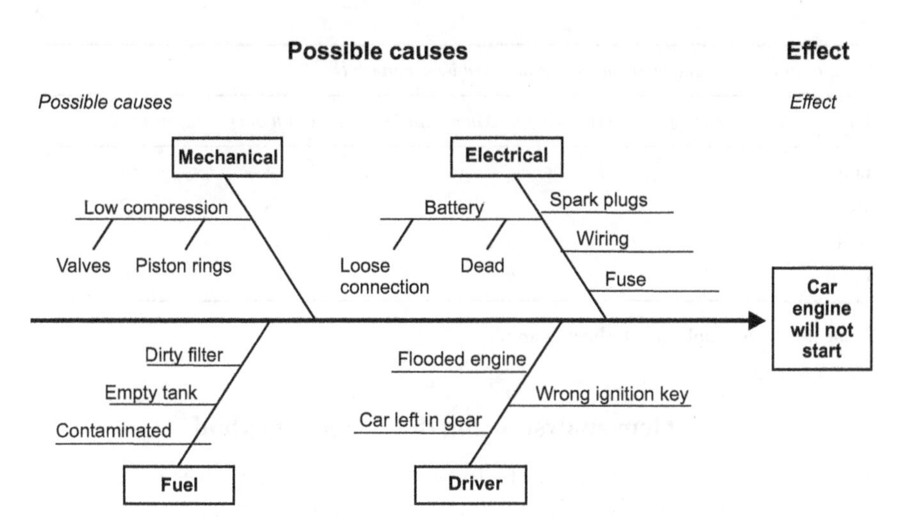

Figure 26.1 A cause and effect diagram. This Ishikawa fishbone diagram example analyses possible reasons for a car failing to start.

A fishbone diagram example is given in Figure 26.1, where the problem is a car engine that refuses to start.

Check sheets

A check sheet is a structured format that makes it easy to record and analyse data. Tally marks are made to indicate how often something occurs. That displays gathered data in a clear format allows easy analysis. Check sheets can identify and quantify possible causes, types and locations of problems. They are also useful for standardizing data collection for a process, ensuring that everyone collects comparable data in the same format. They can be used during the first stage of problem-solving to identify a problem. In the second stage they can help to identify the root causes of problems. The process for designing and using a check sheet is as follows:

- Decide which data are needed.
- Decide how the results will be analysed and used.
- Keep the form simple and easy to understand.
- Include only information that you intend to use.
- Make sure that everyone will interpret the categories in the same way.
- Compile separate check sheets for different days.
- Try out the form in advance and validate it by testing it on someone not involved in its design.
- Gather the actual data.

A check sheet example is shown in Figure 26.2.

Receipts for missing expense claim receipts analysed by department					
Type	*Marketing*	*HR*	*Manufacturing*	*Purchasing*	*Totals*
Taxis	10	-	-	4	14
Meals	12	2	-	8	22
Fuel	2	6	1	1	10
Total	14	8	1	13	46

Figure 26.2 A simple check sheet example.

Problem analysis using the '5 Whys' method

The '5 Whys' process was developed by Toyota to help employees trace problems back to their root causes. It allows the practitioner to deal with the real problem rather than just the observed symptoms. Sometimes the solution to a problem involves correcting a single, immediate cause but at other times the problems and permanent solutions will lie much deeper. This tool can be used on any problem at any level. The method begins by naming the problem in a concise definition statement (in similar fashion to the Ishikawa fishbone method). It then asks the question 'Why?' repeatedly, progressing from one possible cause to the next until the ultimate source of the problem is determined. The question 'Why?' must be asked as many times as possible to establish the root cause of the problem. This method (illustrated in Figure 26.3) focuses attention on the root cause of a problem rather than just the symptoms.

Concentration diagrams

A concentration diagram is a diagram or map on which a particular type of problem is marked wherever it occurs. This highlights the places where problems occur and whether or not they cluster. For example, the occurrence of serious accidents might be marked with crosses on a city road map (in which case the diagram might be called an *incident map*). In a book about the murders of Jack the Ripper one would expect to find a Whitechapel street map marked with crosses showing the scenes of crime. Here are the necessary steps:

1 Display a plan of the object under investigation on a table or wallboard.
2 Whenever the unwanted event occurs, mark its position on the plan (using a pin, a pen or some other method).
3 Ensure that each event can be identified clearly and separately.
4 Continue until enough data have been collected to show one or more significant clusters of events that will enable a realistic decision to be made on the main problem areas.

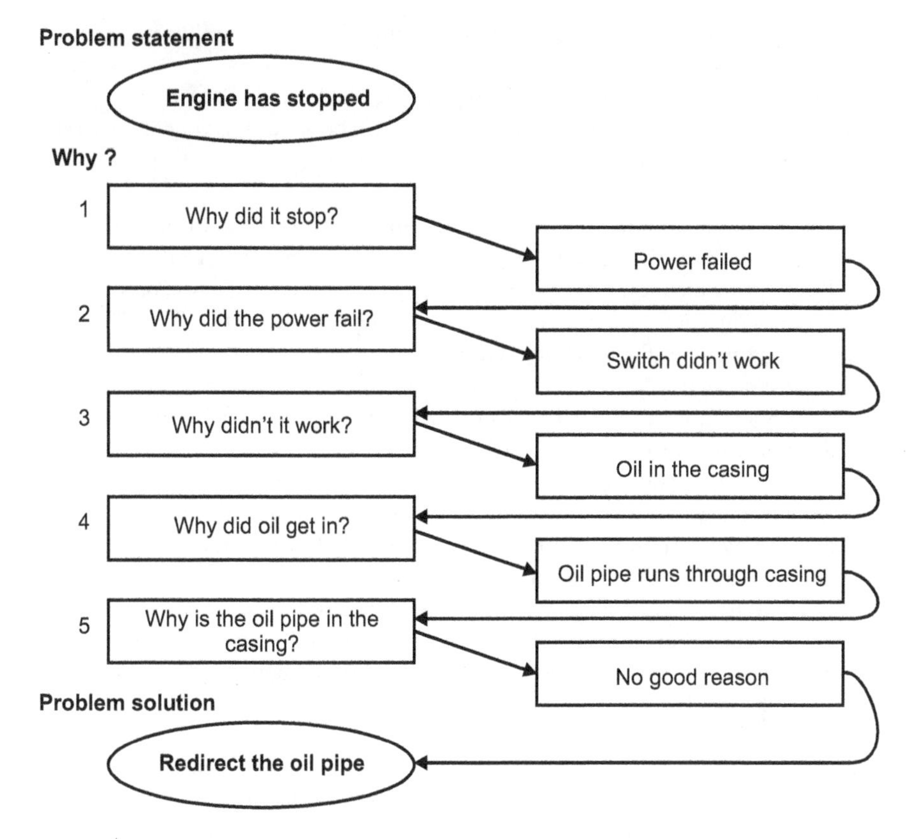

Figure 26.3 The 'Five Whys' method of cause and effect analysis. The problem in this example is that a machine has stopped working.

Pareto charts

In the late 1800s, the Italian economist Vilfredo Pareto found that 80 per cent of the land in Italy was owned by only 20 per cent of the population. He also found that this proportion applied to other situations, such as in his garden where 80 per cent of the peas harvested came from only 20 per cent of the pods. Now widely known as the Pareto Principle or the 80:20 rule, this ratio has been found to apply in many areas. Here are some examples:

- It is commonly found that 80 per cent of a company's stock value typically comes from only 20 per cent of all the items stored.
- Roughly 80 per cent of difficult management problems come from only 20 per cent of all the problems encountered.
- 80 per cent of measurable results and progress will come from just 20 per cent of the items on a manager's daily 'to-do' list.

- 80 per cent of a manager's interruptions come from the same 20 per cent of people.
- Roughly 20 per cent of input errors typically cause the lion's share of defects.
- Roughly 80 per cent of customer complaints will be about the same 20 per cent of your projects, products or services.

A Pareto chart communicates the results of an analysis that aims to narrow down the sources of trouble by ranking or prioritizing data in their order of importance. The Pareto principle means that, for nearly every event or consequence, only a small number of all the contributing factors will account for the bulk of the effect. Drawing a Pareto chart requires the following steps:

1 Classify and group data based on shared characteristics (such as non-conformances, downtime or customer complaints).
2 Organize the data in order of their magnitude on the vertical axis.
3 Arrange the groups along the horizontal axis. Begin at the extreme left with the one costing the most or occurring most frequently. Proceed from left to right in descending order of value or frequency.
4 Draw vertical bars scaled to their values on the vertical axis.
5 Draw a curve from left to right across the top of the bars to fit the cumulative item totals. This curve will end at 100 per cent in the upper right-hand corner and will complete the Pareto analysis.

A Pareto chart example is shown in Figure 26.4. To interpret the results, compare the height of the bars to evaluate the relative importance of the problems. Thus Pareto analysis can show at a glance which problems, options or possible causes should be given top priority.

In problem solving, Pareto analysis can be used when defining the situation to help narrow the scope of a problem. When identifying the root causes Pareto analysis can also be used to rank problems and look more closely at non-conformance data. Best results come from tackling the top priority problems and causes first. The causes of non-conformances can be plotted against the number of event occurrences. Pareto analysis is often used in conjunction with other techniques such as brainstorming and cause and effect diagrams.

Risk breakdown structures

Previous sections described risk identification and quantification. Although those are all valid methods, their outcomes tend to be linear and isolated representations of each risk. That does not really help with identifying the areas within a project that are most susceptible to risk. In the same way that a WBS is recognized as a way of presenting project work in a hieratical, structured and manageable format, risk data can be organized in a risk breakdown structure (RBS). That will provide a standard presentation of risks that can facilitate understanding, communication and management of risks.

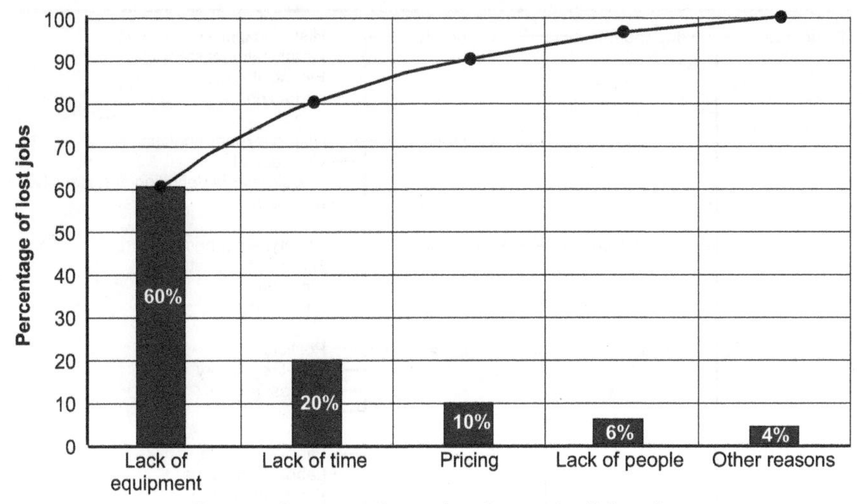

Figure 26.4 Analysis of reasons for lost jobs. This graph demonstrates the Pareto principle where, in this case, 60 per cent of potential orders from customers were lost through only about 20 per cent of all the possible reasons.

Following the pattern of a WBS, the RBS should be defined to provide a source-orientated grouping of risks that defines the total risk exposure of the project. Each descending level should represent an increased detail definition of the sources of risk. The RBS is therefore a hierarchical structure of the potential risk sources. A generic RBS is illustrated in Figure 26.5.

Just as the WBS provides a basis for many aspects of project management, the RBS can be used to structure and guide the risk management process. Each RBS element can be brainstormed, thus encouraging participants to identify risks under each of the RBS levels. Structured risk analysis can be achieved by interviewing the individuals responsible for each of the main RBS areas.

As with the WBS, the RBS is a useful tool for identifying items that might otherwise have been forgotten. Any risk uncovered by this method should be placed within its relevant risk area on the RBS. If a risk is identified that has no obvious home on the RBS it is possible that a key area of project risk has been overlooked. Then, of course, the newly identified risk category must be added to the RBS. The RBS can also be used in the opposite manner by reviewing which risks exist within each of the RBS categories. That can reveal possible gaps or blind spots within the risk identification process. Following these procedures should then provide an assurance that all the common risk sources have been explored.

Risk assessment and reporting using a risk breakdown structure

Identified risks can then be categorized according to their source by allocating them to the various elements of the RBS. This then allows areas of and concentration of

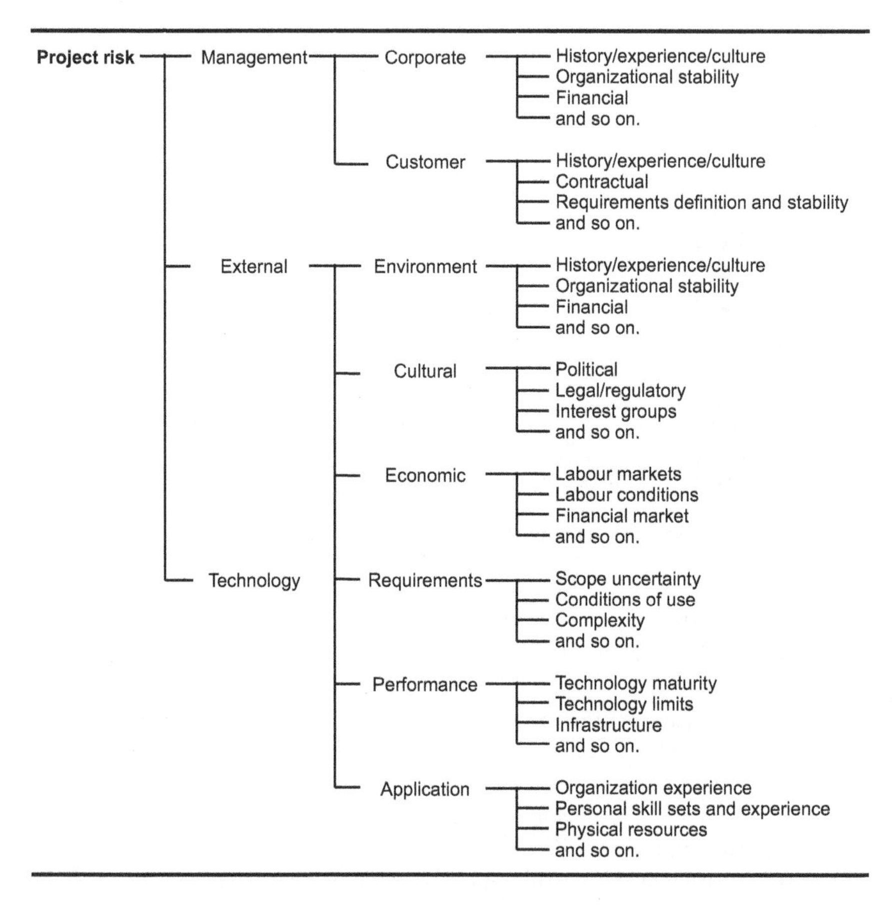

Figure 26.5 Format of a project risk breakdown structure.

risk within the RBS to be identified, potentially identifying the most significant risk sources. That could be done simply by adding up the number of risks within each RBS area, but that might be misleading because it does not consider the severity of each risk. A more effective measure is to allocate a score to each risk based on its exposure to the project. That will then give a quantifiable value to each of the risk sources. In addition, a percentage of the total can also be used to allow a comparison within each area.

Figure 26.6 shows a part of an RBS in expanded detail. In this example each item has been given a risk severity ranking or score. When the possible risk categories are displayed it appears that there are more risks associated with the technical aspects of this project. However, when each risk is assessed and rated according to its severity or impact, the potential severity is greater for organization and people.

The RBS can also be used to roll-up risk information on an individual project to a higher level for reporting to senior management, as well as drilling down into the

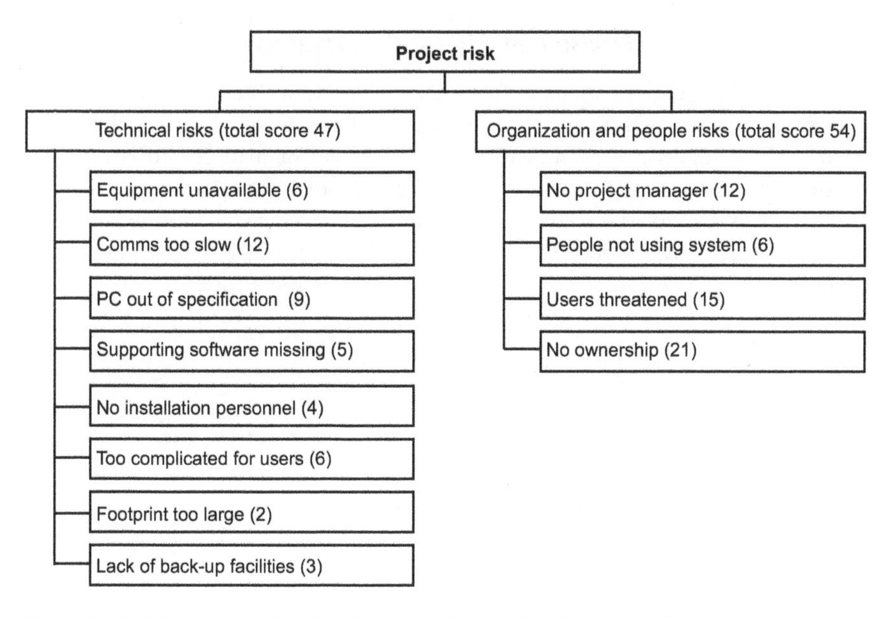

Figure 26.6 Use of the project RBS to score the severity of a potential risk. In this extract from a project RBS each risk item has been given a risk impact or severity score (shown in brackets). Although there are more possible technical risks, the risks associated with organization and people are collectively ranked higher.

detail required to report on project team actions. Reports to senior management may include total numbers of risks or total risk score in each of the higher-level RBS areas. Project teams can also be notified of risks within their part of the project by selecting relevant RBS areas for each team member.

One of the most important aspects of the RBS is that it can provide a consistent retrospective insight to project risk events after the project is completed. An RBS analysis can reveal risks that occur frequently, allowing generic risks to be recorded for future reference, together with effective responses. When routine analyses of post-project reviews indicate that a particular risk occurs repeatedly, preventative responses can be developed and implemented.

Decision-making risks

Decision-making falls into two categories, namely those made with certainty and decisions taken with risk. Decision-making under certainty is the best and easiest case to work with. With certainty, we assume that all the necessary information is available to assist us in making the right decision, and we can predict the outcome with perhaps 100 per cent confidence. But as we progress from certainty to risk, clearly potential damage to the project increases.

Decision making under certainty

Decision making under certainty implies that we know with perfect accuracy what the states of nature will be and what the expected payoffs will be for each state of nature. These results can be shown with payoff tables or matrices. To construct a payoff matrix, we must identify (or select) the states of nature over which we have no control. We then choose our own action to be taken for each state of nature. Our actions are called strategies, which are the risks that we are willing to take. The elements in the payoff matrix are consequences or outcomes for each risk. A payoff matrix based on decision-making under certainty has the following two controlling features:

1. Regardless of which state of nature exists, one dominant strategy or risk will produce larger gains or smaller losses than other strategies or risks for all the states of nature.
2. There are no probabilities assigned to each state of nature. In other words, each state of nature has an equal likelihood of occurring.

Case example

Consider a company wishing to invest £50 million to develop a new product. The company decides that three 'states of nature' exist for the predicted market demand. These three states of nature are represented on the payoff matrix as follows:

N_1 = strong market demand.
N_2 = an even market.
N_3 = a low market demand.

The company has narrowed its choices for developing the product to three, which have been labelled S_1, S_2 and S_3. There is also a possible strategy S_4, which would be not to develop the product at all but in that case there would be neither profit nor loss so we can ignore it. For this case example assume that the decision is made to develop the product. The payoff matrix for this example is shown in Figure 26.7(a). Looking for the controlling features, it can be seen that regardless of how the market reacts strategy S_3 will always yield larger profits than the other two strategies. So the project manager would choose strategy S_3 as the best option.

Decision-making under risk or uncertainty

In practice one dominant strategy will not exist for all states of nature. In a realistic situation, higher profits are usually accompanied by higher risks and therefore higher probable losses. When there is no dominant strategy, a probability value must be assigned to the occurrence of each state of nature.

(a)

Strategy	States of nature		
	N_1 = up	N_2 = even	N_3 = low
S_1 = A	£50m	£40m	£(50)m
S_2 = B	£50m	£50m	£60m
S_3 = C	£100m	£80m	£90m

(b)

Strategy	States of nature			
	N_1 = 0.25	N_2 = 0.25	N_3 = 0.50	Expected value
S_1 = A	£50m	£40m	£90m	£67.5m
S_2 = B	£50m	£50m	£60m	£55.0m
S_3 = C	£100m	£80m	£(50)m	£20.0m

Figure 26.7 Pay-off matrix examples.

Case example

Figure 26.7 (b) shows a case for the same company as 25(a). Here the payoffs for strategies 1 and 3 of 26.7(a) are interchanged for the state of nature N_3. Here it is obvious that there is no dominant strategy. When this happens probabilities must be assigned to the possibility of each state of nature occurring. The best choice is therefore the strategy with the largest expected value, where this is the summation of the payoff times and the probability of occurrence of the payoff for each state of nature. The expected value is given by:

$$E_i = \sum_{J=1}^{N} P_{ij} p_i$$

where E_i is the expected payoff for strategy *i*, P_{ij} is the payoff element, and p_j is the probability of each state of nature occurring.

For example, looking at Figure 26.7(b) the expected value E_1 for strategy S_1 is therefore given by:

$$E_1 = (50 \times 0.25) + (40 \times 0.25) + (90 \times 0.50) = 67.50$$

The expected value can be interpreted as the average value that the project manager can expect if this effort is performed 100 times. Repeating the procedure for strategies 2 and 3, we find that $E_2 = 55$ and $E_3 = 20$. Therefore, based on the expected value, the project manager should choose strategy S_1. Should two strategies produce equal expected values, the decision can be made arbitrarily.

To quantify risk management, we must identify the following:

- The risk we are willing to take (that is the strategy).
- The expected outcome (element of the pay-off table), and;
- The probability that the outcome will happen.

In the example of Figure 26.7(b) we should accept the risk associated with strategy S_1, since this gives the greatest expected value. A positive expected value suggests that taking the risk should be considered. If the expected value is negative the risk should be avoided.

The key factor in decision-making under risk is assigning the probabilities for each state of nature. Clearly if wrong choices are made different expected values will result, most probably giving a wrong indication of the best risk to take.

Decision tree analysis

A decision tree is an excellent tool when the need arises to choose between several courses of action. The method provides a highly effective structure within which one can lay out various options and investigate their possible outcomes. Decision trees can also help to form a balanced picture of the risks and rewards associated with each possible course of action.

Suppose that you need to make a difficult decision and decide to use a decision tree. Begin the tree by drawing a small square (decision box) towards the left-hand side of a large piece of paper. Now draw a line out from this decision box towards the right for every possible solution that you can imagine. Write a short description of each solution along its line. Keep the lines apart as far as possible so that you can expand your thoughts. At the end of each line, consider the results. If the result of taking that decision is uncertain, draw a small circle. If the result is that you need to make another decision, draw another square. Squares represent decisions, and circles represent uncertain outcomes. Write the decision or factor above the square or circle. If you have arrived at a solution when you reach the end of the line, just leave it blank.

Starting from each new decision square on your diagram, draw out more lines to the right representing the options that you could choose. From each circle draw lines representing possible outcomes. Again, make a brief note along each line to indicate what it means. Keep doing this until you have drawn out as many of the possible outcomes and decisions as you can envisage leading on from the original decisions.

Now review your tree diagram. Challenge each square and circle to see if there are any solutions or outcomes you have not considered. If there are, draw those in. If necessary, redraft the tree if parts of it are too congested or untidy. You should now have a good understanding of the range of outcomes possible from your decisions. This process is illustrated in Figure 26.8(a).

Evaluating the decision tree

Now you are ready to begin evaluating the decision tree and work out which option is shown to have the greatest worth. Begin by assigning a cash value or score to each possible outcome. In Figure 28(b) scores have been allocated in decimal steps over a range from zero to 1.0. Next, look at the circles (each of which represents an uncertainty point) and estimate the probability of each outcome. If you use percentages, the total must come to 100 at each circle. If you use decimals (as here) these must add up to 1.0 (working from right to left). If you have data from past events you might be able to make good estimates of the probabilities. Otherwise write down your best guesses.

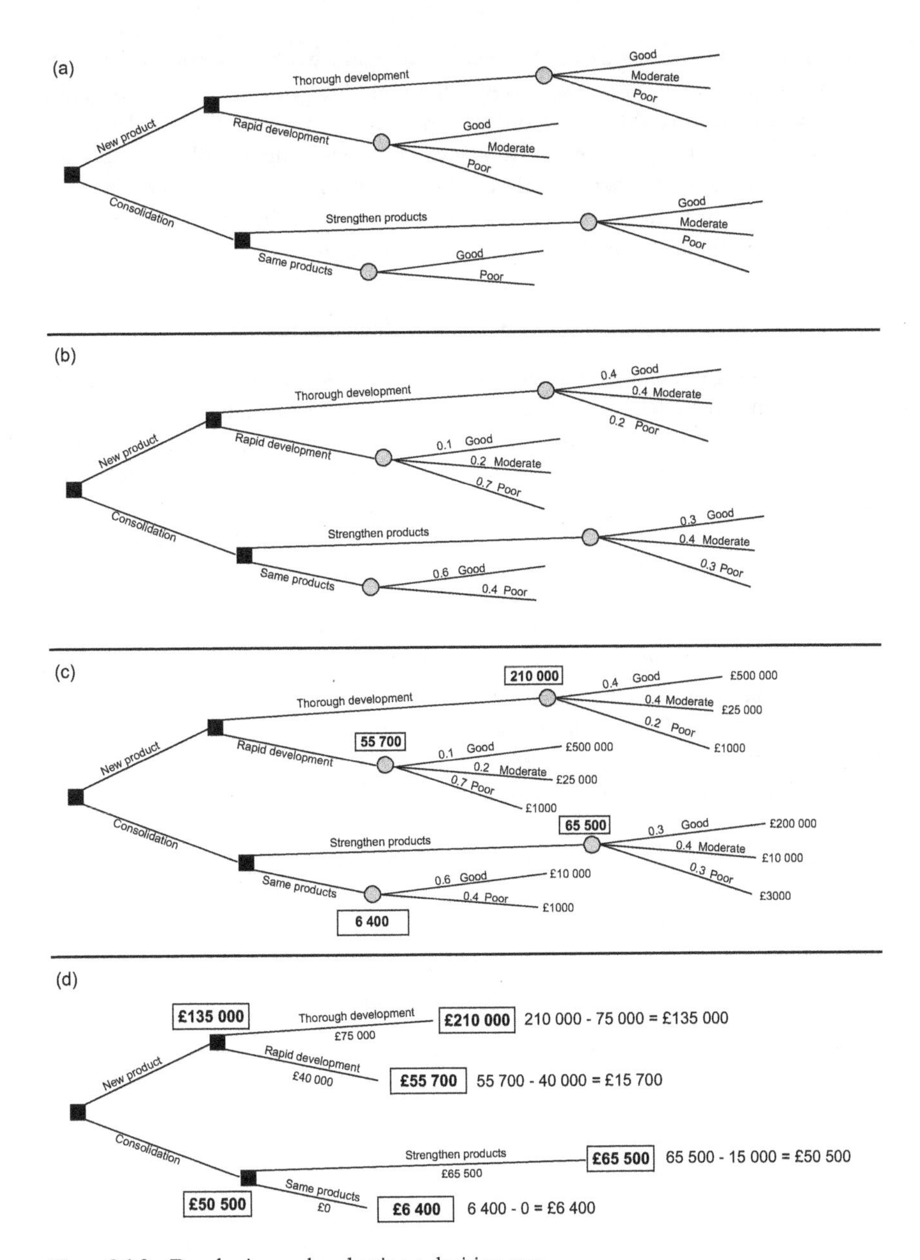

Figure 26.8 Developing and evaluating a decision tree.

Calculating tree values

When you have worked out the value of the outcomes and assessed the probability of the outcomes of uncertainty, it is time to start calculating the values that will help you make your decision. This stage of the process is illustrated in Figure 26.8(c). Again begin at the right-hand side of the decision tree and work back towards the left. As you complete each set of calculations on a node (decision square or uncertainty circle), all you need to do is to record the result. You can ignore all the calculations that lead to that result from now on. Where you are calculating the value of uncertain outcomes (circles on the diagram), do this by multiplying the value of the outcomes by their probability. The total for each node of the tree is the sum of these values. Note that the value calculated for each node is shown in a box.

Calculating the values of decision nodes

When evaluating each decision node (represented by a square on the diagram), write down the cost of each option along its decision line. Then subtract the cost from the outcome value that you have already calculated. This will give you a value that represents the benefit of that decision. Note that amounts already spent do not count for this analysis: these are sunk costs and (despite emotional counter-arguments) should not be factored into the decision. When you have calculated these decision benefits, choose the option that has the largest benefit, and take that as the decision made. This is the value of that decision node.

These calculations are illustrated in Figure 26.8(d). For example, the gross benefit previously calculated for 'new product, thorough development' was £210,000 (shown in the box). The estimated future cost of this approach is £75,000, leaving a net benefit of £135,000. Following the same calculation steps, the net benefit of 'new product, rapid development' is far less at £15,700. So if we develop a new product it is worth much more to us to take our time and get the product right, rather than rush the product to market.

Suppose the choice was to lie between improving (consolidating) our existing product or to risk developing a new product rapidly (probably botching the process), consolidation is the indicated decision (£50,500 against £15,700).

Key points about decision trees

Decision trees provide an effective method for decision-making because they:

- Lay out the problem clearly, allowing all options to be challenged.
- Allow full analysis of the possible consequences of each decision.
- Provide a framework for quantifying the values of outcomes and the probabilities of achieving them.
- Help us to make the best decisions based on existing information and best guesses.

Delphi technique

The Delphi technique relies upon expert judgement to determine the possible risk events in a project. The technique is used when embarking upon a project where there is little or no experience held within the organization on projects of that type. The expertise needed will not usually be found within the project organization so external consultants will have to be used. However, these consultants can operate remotely and will not have to work among the project staff. They will need to be informed about the proposed scope of the project and will be asked to identify and list the consequent risks.

There will normally be several iterations as the analysis becomes increasingly detailed, requiring much interchange with the consultants. However, these reiterations can be time-consuming. Care is therefore needed to ensure that too many iterations do not become necessary, which could lead to drop-out from the experts.

Strengths, weaknesses, opportunities and threats (SWOT) analysis

SWOT analysis is a good method for identifying the areas of strength and weakness within a project and, from those, the opportunities and threats that might affect the project (in other words, things that could expose the project to risk). The main advantage of using SWOT analysis is that it is likely to be used in other areas of the organization and is often used when dealing with problems. It is therefore a technique that can be integrated with other business functions such as business development and marketing, both of which will frequently be involved in the early project phases.

Figure 26.9 is an example of a SWOT analysis grid. If you look at the bottom half of the grid you will notice that some entries appear in both segments. For example,

	Positive	*Negative*
Internal factors	*Strengths* Technological skills Leading brands Distribution channels Customer loyalty/ relationships Production quality Management	*Weaknesses* Absence of opportunities Weak brands Poor access to distribution Variable product/service Sub-scale Management
External factors	*Opportunities* Changing customer tastes Liberalization of geographical markets Technological advances Changes in government politics Lower personal taxes Change in population age structure New distribution channels	*Threats* Changing customer tastes Closing of geographical markets Technological advances Changes in government politics Tax increases Change in population age structure New distribution channels

Figure 26.9 A SWOT analysis example.

the entry 'new distribution channels' appears as both an opportunity and a threat. This highlights the fact that even if there is an opportunity, it could also have a negative impact as well.

References and further reading

Crosby, P.E. (1978), *Quality is Free, the Art of Making Quality Certain*, Maidenhead: McGraw-Hill.
Deming, W.E. (2000), *Out of the Crisis*, Cambridge, MA: MIT Press.
De Feo, J.A. (2017), *Quality Handbook*, 7th edn, New York: McGRaw-Hill.
Seaver, M. (2003), *Gower Handbook of Quality Management*, 3rd edn, Aldershot: Gower.
Smith, D.J. (2011), *Reliability, Maintenance and Risk, Practical Methods for Engineers*, 8th edn, Oxford: Butterworth-Heinemann.

PART VI

Purchasing and contracts

27
CONTROLLING PURCHASING

Dennis Lock

The purchasing function typically devours well over half the costs of a project. If vital materials cannot be obtained on schedule, then your project will run late. Purchasing is a central and essential project function for obtaining goods and services. Yet, astonishingly, it is often taken for granted or ignored completely in project management literature. This chapter explains procedures intended to achieve the 'rights' of purchasing: getting the right goods of the right quality to the right place at the right time at the right price.

Introduction to the law of contract

Let's begin with the legal stuff. Anyone buying project goods and services commits their organization to a contract when they do so. It is therefore advisable to have some knowledge of the law of contract when placing a purchase order. A valid contract has four elements, which are as follows:

Offer and acceptance. The seller makes an offer to provide goods or services and the buyer accepts the offer.

Consideration. In this case consideration has nothing to do with politeness or seeing the other person's point of view. Instead it means that in exchange for the goods and services provided the buyer must offer in exchange something of value (which usually means a cash sum) to the seller.

Intention to be legally bound. This condition should need no further explanation here.

Capacity. The parties must be legally able or authorized to enter into the contract. A company cannot enter into a contract to supply goods or services that fall outside the scope defined in the objects clause of its memorandum of association. Thus a fashion knitwear company could not enter into a legal contract to supply a project company with 100 tonnes of steel rebar.

The often quoted sentence 'A verbal contract isn't worth the paper it's written on' has been attributed to various sources, including particularly the film producer Sam Goldwyn but a company can be committed to a verbal contract if there are reliable witnesses to prove that the relevant conversation took place. In all project transactions it is essential to have the contract enshrined in a document, and any variation to that contract should also be documented. The usual document for a purchasing contract is a purchase order, such as that shown later in this chapter in Figure 27.2, but items bought on line can blur this issue. Changes to a purchase order after issue must be documented and that is done by agreeing and issuing a purchase order amendment for every change.

Local small value purchases made informally

The very simplest form of a project purchase happens when a member of staff is despatched to buy some small item for a project in a cash or credit card transaction. The goods are then used on the job or project to keep it moving and the buyer is repaid using a petty cash voucher. So, can we ignore this simple process here in the context of project controls? Well, actually, no, we can't. As you will be able to read in the section on fraud prevention later in this chapter, even this simple procedure can lead to criminal activity and some financial loss for your organization.

Enquiry and purchase schedules

Suppose that you are the member of a project team and one of your responsibilities is to ensure that all goods and services are bought so they will be available when needed for use on the project. Your first step will be to compile a shopping list. In small manufacturing jobs that list would be known as a bill of materials. For a project the control document will probably be known as a purchase schedule.

Project purchase schedules begin their lives as enquiry schedules, in which every foreseen purchase of significant value is listed. The enquiry schedules are the initial control documents for scheduling and recording progress on individual high-value purchases. When the supplier for each item has been decided, data for that item can be entered in the project's purchasing schedule, probably listed under the following column headings:

- design specification number (which identifies the item being purchased);
- revision number (which allows for design modifications);
- description or name of the item to be purchased;
- quantity (the number or amount of the item required for the project);
- supplier;
- purchase order or contract number;
- amendment or revision number (which establishes the current design or contract state);
- date needed;
- cost, which should be the total cost delivered to the project stores or site.

Purchasing lifecycles

Figure 27.1 shows two purchasing lifecycles. The upper part of the figure is the lifecycle for the routine purchase of project goods that are available from suppliers' catalogues or are otherwise not complex. The steps in this lifecycle should be familiar and straightforward to all readers and the diagram is self-explanatory. The usual contract document for each straightforward purchase is a purchase order form, an example of which is shown in Figure 27.2.

However, most suppliers state in their terms and conditions that they have the right to vary or modify their product design. It could be that some seemingly insignificant feature of the product design as offered might be of crucial importance for the particular project application. This could be something as simple as the colour of the item, or whether or not a valve turns anticlockwise to open. So prudent project purchasers send their own description, sketch, drawing or specification to accompany each purchase order, just to ensure that the goods received will be as expected and fit for use on the project.

Lifecycle for a high value project purchase or contract

The lifecycle shown in the lower half of Figure 27.1 applies to higher value purchases and contracts, including equipment that is custom-designed and built specially for the project. This lifecycle begins with the recognition that a particular purchase or contract for services will be needed for the project. Then a detailed description has to be written, which at this early stage will be known as an enquiry specification. Potential suppliers have to be found who can be invited to bid for the contract.

Within the EU, regulations require that some higher value purchases for goods and services must be advertised in *The European Journal*, thus opening the bidding opportunity to suppliers who might be unknown (or even unwelcome) to the purchaser and that can add at least six months to the lifecycle.

The project purchaser will compile an invitation to tender (ITT) that describes both the commercial and technical requirements of the project. An ITT must then be sent to each potential supplier. To ensure that bidders are given a level playing field, if one bidder has a question, both the question and its answer must be supplied to all the bidders.

Bidders are given a common deadline date for reply and all will be asked to submit their tenders in two separate sealed packages on or before that deadline date. One package must give the technical proposal, with drawings and specifications. The other sealed pack must state the bidder's commercial offer including the price and terms of payment.

Bids received later than the specified closing date will automatically be rejected. On that closing date the technical packs containing the bidders' proposed technical designs and specifications will be opened and examined by the project engineers, who will reject any proposal that is not technically sound. The sealed commercial bids associated with the bids that were rejected on technical grounds must now be scrapped or returned to the bidders unopened.

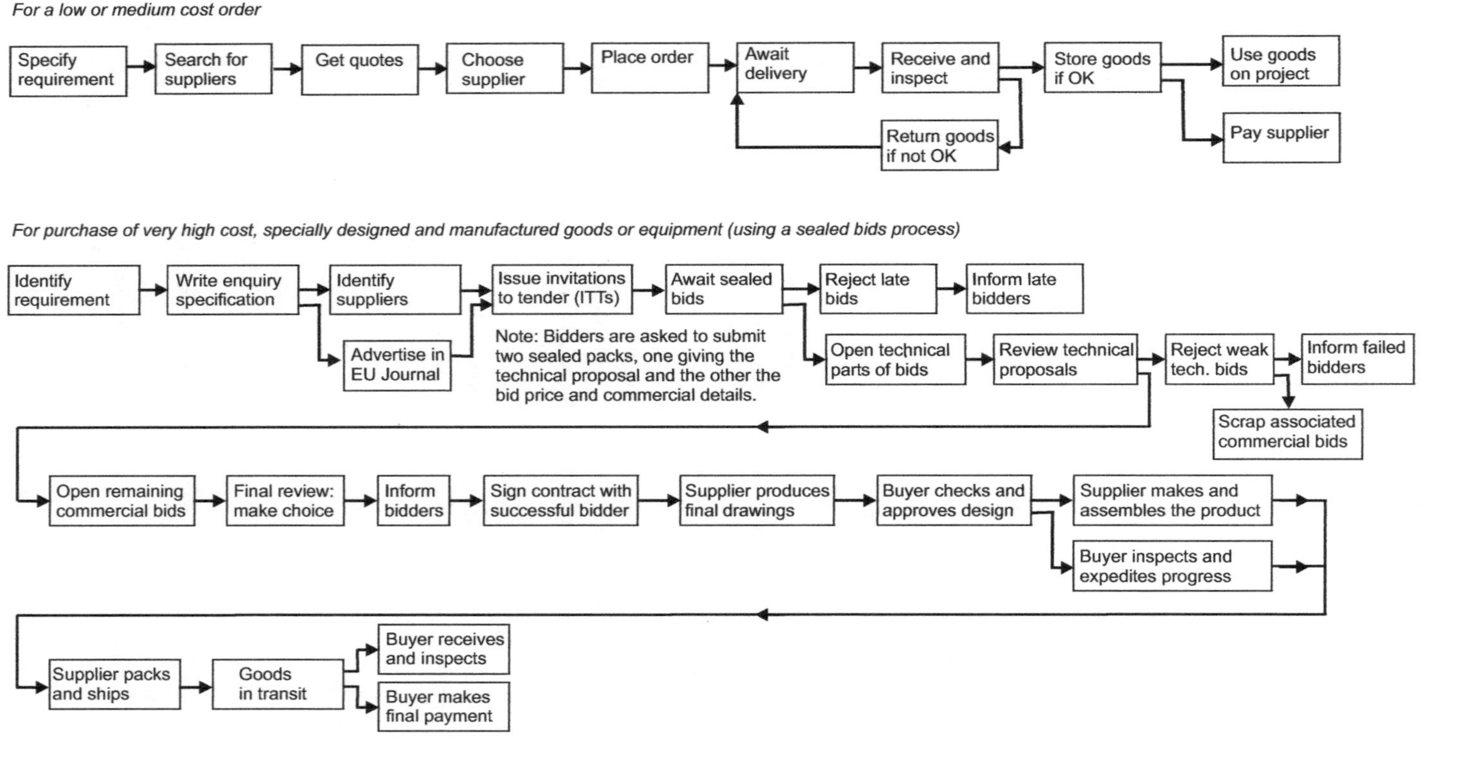

Figure 27.1 Examples of project purchasing lifecycles.

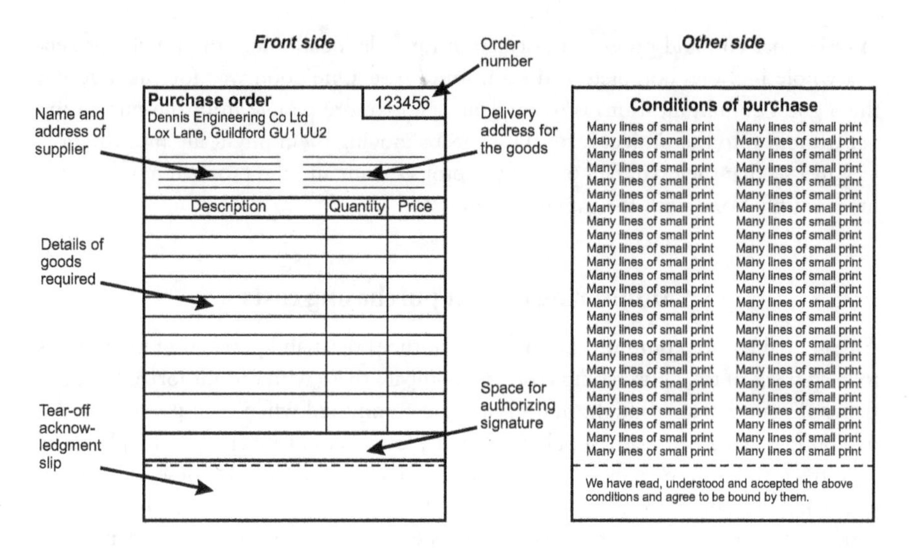

Figure 27.2 Elements of a typical purchase order.
Source: Lock (2013).

Now the surviving commercial packs associated with the successful technical bidders can be opened and examined, probably by a committee comprising both purchasing and technical people from the project organization. Some companies log the salient parts of these proceedings and the technical and commercial data in a bid summary document, to allow easier comparison of all the results.

At the end of this process the favourite bidder is congratulated and notified. Once the deal is done, the remaining bidders are informed. Project control actions necessary during the resulting purchasing lifecycle are illustrated in Figure 27.1 (above) and are described later in this chapter.

Communications between the project engineering and purchasing functions

The project manager and the PMO must have efficient and friendly communications with the commercial departments of their organization. Those commercial departments include the legal, financial and (especially in the context of this chapter) purchasing people.

It is not unknown for bad relations to exist between buyers and engineers. One of the principal causes for this hostility stems from the predisposition of some designers and engineers to talk to potential suppliers directly without involving anyone from the purchasing department. That can lead to the project organization being committed to verbal contracts without the essential commercial input from professional people in the purchasing department.

In the context of project controls, it is essential to have good day-by-day (even hour-by-hour) communication between the PMO and the purchasing department in order that cost and progress information for individual orders and for the project as a whole is always complete and right up to date. One good way for engendering these good communications is to implant one or more people from the purchasing department into the project team. That means moving them physically into the project area so that they will feel part of the project team and the project engineers can more readily bond with the purchasing function.

Monitoring project purchasing costs

An essential part of communications between the project and purchasing functions is to keep track of total purchasing costs and compare these with the authorized budget. Costs of materials and bought out services are committed when each purchase order or contract is agreed with suppliers. When plotted cumulatively as a graph, these costs trace an S-curve, as shown in Figure 27.3. Early indications of these costs are an essential requirement for budgetary control. If an overspending trend can be detected early, that gives the project manager the best chance of recognizing the need to take extra care. That early warning depends greatly on receiving feedback from the purchasing function and that, in turn, means having very good communications as outlined above.

The delayed S-curve shown in Figure 27.3 relates to the cash flows, when the suppliers' invoices are actually paid. This information comes not from the purchasing department, but from the company's cost and management accountants. So, although of less immediate importance in the context of purchasing controls, good communications between the project manager or PMO and the finance department are essential. Good communications in this respect are also necessary for controlling other project costs (particularly labour).

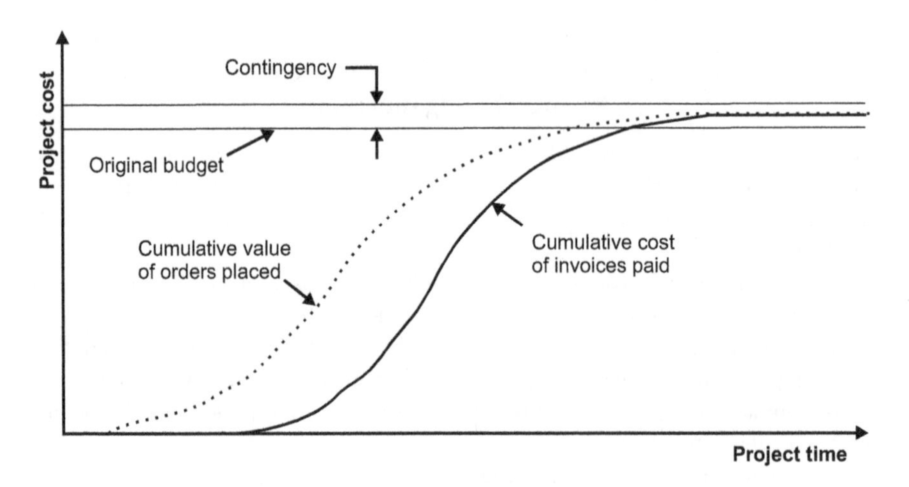

Figure 27.3 Ideal relationship between purchasing costs and project time.

I have labelled these S-curves as ideal relationships between purchasing costs and time for two reasons, indicated by the shapes and timing of the graphs:

1 The total costs in this example lie within budget.
2 Everything is settled and paid for without undue delay, indicating that there were no disputes.

Shortage lists

Shortage lists name goods that are desperately required for project use and which (for a variety of reasons) are not available. Needless to say, the purchasing department will be expected to take urgent action to clear shortages. These lists will typically be compiled as a result of feedback from the project manufacturing or building site to report that progress will be held up unless the particular goods are not delivered very quickly.

Shortage lists are a good example of management by exception, which means that they concentrate management effort on things that are going badly rather than wasting valuable time in self-congratulation for the activities that are doing well.

Inspection and expediting

Once an order has been placed, it would be wonderful if every company engaged in projects could simply wait and rely on all their suppliers to fulfill their obligations. However, there are many reasons why suppliers can fail in this respect. They might be suffering from machinery breakdowns, industrial action, overload, or could simply be giving higher priority to orders from other customers. So a prudent project manager will expect the purchasing department to monitor progress, rather than wait to be told on the due completion date that an order will be delivered late.

For specially manufactured goods and equipment the supplier might wish to send one or more of its own staff to the supplier's premises at intervals to see and check progress at first hand. Such visits can also include the witnessing of final inspection and testing. In some cases the supplier will be obligated to allow such visits by means of a special clause in the purchase order or contract.

Transport arrangements

The purchasing function involves not only the manufacture or supply of goods and equipment but also ensuring that those supplies are made available to the project site organization when they are needed. That can be as simple as arranging a local delivery but could instead involve the shipping of goods to a far corner of the world.

Incoterms

A purchaser needs to know, and agree with the seller, exactly when responsibility transfers from one party to the other when goods are transported. For local purchases this might have very little financial effect but international shipping is a different matter and it is important that the boundaries of responsibility for transportation are clearly defined in proposals, contracts and on purchase orders. Incoterms, defined and published by the International Chamber of Commerce (ICC), are accepted worldwide as the succinct and definitive method for setting out these boundaries. Incoterm is a registered trademark of the ICC. Some Incoterms are listed below, in ascending order of the sender's scope of responsibility.

- EXW (Ex works). All risk for carriage and insurance is assumed by the buyer.
- FCA (Free carrier).
- CPT (Carriage paid to nominated place).
- CIP (Carriage and insurance paid to nominated place).
- DAT (Delivered at terminal. Buyer will have to arrange carriage, customs and insurance after picking up from the terminal).
- DAP (Delivered at nominated place).
- DDP (Delivered duty paid).

And, for transport by sea or canal:

- FAS (Free alongside ship).
- FOB (Free on board).
- CFR (Cost and freight).
- CIF (Cost, insurance and freight).

Incoterms are subject to review by the ICC from time to time and the list here is presented only as an approximate guide.

Agents for overseas purchases

Any company that regularly imports project materials and equipment from overseas countries, or has to ship goods overseas, should appoint a professional purchasing agent who operates in each relevant global area. That does not mean having an agent in every possible nation, but it could mean (for example) having an African agent, a South American agent and so on. These agents are usually independent people or companies who will have a number of clients in addition to your own project organization.

Local knowledge is very important. For example, when ordering lengths of steel pipe one company was able to avoid an expensive mistake because the local agent advised that rail transport to the project site involved a tunnel with a sharp bend, so that the maximum load length had to be restricted. So the local rail company had no long wagons and, even if they had, the steel pipes would have hit the tunnel walls.

Overseas agents can arrange for inspection and expediting visits, saving the project organization from having to send people from the home organization. They will also have vital experience in local export/import regulations, including the preparation of export invoices (which can be very complicated).

Freight forwarding agents

Freight forwarders are specialists in the transport of goods right across the globe. They have specialist knowledge of all possible means of transport and alternative routes. They can also achieve economies of scale by ensuring that your goods are 'married' with other shipments so that maximum advantage is taken of space in shipping containers.

One great service provided by freight forwarders is to share their knowledge with your organization concerning documentation. Some nations have very complex requirements including the provision of special export invoices (in some cases requiring an extravagant number of copies, using forms of special designs). Freight forwarders also have very good communication networks, so that your goods can be tracked through all stages of their journeys.

Couriers

Occasionally a project organization will have the need to transport a small component or a signed document that has to be delivered in person to an overseas destination. Some agents can provide a courier service, where one of their employees will personally pick up your vital item and ensure that it is carried personally and safely to its intended overseas destination.

Audit and fraud prevention

The power and authority granted to you as an individual to commit your organization to significant sums of money in purchasing and contracts can be associated with the temptation to commit criminal fraud activity. So it is good practice to have your purchasing and contract procedures audited occasionally in order that potential risks can be spotted.

A major fraud risk

In one organization, where I was administration manager for an office complex in London, we were visited by auditors from our head office. I was not present at the discussions but the following conversation was reported to me (with some amusement) by the director to whom I reported.

Auditor: Who is responsible for the purchasing function here?
Director: Dennis Lock.
Auditor: Who signs your purchase orders?

Director: Dennis Lock.
Auditor: What is his limit of authority?
Director: He has no limit.
Auditor (by now becoming alarmed): Who clears incoming invoices for payment?
Director: Dennis Lock.
Auditor: What limit of authority does he have on those?
Director: None.

Well, here was a glaring opportunity for fraud. The outcome was that limits were imposed on clearing incoming invoices over a certain (high) value. I fully understood the reasons for these questions and their result. Every company should apply such precautions and ensure temptation is not put in the way of those able to sign purchase orders and contracts.

A petty fraud event

I want to end this chapter with an amusing little tale of a petty fraud that actually happened when I was a manager in a project engineering department. It is a warning that every purchasing transaction may not be what it seems, and a fraudster could always be lurking in the background, looking for an opportunity.

Our purchasing manager employed one woman as his clerk and assistant, and she made sure that every procedure was followed correctly. Let's call her Janet. She exuded an air of authority and, to be honest, we were all a little afraid of her.

In those days design offices did not have the advantages of computers, which allowed easy drawing changes and frequent use was made of a commodity called Cow Gum that allowed our marketing designers to position, then reposition, bits of a design on the page. My department used Cow Gum in small quantities and we bought it using petty cash vouchers from a local office stationery company.

On one busy morning Janet accosted me in the corridor outside the purchasing office and asked me to sign a petty cash voucher for payment. The items on the voucher were of low value, including one small can of Cow Gum, and I signed the voucher.

One evening, a month or so later, when most other people had gone home, I was walking past an open office door when two voices hailed me. The voices belonged to the company accountant and the HR manager. They thrust a petty cash voucher in front of me and asked, with some hostility, 'Is this your signature?' Undoubtedly it was. 'Then why did you authorize this purchase?' pointing to a line on the petty cash voucher authorizing the purchase of an enormous quantity of Cow Gum. Suitably humiliated I could offer no explanation. But my embarrassment was short-lived because both of my inquisitors suddenly burst out laughing, saying, 'Don't worry Dennis, she has done this to everyone'.

For some time Janet had systematically been changing details on petty cash vouchers and purchase orders retrospectively, so that by some clever trick the extra cash found its way into her purse. Now Janet was in police custody, never to return to her job. I learned a lesson never to trust anyone where company money and commerce was involved.

References and further reading

Lock, D. (2013), *Project Management*, 10th edn, Farnham: Gower.

Johnsen, T.E., Howard, M. and Miemczyk, J. (2018), *Purchasing and Supply Chain Management: a Sustainability Perspective*, 2nd edn, Abingdon: Routledge.

Whatever happened to Cow Gum? at hullabaloo.co.uk/blog/whatever-happened-cow-gum

28
PROJECT CONTRACTS

Shane Forth and Dennis Lock

The previous chapter described project purchasing, which principally involves contracts relating to the supply of goods. This chapter is also associated with project purchasing in many respects but is more concerned with contracts for the provision of services that can range from small specialist jobs to heavy engineering and construction activities costing billions of pounds. Purchases tend to have finite and easily controllable lifecycles. Contracts for project work and services usually come with complex conditions, less easily controllable lifecycles, more complicated terms of payment and need special control measures.

Maintenance and service contracts

All companies (not only those engaged in projects) have to manage contracts. Such contracts include (for example) staff employment contracts. They also include service contracts for the maintenance of offices and office equipment in and around the business premises. Most maintenance and services contracts are part of normal business housekeeping and project managers only become aware of them when something goes badly wrong.

Maintenance and service contracts cover everything from office and window cleaning to the supply of catering services. They also include security services, ranging from alarm systems to the provision of on-site guards and guard dogs. Lifts and hoists are all subject to compulsory service contracts. I could list many more. All of these services are intended to run like clockwork in the background. They are the concern of the organization's administration management and might be under the control of someone bearing a title such as 'facilities manager' or 'maintenance manager'.

Service contracts attract regular monthly payments, triggered by invoices from the accounts departments of the service providers. The receipt of such invoices can be relied upon because they will be generated automatically by the service providers on or just before their due dates. Service providers are very proficient in the provision of invoices. Unfortunately, provision of the associated services cannot always be relied upon and (unlike the invoices) the contracted service and maintenance work will not always be generated automatically. Thus windows might go uncleaned, only one security guard will be on site when the contract specifies three, or the security guard does turn up but then spends all his nocturnal duty hours studying for his examinations. Air conditioning plant breaks down because fan bearings have not been lubricated as stipulated in the contract. I write from experience, when my duties were extended to take over responsibility for facilities management from another manager who had been failing in his job for some time.

Failures in service contracts are common in premises where vigilance is not maintained. They can go unnoticed by project staff until something goes wrong. When everyone has to be sent home on a cold winter day because the offices have no heating, or on a hot summer day when the air conditioning has suffered a catastrophic failure and office temperatures rise to unbearable heights, project managers will suddenly become very interested. So, although this subject will not be discussed further here, we should all be aware of these background maintenance and service contracts, without which company operations (including project design and management) could not operate.

Composition of contract documents for project operations

A contract document for project works or services will begin by naming the parties to the contract, together with their addresses and the date from which the contract will operate.

Schedule of work

Next comes the schedule of work to be performed under the contract, specified as accurately as possible by direct reference to technical drawings and specifications, site plans and so on, each such document defined specifically by its title, identification number, revision number and date of issue. This schedule of work will also include the date when the work shall be completed or, alternatively, when specified phases of the work shall be completed (often with reference to milestones).

Obligations of the parties

There might be a paragraph or two included here about the purchaser's obligations, such as undertaking to ensure that certain essential facilities, assistance or means of access will be provided to the contractor.

Especially for construction and other contracts involving working on exterior sites, environmental considerations will need to be listed, such as not causing inconvenience of any kind to nearby businesses and residents, provision for waste disposal, avoidance of pollution, observance of local bye-laws and so on.

Contractors will usually be asked to agree to allowing inspection and expediting visits to the factory or project site made on behalf of the purchaser. For construction sites quantity surveyors and inspectors or other representatives from the local authority will all be expected to have access.

Terms of payment

This part of the contract document will define arrangements by which the purchaser is obligated to pay the contractor. Now things become somewhat complicated because, especially for construction contracts, there are several clearly definable and quite different contract types, identifiable by their terms of payment. These contract types are outlined below in the section headed Contract Payment Structures.

Model forms of contract

Some companies have legal departments that spend many hours perusing potential project contracts, which are then subjected to discussion rounds between the purchaser, the legal department and the contractor before agreement and signature. However, several professional organizations such as the Institution of Civil Engineers (ICE) offer model forms of contract, making the composition of a formal contract document far easier and less risk-prone than having to develop one from scratch. NEC is a respected organization that has developed the National Engineering Contract, a model used by several large organizations. Contact details are:

NEC,
One Great George Street,
London,
SW1P 3AA
Telephone: 020 7665 2446
Email: info@neccontract.com

There are also standard forms of contract that are designed not by professional organizations, but by bodies representing various specialist trades. For example, when I once had to place an order for upgrading a passenger elevator and installing fire-resistant doors at all levels (which was completed successfully), the chosen contractor waved away my carefully prepared purchase order and placed a trade-standard contract document in front of me for signature that specified the following series of progress payments:

- on signing the contract, 10 per cent of the total contract price;
- on approval of the contractor's drawings, a further 20 per cent;
- on delivery of the main bulk of materials to the project site, 30 per cent;
- on completion of the work and handover for use, 30 per cent;
- after six months of satisfactory operation a final retention payment of 10 per cent.

Contract terms and payment structures

Parts of this section are adapted from Lock (2013). Contracts, particularly for construction projects, are often classified according to the way in which their terms of payment are defined. The preferred payment terms will typically depend on the following factors:

- Risk, uncertainty, and any other factor affecting the accuracy with which the project's scope of work can be defined, estimated and budgeted.
- The customer's intention to set the contractor performance incentives. These incentives are usually aimed either at completion on time or at completion below budget, but they can also have a bearing on the standard of workmanship and quality.
- A penalty clause may be included in the contract as an attempt to limit failure in performance, the most common form being a penalty payment calculated according to the number of days, weeks or other stated periods by which the contractor is late in successfully completing the project.

Fixed (or firm) price contracts

With a fixed price contract the project purchaser understands that the contractor cannot, in normal circumstances, increase the price quoted. The offer to carry out a project for a fixed price demonstrates the contractor's confidence in being able to complete the specified project without spending more than its estimated costs.

In practice there are sometimes clauses, even in so-called fixed price contracts, which allow limited price renegotiation or additional charges in the event of specified circumstances that may arise outside the contractor's control (national industry wage awards were once a common cause).

Cost reimbursable contracts

Many contracts cannot start with the inclusion of a known total fixed price because the scope of work cannot be accurately defined. Most of these are 'cost reimbursable' contracts where the customer agrees to repay the contractor for work done against a prearranged scale of charges. These charges might be for certified quantities of work completed or for reimbursing the documented costs of labour time and materials used.

Fixed price contracts are usually avoided by contractors in cases where the final scope of a project cannot be predicted with sufficient accuracy when the contract is signed, or where the work is to be carried out under conditions of high risk. Many construction contracts for major capital works or process plants are subject to high risk, owing to site conditions, the weather, or to political and economic factors outside the contractor's control. Projects for pure scientific research (where the amount of work needed and the possible results are completely unpredictable) would obviously be unsuitable for fixed price quotations.

Even in cost reimbursable contracts, with no fixed prices to bid, managers have to decide the levels at which to set the various charging rates. It cannot be assumed that a contractor will charge all customers the same rates. Some customers will demand details of how the direct and overhead charges are built up, and will expect to negotiate the final rates before agreement can be reached.

In any contract where payment is related to agreed rates of working the customer will want to be assured of the veracity of the contractor's claims for payment. This might even entail access to the contractor's books of account by the customer or by auditors acting for the customer. In construction contracts based on payment by quantities, independent quantity surveyors can act for the customer by certifying the contractors' claims to verify that the work being billed has in fact been done.

Estimating accuracy might seem less important where there are no fixed prices operating but tenders for contracts with no fixed prices might have to contain advisory budgetary estimates of the total cost to the customer. If these are set too high, they can frighten a potential customer away, and the contractor stands to lose the contract to competitors. If the estimates are set too low, all kinds of problems could arise during the execution of the work, not least of which might be the customer running out of funds with which to pay the contractor. Any contractor wishing to retain a reputation for fair dealing will want to avoid the trap of setting budgetary estimates too low, especially where this is done deliberately in pursuit of an order. In any case, avoidable estimating inaccuracies must prejudice subsequent attempts at planning, scheduling, budgeting and management control.

Summary of contract types

Quoted prices or rates do not always fall entirely into the clear category of fixed price or cost reimbursable, often because one of the parties wishes to introduce an element of performance incentive or risk protection. Some contracts (compound contracts) incorporate a mix of these arrangements. Others (convertible contracts) allow for a change from a reimbursable contract to a fixed price arrangement at some pre-agreed stage in the project when it becomes possible to define adequately the total scope of work and probable final costs. Some well-known options are summarized below.

Fixed price

A price is quoted by the contractor and accepted by the customer for the work specified in the contract. The price will only be varied if the customer varies the contract, or if the contract conditions allow for a price increase to be negotiated under particular circumstances (for example, a nationwide wage award in the particular industry). Thus the contractor accepts the risks and, provided that the customer does not make changes to the scope or specification, the contractor must pay for any excess costs arising from underestimates, technical difficulties or other causes. Should the contractor realize, at some stage, that costs are rising well above budget, it can be argued that there might be a temptation to cut costs by skimping work and thus putting quality at risk.

Target price

Target price contracts are used where there is some justifiable uncertainty about the likely costs for carrying out the project as it has been defined. The contract will allow for price adjustment if the audited final project costs either exceed estimates or show a saving, so that the risks and benefits can be shared to some extent between the customer and the contractor through what are known as 'painshare/gainshare' incentives. These arrangements are typically used when a client and the contractor(s) recognize the need for genuine collaboration between them to deliver a project and form a joint venture or consortium, with the incentives 'painshare/gainshare' further extended through the supply chain contractors.

Guaranteed maximum price

A guaranteed maximum price arrangement is a target price contract in which, although cost savings can be shared, the contractor is limited in the extent to which excess costs may be added to the target price.

Cost reimbursable

A simple cost reimbursable arrangement means that the contractor is reimbursed for costs and expenses, but makes no profit. This type of payment sometimes occurs when work is performed by a company for its parent company, or for another company that is wholly owned within the same group of companies. A formal contract might not be used in such cases.

Cost-plus reimbursable

Cost-plus is a common form of reimbursable contract. As in simple reimbursable contracts, the contractor charges for materials used and for time recorded against the project on timesheets. But the charging rates agreed with the customer are set at levels that are intended to recover not only the direct costs and overheads but are marked up to yield profit.

Schedule of rates

Contracts with scheduled rates are reimbursable contracts (usually cost-plus), charged according to the number of work units performed. A specific work unit charging rate will be agreed with the customer beforehand for each trade or type of work involved.

Reimbursable plus management fee

This is a form of reimbursable contract in which the contractor's profit element is charged as a fixed fee, instead of being built in as a 'plus' element in the agreed rates. Unlike cost-plus, the contractor's profit revenue does not increase with cost but instead decreases proportionally as total project costs rise, arguably providing an incentive for the contractor to keep costs low.

Bill of quantities with scheduled rates

A bill of quantities contract is reimbursable, operating with an agreed schedule of rates, but the total number of work units expected in each trade or type of work is estimated and quoted beforehand.

Risks associated with contract types

Project contract arrangements are normally determined by the purchaser (client). Contractors bidding for the work are usually limited in the extent to which they can negotiate amendments. For this reason contractors need to study the conditions in the ITT with great care and understand their exposure to commercial risk. Although important for any project, the need for good project controls increases with the risks involved.

The national economy typically goes through boom and bust cycles. During economic downturns (especially for longer term projects with high capital investment) there is a strong tendency and internal pressure for clients to protect their interests, with limited appreciation of the contractors' needs to achieve fair business profits. Clients will thus prefer those forms of contract that place the ownership and management of risks in the hands of the contractor. This can happen even where the client is in a far better position to manage the risks. This state of affairs has been identified as a common cause of project failures in several major studies since the 1980s. See, for example, Latham (1994) and Egan (1998). In the short periods of enlightenment that follow such studies, the collaborative approach in the use of target price contracts comes to the fore. Then, that provides a balanced sharing of the financial risks and rewards between the parties involved. Figure 28.1 illustrates the typical allocation of risks between client and contractor for the different contract types.

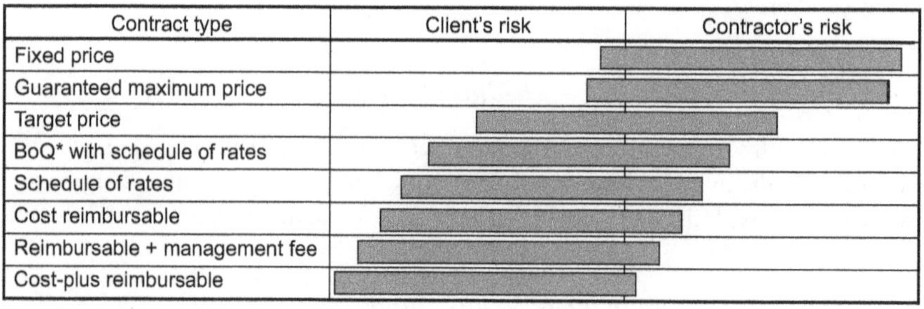

Contract type	Client's risk	Contractor's risk
Fixed price		
Guaranteed maximum price		
Target price		
BoQ* with schedule of rates		
Schedule of rates		
Cost reimbursable		
Reimbursable + management fee		
Cost-plus reimbursable		

* BoQ = bill of quantities

Figure 28.1 Apportionment of risk between client (purchaser) and contractor according to contract type.

At the beginning of my career (SF) in the mid- to late-1970s there were many major engineering and construction projects active in the UK. Relationships between clients and contractors were adversarial. To some extent the contractors held sway as the dominant force, owing to the demand and supply situation. Some behaviour was so aggressive and immature that I thought I had returned to the environment of my school playground. Since then I have seen many (but by no means all) projects succeed with target price arrangements and the 'alliance and partnering' approach, owing to the more collaborative working environment and integrated team approach (please see Chapter 7 for example). Provided that:

- the target price is realistic and aligned with the levels of project definition, risk and uncertainty when the contract is placed;
- the 'painshare/gainshare' initiatives are well-balanced between the parties, and;
- the parties are committed to this approach.

the client and contractor will both be driven to consider what is best for the project rather than adopting traditional parochial – even adversarial – positions.

Abbreviations

The following are commonly used abbreviations to define the division of responsibilities between customer and contractor in construction contracts:

BOOM	Build-own-operate-maintain
BOOT	Build-own-operate-transfer ownership
BOTO	Build-own-train-operate
OMT	Operate-maintain-train
TK	Turnkey

Contract variations

For many reasons, some beyond our control and others that are not, contract works can be subject to changes after the contract has been signed. For a number of important reasons every change to a contract must be documented (and also, clearly, agreed between the parties). Variations that affect the project design (as recorded in specifications and drawings) must also be subjected to scrutiny, approval and administration even if the contract cost will not be affected but that aspect of change is described in Chapters 36 and 37.

Imagine a contract that was placed with a contractor for a fixed price, when several competing contractors were keen to have the work and all the competitors were tendering prices that were as low as they dared to go without excessive risk. Now that your contract has been placed and your work has been assigned to one contractor who no longer has competitors, the contractor can potentially charge very highly for any work that you add to the contract. I have even heard contractors say that it is only through contract variations that they expect to make any profit. So contract changes (and project scope creep) are dangers to be avoided as far as humanly possible.

Where a significant change is inevitable post contract signature, the administrative mechanism relies on documents known as contract variations or contract variation orders. These documents have to be serially numbered for easy identification and when issued become part of the contract. Each variation order must define the exact nature of the change, the agreed decrease or (usually) increase in price, any agreed extension to the completion date and so on. When the project is finished, variation orders will become part of the archived project definition. In a few unfortunate cases they might also be needed as evidence in the event of legal action through the courts to settle disputes.

Dayworks orders

Imagine, if you can, a project for construction works that is well under way. A new building is taking shape alongside existing premises - an extension of which the company owners are justly proud because it is a sign of success for their family business. Then one day the CEO's wife notices that the paintwork on the new building clashes with the colours of the older building. She complains to her husband, he summons his project manager to his office, the project manager visits the contractor's site foreman and asks 'Please do you think these new doors could be repainted to match those on the old building?' The site foreman answers, no problem and asks the project manager to authorize this trivial change on a dayworks order form.

Dayworks forms are used commonly on project sites to authorize trivial, inexpensive changes. Sometimes those changes will be essential for health and safety or for unforeseen practical reasons. The procedure is extremely simple and devoid of red tape. The contractor's site manager or foreman is equipped with a pad of NCR (no carbon required) serially numbered pages (probably only A5 in size). So 'change door colour' would be written on one pair of these sheets and signed by both managers on the spot. In time the site hut would accumulate a pile of such sheets, probably impaled on a spike. The project owners' matching collection should end up in the project manager's office or with the PMO clerk.

This is a common sense trivial procedure for correspondingly trivial amounts of work. But suppose your project lasts for several years, costs many millions of pounds, and occasions the use of dayworks orders frequently on a large project site. The principal contract payments are claimed and paid according to the contract terms and conditions, but those dayworks sheets can remain neglected until the time comes to dismantle the site organization and call the project finished. Now the two parties to the original contract find that they have to reconcile hundreds of dayworks orders and agree the outstanding cash amount.

If that sounds unbelievable, please do believe that it can and does happen. It happened to me (DL) when my department took over responsibility for buildings maintenance works at our London offices. That contract was coming to the end of its two-year life and all the time there had been no problem with paying the contractor's main invoices. But we had to employ an independent quantity surveyor for a few weeks before we could agree with the contractor how much extra was owed for the dayworks. It was a considerable sum of money.

Contract administration

From the accounts of project variations and dayworks orders in the previous section it should be apparent that contracted works, especially when they take place over more than a few weeks, need particular attention to ensure that all the design facts and the project accounts figures are kept up to date and accurate. This important but often undervalued process is universally known as contract administration.

Provided that all contract variations and dayworks orders are documented accurately when they take place, contract administration is not a difficult task. It is a straightforward job that can be undertaken, for example, by a technical clerk or a records clerk who would probably be located near the project manager's office or in a PMO. That same person is also the logical owner of the task to maintain other registers, such as the register of project changes and design modifications.

References and further reading

Boundry, C. (2010), *Business Contracts Handbook*, Farnham: Gower.
Duxbury, R. (2012), *Nutshells Contract Law*, 9th edn, London: Sweet & Maxwell.
Egan, J. (1998), *Rethinking Construction*, London: Department of Trade and Industry.
Hornby, R. (2017), *Commercial Project Management*, Abingdon: Routledge.
Lock, D. (2013), *Project Management*, 10th edn, Farnham: Gower.
Latham, M. (1994), *Constructing the Team*, London: TSO.
Marsh, P.D.V. (2001), *Contract Negotiation Handbook*, 3rd edn, Aldershot: Gower.
Peel, E. (2007), *Treitel on the Law of Contract*, London: Sweet & Maxwell.
Stone, R. and Devenney, J. (2019), *The Modern Law of Contract*, 13th edn, Abingdon: Routledge.
Trebes, B. and Mitchell, B. (2012), *NEC Managing Reality: A Practical Guide to Applying NEC*, London: Thomas Telford/Institution of Civil Engineers.
Turner, J.R. (1995), *The Commercial Project Manager*, Maidenhead: McGraw-Hill.
Turner, J.R. (ed.) (2003), *Contracting for Project Management*, Aldershot: Gower.
Uff, J. (1991), *Construction Law*, London: Sweet & Maxwell.
Wright, D. (2004), *Law for Project Managers*, Aldershot: Gower.

PART VII

Monitoring and measuring for control

29

THE INTEGRATED PROJECT BASELINE

Alan McDougald

Integrated baseline plans express the initial expectations of the principal project stakeholders. They also set parameters that are essential throughout the subsequent monitoring of progress and performance. Approved changes during the project lifecycle will usually cause divergences from the original baseline. However, unauthorized divergences can signal potential project control failures; they identify the priorities for targeted intervention and corrective actions needed to keep the project on track.

Individual project baselines

Project baselines are determined at or immediately following the time of project authorization. Their source depends somewhat upon whether the projects are conducted for external clients, or if they are internal change projects (such as organizational change or IT projects). For external projects, the baseline components will be dependent upon the agreed contract documents. The baseline components of an internal project will usually be derived from the approved business plan for the project.

Principal baseline elements can be listed under scope (project definition), schedule or cost. Each of these elements can usually be subdivided into a number of subcategories. The essential starting point for establishing project baselines is the project description and scope, as documented in the internal project authorization document or client contract. In more detail, baseline components include or are derived from the following:

- Scope (project definition): the clarity of work that is included (and importantly that is not included) in schedules and cost plans. See Figure 29.1.

- WBS. This essential document expands and details the project scope description. The importance and essential nature of the WBS is described in Chapter 5. Dennis Lock claims that the WBS is to the project what the skeleton is to the human body.
- The cost and budget baseline is derived from the cost estimates. Figure 29.2 outlines the process.
- Schedule, preferably using the critical path method (see Chapters 15 to 20). The baseline schedule should include all milestones that define the timing of targeted deliverable dates. Milestone achievements are often tied to sales contracts and the issue of invoices and other claims for payment. These issues are a common source of legal battles, and deviations from promised deliveries can result in claims for payment of damages. The schedule must be constructed or adjusted according to the available resources. Planning and scheduling methods are described in Chapters 15 to 22. Steps leading to the creation of a schedule baseline are summarized in Figure 29.3.
- A time-phased plan for expenditure, compared principally against cash receipts (from clients or other sources) is the basis for cash flow management. The essential elements needed to prepare a net cash flow baseline schedule are shown in Figure 29.4.

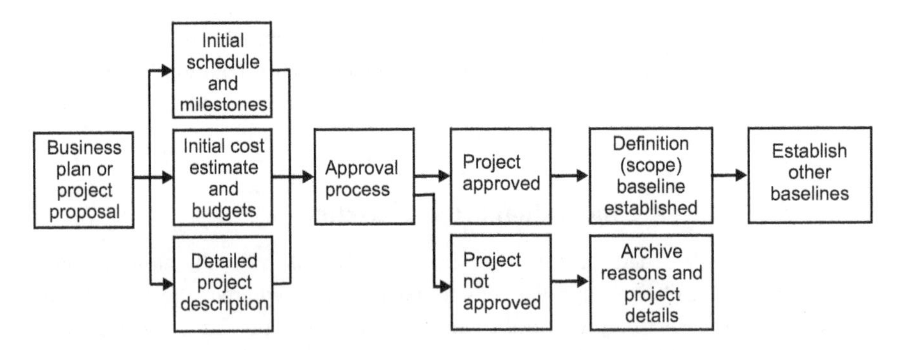

Figure 29.1 Establishing the initial project scope or definition baseline

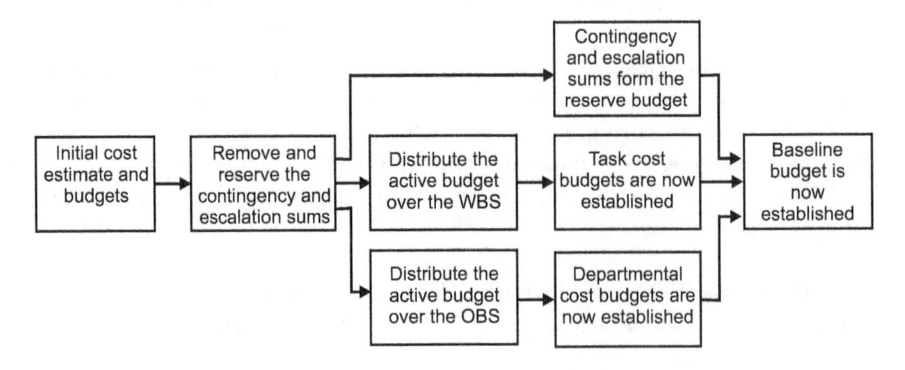

Figure 29.2 Establishing the initial project cost or budget baseline

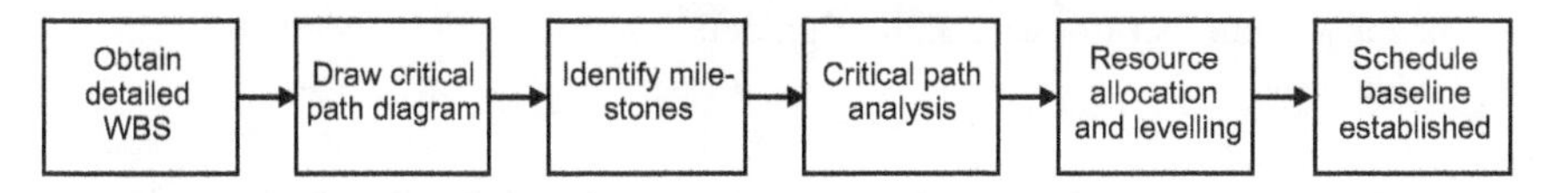

Figure 29.3 Establishing the project schedule baseline

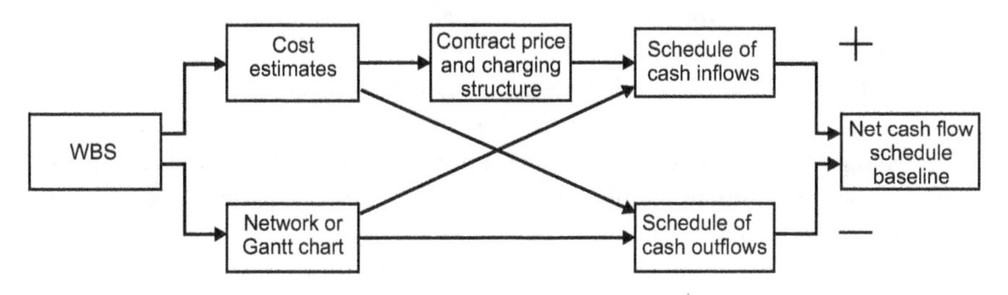

Figure 29.4 Establishing the net cash flow schedule baseline for a commercial project.
Source: Adapted from Lock (2013), used with permission.

- Responsibility, which means specifying who is responsible for executing each of the different project activities. This means having a clear understanding of the project organization (as defined in an OBS). Some organization structures, particularly of the balanced matrix kind, can blur the responsibility picture. Organization structures are described in Chapters 6 and 7.
- Risks associated with the project. These include potential constraints created by lack of funding. Risk treatment (mitigation plans) should be built into the baseline schedule and cost plans. Risks are discussed in Chapters 23, 24, 25 and 26.

Creating initial baselines

Once the approval to proceed with a project has been given, a baseline schedule defining details of work to be completed can be made. A baseline schedule is started by creating a WBS and then preparing a zoned timeline (sorted by work category) of activities and milestones. The principal milestones and top-level activities will often have been identified during pre-authorization studies. Critical path logic is applied to the timeline for top level and level 2 activities and beyond. This process defines dependencies, predecessors and successors of all work activities. After this logic has been defined, task durations must be estimated and WBS level 3 activities are added as required to help determine the critical path. Every activity must be logically linked to all other activities. The logical path with the longest duration is identified as the critical path, using the methods described in Chapter 16.

Resources required to complete all tasks must also be defined. For each activity or task this includes the following:

- materials;
- equipment;

- manpower number (for each skill or grade);
- budget cost.

In some projects it might be necessary to include subcontractors' work in the above list.

Milestones for starts and completions where needed for particular tasks should be integrated into the network diagram and critical path analysis. Now the critical path logic will establish the planned start and completion dates for all tasks. After checking the critical path results for any logic errors, a risk assessment should be performed. Resource schedules should be levelled to minimize unwanted peaks and troughs (see Chapters 18 and 19). If resource shortages would impact the critical path, provision of additional resources should be considered, taking into account any additional expenditure that would result.

All subcontractor information should be broken down to show labour (by skillset or craft, equipment provided and materials. Each task should be loaded with the resources required to do the job in a timely manner based on availability and cost. All equipment and materials should be quantified by activity and loaded as a resource. Resourcing loading and levelling can be accomplished with Primavera or SAP software but some popular and widely available software will not be suitable. The original estimate should be detailed to provide all resource requirements.

As each item of the required resources has been identified and loaded on to the project plan, a new determination (schedule calculation) has to be made to ensure that the critical path remains fully and accurately identified.

Scope and WBS

A WBS identifying specific work activities to reflect the project scope is essential. It should be sufficiently detailed. The following is a checklist of topics (not all of which will apply to every project). These are listed in alphabetical order:

- approvals, licences and permits (naming the responsible parties);
- buildings and infrastructure;
- civil works, to include all underground work for electrical and piping;
- commissioning and start-up;
- construction equipment (by size and type);
- controls and instrumentation;
- cost contingency;
- delivery of products;
- electrical (equipment and materials);
- engineering, including design/studies;
- health, safety and environmental issues;
- major mechanical equipment (all equipment by category);
- materials (bulk commodities);

- piping (by size and type);
- storage facilities:
- structural and steel works.

Cost

All costs can be managed as resources. Incorporation of labour and materials costs (including subcontractor costs) can be reported by selecting and filtering against specific data. Management cost performance can also be measured by selecting data according to the organizational breakdown structure (OBS).

Cost factors and performance can be monitored, for example, by loading planned labour-hours as resources. Some software allows estimated cost and labour hours to be stored as resources and compares planned versus actual hours, thus identifying labour efficiencies.

The original estimate can be compared with the current (revised) estimate, allowing cost factors of over- or under-runs and trends to be calculated.

Responsibilities

Responsibility for each task should be indicated by means of a simple (two-character) departmental code. These codes can identify either the department or (preferably) the relevant departmental manager. Such codes will be found invaluable for sorting, filtering and editing the data so that each manager can be given regularly updated schedules of forthcoming tasks for which he or she is responsible.

Risk

The initial risk assessment should be performed to identify all project risk. Actions for risk mitigation and improvement should always be taken after critical path analysis has identified the available float of all activities, and critical activities have been identified. Risk management procedures are described in Chapters 23, 24, 25 and 26.

The integrated baseline

As its name implies, the integrated baseline combines all the separate baselines described above to give a comprehensive baseline for the entire project. Figure 29.5 gives an idea of the result when all the individual baselines described above are integrated over the project.

If the integrated project baseline is to be the blueprint for project success it must be logically organized. Then measurement and comparison of actual costs and progress against the integrated baseline will highlight areas of cost and schedule concerns and identify associated threats and opportunities (risks). The results will allow management to act quickly to reverse or minimize unwanted baseline deviations.

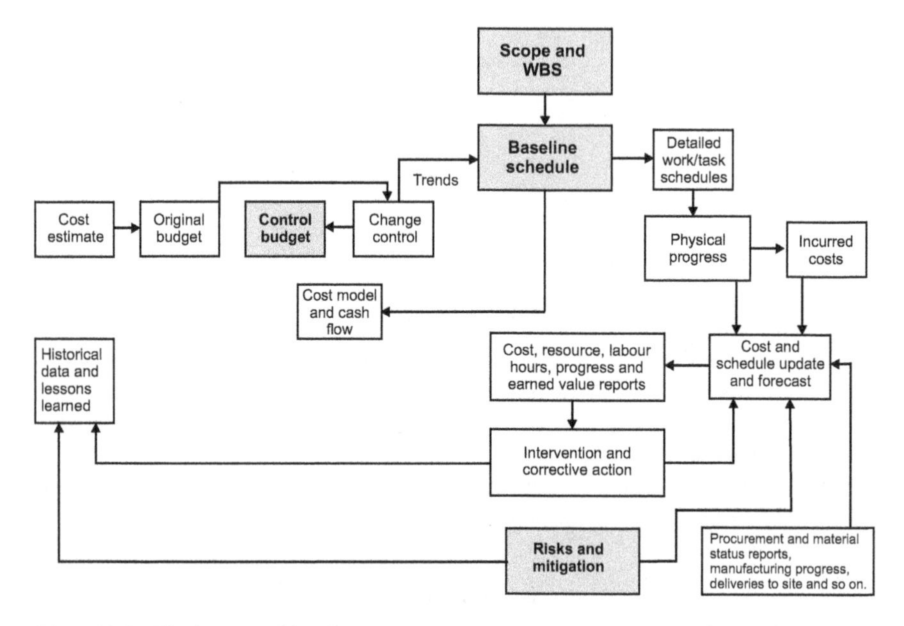

Figure 29.5 The integrated baseline concept.

Software applications for the integrated baseline

The practice of project controls has moved on from its origins in the analogue age of wall-mounted charts, typewriters, drawing boards and mainframe computers and is now firmly entrenched in the digital age. Today, with business intelligence tools and sophisticated software integration, information can move from the project to top management levels and across continents, providing high visibility of project objectives, progress and performance at a computer key stroke. There are many good software packages that can produce an integrated project baseline. Some can provide amazing graphic charts and views showing (for example) planned versus actual time and money data.

Of course, as with most other things, the data used to make good decisions must be accurate and timely. Bad (illogical) plans and inaccurate cost information rob managers of the tools needed to make good decisions. With good data and competent software everyone is kept in the loop, with potential threats identified accurately and quickly. Logistics and uncertain delivery times are therefore less of an issue but still they are potential threats that must be addressed as soon as possible.

Red flags

Most managers have limited time for viewing presentations. So presentations must be clear, timely and accurate. Managers should spend most of their time in fixing problems, not in identifying them. Project controls people (and the procedures they use) should not raise red flags unless there is complete confidence that the reports of time, money and quality performance are accurate.

Application of the integrated baseline for project control is described in the following chapter.

References and further reading

Devaux, S.A. (1999), *Total Project Control: A Manager's Guide to Integrated Project Planning, Measuring and Tracking*, New York: Wiley.

Lock, D. (2013), *Project Management*, 10th edn, Farnham: Gower.

30

REVIEWING THE INTEGRATED PROJECT BASELINE

Shane Forth

Chapter 29 described the process of establishing an integrated project baseline plan that defines the initial expectations of the principal project stakeholders and sets the essential parameters for project control. This chapter describes the integrated baseline review (IBR) procedure, which is necessary to establish whether or not the processes used to produce the integrated baseline have been followed sufficiently well. The IBR should give the principal project stakeholders confidence that they have a robust, realistic integrated baseline plan that is fit for purpose.

Preparation

Dependent on the organizational arrangements for the project, the IBR team may come from an external client's organization, the owner's engineer or from a main contractor. Ideally an IBR team acts as a fresh pair of eyes, independent from the project. A typical team will be led by an experienced project controls specialist, supported by a small number of functional leads who act as subject matter experts for various aspects of the review. Provided that it is independent of the project, a PMO is ideally placed to play a significant part in the IBR team.

It is common on major projects for the IBR to take at least a full week. Then, a formally-issued report of the findings and any recommended actions will follow. In accordance with predetermined timescales and in advance of the IBR actually taking place the following steps are required:

1 The leader of the review team should establish a date, timescale, agenda and proposed list of participants for the IBR. This must allow sufficient time following the review for actions and recommendations to be implemented before the integrated baseline is authorized.
2 The proposed project team members required to participate in the IBR should be agreed with the review team. For more complex project organizations, IBR

team members might come from multiple sub-organizations involved in the project (such as the owner's engineer, joint venture and consortia representatives, the main contractor (or contractors), principal subcontractors and vendors).

3 The project team (via a nominated project manager or director as a focal point) should provide the review team with a project overview and key project documents so that the team members can familiarize themselves with the project.

Overview of the IBR process

As a preliminary step in a typical IBR process the project team must present a detailed project description to the review team. The review team will then conduct the following steps:

1. Summarize their understanding of the project's purpose and objectives.
2. Describe the IBR process in outline to the project team.
3. Hold a series of structured and semi-structured interviews with nominated members of the project team to discuss the initial baselines and their integration.
4. Observe a series of data traces through the initial baselines and their integration by the responsible members of the project team. This process will show the relevant review team members the sources, pathways and flows of project control data within and between the project scope and WBS, cost and budget, schedule and cash flow. All of these must include consideration of resources. This observation will also determine whether or not the data have been checked and validated as being relevant and appropriate.
5. On completing the IBR, bring it to a close by debriefing and summarizing the findings, making recommendations and listing any necessary actions identified by the review, together with dates for those actions to be completed.
6. If considered necessary, arrange a follow-up meeting with the project team to present the formalized findings, recommendations and actions required.

During the IBR the lead person from the review team will conduct the interviews, supported by the subject matter experts (who take notes and can ask supplementary questions). A similar process will apply to the data trace discussions but the review team roles might be reversed. Responses to questions from structured or semi-structured interviews and data traces can be presented in a number of ways. Typically, a brief descriptive narrative will be supplemented by a visual presentation method, of which the following are a few examples.

- A numerical scoring mechanism (perhaps with a rating scale of 1 for best to 5 for terrible).
- Traffic light or RAG indicators (with red, amber or green signals indicating bad, mediocre or good performance, respectively).
- A Likert-type scale. This method will be familiar (for example) to anyone who has filled in a satisfaction survey questionnaire for a travel operator (see Figure 30.1).

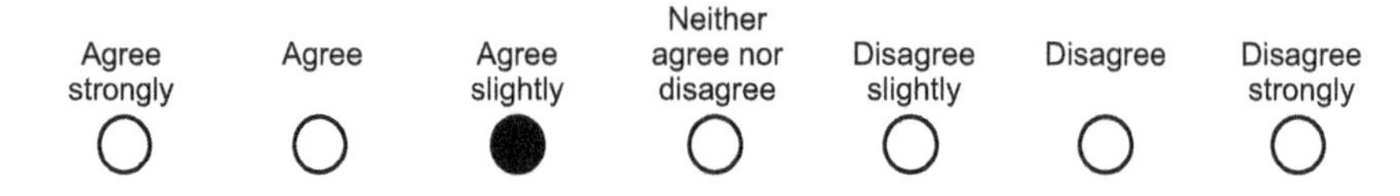

Figure 30.1 The Likert scale method used in the context of an integrated baseline review.

Visual presentations of project data are described in greater detail in Chapter 38.

Typical IBR questions in structured and semi-structured interviews

Project scope

The aim of IBR lines of enquiry for scope is to establish where the scope definition is clear, where there are areas of uncertainty and what (if any) assumptions have been made to compensate. This will indicate whether or not the scope definition is adequate for developing the project schedule, cost estimate and budget and the degree of potential scope changes that might occur during project execution. The following is a checklist of relevant questions:

- Which scope areas are well-defined?
- Which scope areas are not well-defined, lack clarity or contain ambiguity?
- What assumptions have been made?
- What is excluded from the scope?
- What is the probability that the scope will change significantly?
- Is the scope sufficiently defined to allow change to be identified clearly?
- Which areas are most liable to potential significant change?
- What potential is there for scope creep?
- Are the conditions of contract clear about what constitutes change and how resulting payment will be made?

The WBS and responsibilities (OBS)

The WBS and OBS bring scope, work and responsibilities together. Typical IBR lines of enquiry will aim to establish whether or not the WBS effectively captures the entire project scope. The IBR will also question whether or not the schedules and budgets are owned by the responsible organizations and people at the work package and cost account levels respectively. Here is a checklist of questions that must be asked in this context:

- Is the WBS complete? Does it include the entire scope of work for the project or contract?

- Is the WBS supported by a WBS dictionary that relates it directly to the statement of work for the project?
- How is the WBS aligned with the project execution plan?
- How well are the subcontractors' WBS defined and aligned with the project WBS?
- What is the plan for developing the WBS further as the project scope is defined further?
- Is the WBS product-based? If not, explain the rationale?
- Do you consider the WBS level of detail appropriate and fit for purpose?
- How have the budget holders been assigned to the scope of work? For example, by work packages or by cost accounts and the OBS?
- How well are the work packages or cost accounts described, broken down and budgeted in the WBS?
- How are responsibilities defined in the project execution plan?

Cost, budget baseline and cash flow

Difficulties encountered in cost estimating and budgeting and the risk of inaccurate cost estimates, whether due to poor scope definition, errors and omissions or too much optimism were discussed in Chapters 10, 11 and 12. These difficulties provide good reasons for challenging proposed cost and budget baselines as part of the IBR. Typical IBR lines of enquiry for cost, budget baseline and cash flow should result in a well-considered and independent view, so that these critical components of the integrated baseline achieve a level of confidence that minimizes the risk of unwanted surprises later. Here is a checklist of relevant questions:

- What is the overall class of estimate? Give reasons.
- Which parts of the estimate are least accurate and how is this reflected in the cost and schedule baselines?
- What estimating methods (for example, top-down, bottom up) were used and why?
- Which organizations and people provided input to the estimates?
- Were any inputs missing from the estimate and, if so, which?
- How well are labour, materials and other direct costs separated?
- What estimating norms were used?
- How has this estimate been compared with other external and internal projects?
- What historical data were used for estimate development?
- How do the level of detail, assumptions and exclusions support the declared estimate class?
- What estimating contingency is included for general uncertainty?
- What contingency is included from the risk register?
- How is the level of management reserve related to the risk assessment?
- Have processes been established for authorizing and drawing down sums from the management reserve?

- What are the time-phased cash flow implications in terms of investment/funding and return on investment/profit margin?
- What measures are in place to mitigate any adverse impacts of the project cash flow?

The schedule

Chapter 20 stressed that from the moment a work schedule is first drafted to the time when a project ends any plan is a good breeding ground for multiple mistakes of different kinds. Maintaining schedule integrity should eliminate mistakes and unwise choices, so that the schedule remains valid and practical throughout the entire project lifecycle.

The IBR provides an effective way for ensuring the integrity of the baseline schedule and its integration with the other components. Two different sets of checks should be applied, according to the following lists:

First schedule checklist

The first schedule checklist is directed at the network logic and work content. Here are the checklist questions:

- How does the network logic satisfy project/contract requirements?
- Which organizations and people had input to the schedule development?
- What inputs (if any) were missing from the schedule development?
- How does the schedule logic represent a sound practical approach and sequence for the work?
- Does the schedule contain any open ends other than the project start and finish? If so, how is this justifiable?
- Is the schedule free of imposed dates? If not, are the imposed dates documented and justifiable?
- Is the schedule free of negative lags?
- What milestones does the schedule contain for decision points and key interfaces (for example, access dates)?
- What required on site dates are included for material and equipment deliveries?
- How is the level of schedule detail proportionate to the project scope definition?
- To what extent does the schedule include parallel working or the fast-track overlapping of activities?
- Is the schedule deterministic? Or has schedule risk analysis been built in to confirm achievability?
- Has a schedule technical integrity review been carried out?

Second schedule checklist

A second schedule checklist is necessary to ensure that the estimates for activity durations and resource levels are sound. Typical review questions will include the following:

- In what way are the estimated for activity durations and the associated resources reasonable?
- What multiple methods have been used to determine and validate activity duration and resource requirements for activities that are critical or near-critical?
- How well do the organizations and people scheduled to carry out the work accept (buy-in) to the estimated activity durations and scheduled resource levels?
- How much does the schedule allow for contingencies, such as uncertainty of labour productivity and rework?
- To what extent are the activity durations and schedule influenced by a prescribed project completion date?
- Are the scheduled resources clearly traceable to the cost estimate and any changes agreed up to the project/contract start?
- How does the resource loading and levelling reflect the recruitment and availability of resources?
- How does the resource loading and levelling reflect the access to, and availability of, work faces?
- How does the resource loading and levelling allow for non-productive time, holidays, absenteeism and work patterns?
- What risks have been allowed for with respect to critical and scarce resources?
- What approach to cost loading of the project schedule been made?

Risk

An IBR will determine if risks to the project objectives have been identified and assessed. It will also ensure that mitigation strategies were included when formulating the integrated project baseline.

Typical IBR questioning for project risk is as follows:

- How have risks that might impact cost and schedule been documented in terms of identification, probability, impact and mitigation?
- What do you consider the level of risk is to the cost baseline?
- What do you consider the level of risk is to the schedule baseline?
- Does the team understand which people 'own' the risks in the project management process?
- What organizations and people have provided input to risk analysis and management?
- What inputs are missing from risk analysis and management?
- What is the process for escalation of issues (exception reporting) and intervention?
- How have the results of any cost and schedule risk analysis been used?

Integration and data tracing

Much of the process for integrating separate baseline components into the integrated baseline comprises bringing the cost and schedule baselines together and ensuring that they are prepared in a manner that works with the progress and performance measurement and analysis once the project is under way (a subject that is discussed in Chapter 33). Typical IBR lines of enquiry for integration are as follows:

- How have contractors' costs and schedules been incorporated in the integrated baseline?
- How are the schedule and cost baselines integrated?
- How is the scheduling system integrated with the budget and cost collection systems?
- What structures have been used and how have they been used to integrate the cost and schedule baselines?
- Do the cost and schedule baselines share the same WBS?
- Are the labour hours and costs distributed from and estimates and budgets and assigned to activities appropriately and how is this managed for detailed planning?
- How are the time-phased budgets for work packages or cost accounts aligned with the start and finish dates in the schedule?
- Have appropriate measurement methods been applied to the project scope and work activities?
- How does the integrated baseline support accurate progress and performance measurement (earned value, for example)?

The importance of good data and software for the integrated baseline was mentioned in the previous chapter. Project controllers use a variety of (often complex) software applications to undertake a wide range of project controls tasks. These include identifying the appropriate data for the following tasks;

- setting baseline targets;
- assessing progress and performance;
- forecasting trends;
- identifying, modelling and anticipating deviations from the baseline;
- assessing the impact of changes;
- using insight to recommend early preventative and corrective remedial actions.

Configuring the various software tools correctly is a complex task, requiring detailed knowledge of their internal workings. Complex data flows between different project control software applications can mean that key project control data elements are entered into one software application and then have to be copied and pasted into other applications. Alternatively, data might have to be entered manually more than once. The following data flow patterns are typical:

1 Roll up and summarization (bottom up).
2 Data decomposition (top down).
3 Touch points and interfaces for data between the different software tools.
4 Extraction of data based on search criteria (filtering).

All these data transfers risk that wrong data could be inserted into key project control documents without the knowledge of project team members. Thus it is essential to verify and validate the data. IBR data traces must be conducted to review the complex data flows across the different software applications with the project team and ensure that everything is working correctly. A common approach is to select a number of work packages or control accounts from the project baseline (typically from different parts of the WBS/OBS intersections) and trace the flows of key data. Here are some examples:

1 Budget (a). From the work package or cost account budget origins trace how the data flow down through the software tools to the cost estimate and cost budget baseline and then to the work package or control account and time-phased cash flow.
2 Budget (b). From the actual hours recorded on timesheets, trace how the data flow from the finance or cost accounting system to the cost control and scheduling software applications.
3 Schedule. Trace the origins of baseline schedule dates and note how they flow down through the software tools to the WBS (see Figure 5.1) and planning levels and then to the scheduling system.
4 Schedule for subcontractors. Check how separate subcontractor baseline schedules align with the project baseline schedule. Trace where subcontractor progress does or will originate and how this flows upwards from work management and progress tools to the project scheduling application.

For risks, the data trace should identify risks from the risk register that have a mitigation strategy and then investigate how the mitigation measures relate to the schedule and cost budget baselines.

IBR closeout

The IBR can be closed when its findings, recommendations and actions have been implemented and the integrated baseline has been authorized. The integrated baseline should now express the expectations of the principal project stakeholders and set parameters that are essential for monitoring and controlling all aspects of project performance.

References and further reading

APM (2016), *A Guide to Conducting Integrated Baseline Reviews.* Princes Risborough: Association for Project Management.

NASA (2016), *Integrated Baseline Review (IBR) Handbook.* Washington, DC: National Aeronautics and Space Administration.

NDIA (2015), *Guide to the Integrated Baseline Review (IBR).* Arlington, VA: National Defense Industrial Association, Integrated Program Management Division.

31
MANAGING PROGRESS

Dennis Lock

Management of any work, whether for a single task or for an entire project, depends on adherence to the control cycle plan-do-check-act, which was discussed earlier in Chapter 4 and illustrated in Figures 4.1 and 4.2. This chapter deals in greater depth with the check element of the control cycle and is derived largely from Chapter 22 of Lock (2013).

General observations

One prerequisite for any control system is a method for measuring the effect of any command given. The information so derived can then be fed back to the command source so that any errors can be corrected by modifying the original command.

An artillery commander watches the fall of initial shots through field glasses and uses the results as feedback signals to the gunners, so that they can correct the aim of the guns. In electronic amplifiers, well-established design practice is to arrange for some of the output signal to be fed back to the amplifier's input in opposite (cancelling) polarity. That feedback process tends to correct any distortion of the signal created within the amplifying system.

Thus feedback signals are used to correct errors, whether these happen to be the aim of a cannon or distortion in an amplifier. If the feedback signal is disconnected, the system is said to operate as an open loop. When the feedback signals are applied to the input, the control loop is closed.

Effective project progress control is a form of closed loop feedback system. For every instruction issued, progress is monitored, deviations from plan (errors or variations) are detected and corrective feedback is applied. The competent project manager will ensure that these corrective actions do take place, so that the control loop is closed. This principle operates at the level of each individual task, at higher levels of the WBS and for the entire project. Because of the close electronic analogy this process has been called *cybernetic control.* The principle was shown earlier, in the plan-do-check-act cycle (Figures 4.1 and 4.2).

Relevant management styles

The development of management theory has one thing in common with the fashion industry, namely that this year's technique is next year's history. However, some management theories survive and become invaluable in helping us to understand management better and apply the principles in practice. Over the past years a number of different names have been associated with management styles that all begin with the words 'management by'. Five examples follow.

Management by objectives

Management by objectives (MbO) is based on setting each manager in an organization a set of quantifiable objectives, and then measuring how well or how badly each manager manages to achieve those objectives. Each objective must be quantifiable and measurable. Project management is nothing less, nothing more, than management by objectives. Quantifiable objectives are set when the project is defined, they are distributed over a WBS and OBS, and then managers strive to achieve the set objectives.

Management by the seat of the pants

This style of management derives its name from riding a horse or flying a plane. It refers to a kind of control based on instinctive feelings of whether or not things are going in the right direction. Although a few individuals possess this ability in project management, most of us do not. Therefore more scientific or practical methods are needed, such as those described throughout this book.

Management by walking about

A manager who 'walks about' will visit parts of the project to see for him/herself the progress that is being made, and most important of all, give words of encouragement to everyone working on the project. If some of the activities are at a remote site or overseas, the project manager should make occasional visits. People will respond well if they know that they and their work are appreciated. A small word of encouragement here and there costs little but the return on investment can be enormous. Failure to perform this very simple function of walking about is a great demotivator. No project can be managed effectively by sitting all day behind a desk.

Management by surprise

Management by surprise has its origins in manufacturing, but it has universal application to any cybernetic control system where the loop is not closed. A manager puts work into the front end of a factory and is surprised when it doesn't come out at the other end.

Management by exception

With a cybernetic control system, error signals have greatest significance for control because these generate corrective action. In the management context, divergences from plans and budgets or any unexpected setbacks are called *exceptions*. Cost and management accountants call their exceptions *variances*. Exceptions (variances) can apply to advantageous differences from plan but, for control purposes, it is the negative exceptions and potential failures that demand attention.

Management by exception means concentrating management efforts on the bad variances or exceptions. If something is running early and 5 per cent below its budget, that's fine, leave it alone, don't tell me about it. But if another job is running 5 per cent over budget, concentrate on that and put it right. Similarly, if all jobs except one are running on time, it's that one late job that has to be reported and expedited.

Exception reporting

Exception reports are a product of management by exception. They might be accounts statements that show overspending or lists of jobs that are running late. At the other extreme, an 'exception report' might be the frenzied beating on a senior manager's door by a distraught project manager who feels that his/her project and world have suddenly fallen apart.

Before allowing any exception report to be passed up the organizational hierarchy (escalated) to a more senior manager for action, the project manager must first be certain that some remedy within his/her own control cannot be found. However, once it has been established that events are likely to move out of control, the project manager has a clear duty to give senior management the facts without delay. All of this is, of course, following the management by exception principle that prevents senior managers being bombarded with large volumes of unnecessary information. The intention is to leave executive minds free to concentrate their efforts on matters of higher strategic importance.

Progress measurement methods

The simplest case for progress measurement is clearly a single short-duration task, where the start, progress and finish are all over in a week or so. Progress assessment is often more difficult for tasks that last for several weeks or months. Common methods include:

- Progress assessment by the project manager or a PMO. This subjective method is usually the only method possible in the early project stages. The results (as with cost estimating) will often be optimistic. However, for large projects this might be the only progress assessment method available for several months.

- Counting the numbers of design documents issued. This crude method is sometimes used in the early to mid-stages of a large project to assess design progress. Accuracy and reliability depends on the type of documents being considered. It can work well for drawings such as piping and wiring diagrams, where a project might need a fair number of these and it is easy to count the numbers released. It is clearly far less reliable when considering the progress of general design work.
- Measurement by counting or assessing physical quantities. In construction projects this function usually depends on the observations of an independent professional quantity surveyor (QS), who can assess progress in terms of definable and measurable units. Periodical certificates issued by the QS are usually a prerequisite for a contractor to issue invoices to the client, and for the client to approve and pay those invoices.
- Milestone achievement. This definitive and reliable method is dependent upon having a sufficiently detailed project network plan that identifies milestone events throughout the project lifecycle, which can easily be recognized. This is another process that can trigger claims for progress payments by a contractor from the project client.
- Photographs taken at different time intervals can provide accurate and valuable evidence of progress. This method is applicable mainly to construction and civil engineering projects, but it can also be useful for some larger engineering projects. Photographs can often be produced as evidence of work achieved at regular progress meetings. They can also enhance periodical written progress reports. Obvious requirements are that a reasonably good camera should be used in adequate lighting conditions, and that comparative images should be taken from the same viewpoint. For very large projects the project manager might consider the publicity value that can be achieved by setting up a time-lapse camera at the project site to record progress of the works from start to finish.

Corrective measures

Corrective measures will only be successful if they are taken in time, which means that adequate warning of problems must be given. This will depend on monitoring progress regularly against a well-prepared schedule that is kept up to date.

Working overtime, perhaps over one or two weekends, can sometimes recover time. The project manager will be thankful, on such occasions, that overtime working was held back as a reserve and not built into the schedules as normal practice. Used occasionally, overtime can be an effective help in overcoming delays. When used regularly or too often, the law of diminishing returns will apply, with staff permanently tired and working under pressure, prone to mistakes or (worse) accidents and leaving inadequate reserves for coping with emergencies.

If problems are caused by shortage of resources, perhaps these could be made available from external sources by subcontracting. Or, there might be additional capacity somewhere else in the contractor's own organization that could be mobilized.

The network logic should always be re-examined critically. Can some tasks be overlapped, bypassed or even eliminated? Is there any task that could be delayed until after the main part of the project has been delivered?

If all else fails, try to find out what the customer's reaction would be to late delivery. If, for instance, the project is to supply and install new equipment in the customer's brand new manufacturing plant or offices, it could be that the customer's own building programme is running late, so that a later delivery date can be negotiated with no one being inconvenienced.

When the news is bad

When work runs late, the things to be considered are the effect that this is likely to have on:

- the current project;
- following projects or other work queuing in the pipeline;
- last, but never least, the client or customer.

Sometimes late running might be acceptable and require no action. Much depends on how much float exists in the schedule. Usually some action must be taken. The project manager must then assess the situation, decide the appropriate action and implement it.

Late jobs with float

If late work has enough free float to absorb the total expected delay, then all that needs to be done is to ensure that the work is expedited and finished without further interruption, within the available free float time. Jobs with no free float have to be regarded more seriously because late working early in the programme will rob later tasks of their float. So all jobs without free float should, wherever possible, be expedited to bring them back on schedule.

Purchasing, manufacturing and construction departments have always suffered at the hands of project managers by being expected to perform miracles when all the float in the schedule has been used up by late-running design tasks earlier in the project lifecycle, long before work enters the purchasing and manufacturing or construction phases.

Late tasks with zero or negative float

If critical tasks (tasks with zero or negative float) are late, then special measures must certainly be taken. It might be necessary to accept more expensive working methods to expedite these late jobs and bring them back on schedule.

If a task budgeted to cost £10,000 is in danger of running several weeks late and jeopardizing the handover date of a project worth £10m, then clearly it would be worth spending an additional £10,000 or even more on the problem task if that could rescue the project programme. The project manager must always view the costs of expediting individual activities against the effect on the whole project. In desperate situations an immediate action order can sometimes be an effective remedy. Immediate action orders are described in Chapter 34.

Updating schedules and records

Updating the project schedule

Updating is the process of producing a fresh set of schedules and other reports to take account of one or more of the following:

1 A change to the project parameters, such as an unexpected increase or decrease in the numbers of resources available, changed cost rates, a newly imposed target date or a scope change.
2 A change in the network logic. This could arise, for example as a result of a serious technical problem and consequent change of strategy. A change in the project scope is often another reason.
3 The important need to produce new schedules that take into account progress made to date.

If a schedule has been produced which proves to be practicable in all respects, and if everything goes according to plan, there is no need at all to update the schedule during the life of the project. This could very well be the case for a simple project, or for a schedule covering only a very short total time span. Even a more complex project has been known to run entirely according to its original plan when the plan was well made and the project was effectively managed. However, most projects are always at risk of things not going to plan. Just when everything appears to be on plan, a key supplier fails to deliver vital materials, a design error is discovered that will take weeks to put right, bad weather delays progress, and so forth.

The integrity of a project schedule must be protected at all times. If anything happens that renders a schedule inaccurate it must be updated as soon as possible; otherwise people will lose faith in the plans and disregard them. Updating frequency can be contemplated:

- on an occasional basis, whenever a project change or one or two late jobs require the project to be rescheduled or;
- on a regular basis (which often means monthly).

Regular updating will certainly be needed if the project is complex, with large volumes of progress data to be digested. For some projects, regularly updated schedules and their resulting reports provide an essential part of reporting to the project client.

For the day-to-day management of project work it is clearly the current schedule that counts. Updated schedules simply follow, to make certain that work yet to be done remains sensibly scheduled. If, for instance, only scrap is produced after six weeks of a manufacturing job, or if it is found that 100 metres of trench have been excavated along the wrong side of a road, the manager does not wait for an updated schedule, but takes immediate practical action. If the activity is critical or nearly so, it should be obvious to all those involved that all the stops need to be pulled out at once to put things right. Of course the schedule must be updated, but that is a consequential process which follows once steps have been taken to sort out the immediate problem.

Drawing and purchase schedules

Drawing and purchase schedules contain much information that will change as the project proceeds. They must be kept up to date. When the project is finished the purchase control schedules, because they list all the purchase specifications, become part of the essential documentation that records the 'as-built' state of the project. Similarly, the drawing schedules list all the drawings used on the project, together with their final revision status.

Managing the progress and quality of purchased materials and equipment

This very important subject is also mentioned in Chapter 27.

Expediting

Expediting is not just a process of chasing up late deliveries. It requires communication (including possible visits) with suppliers during the waiting period. This not only safeguards progress, but it also provides an early warning system should a supplier experience any difficulty in meeting an order. The expediter should react promptly and firmly whenever the reply to a routine expediting enquiry is unsatisfactory or vague. The special efforts of the purchasing department should not stop until the supplier has either shown the necessary improvement or has actually delivered the goods.

Sometimes a new method of approach to the supplier can produce the desired result. An offer to arrange collection of the goods from the supplier's premises will, for instance, make the supplier aware that the purchaser is willing to participate constructively and that, on this occasion at least, there is genuine urgency.

The purchaser might wish to arrange visits to a supplier's premises to check on progress, inspect the quality of workmanship or witness performance or safety tests. Such visits are sometimes linked to the certification of suppliers' claims for stage payments. There are several ways in which responsibility for carrying out inspection and expediting visits can be allocated or delegated. Where suitable engineers are available to the purchasing department or agent concerned, it is often convenient for the purchaser to arrange visits that combine the inspection and expediting functions.

Where specialist engineering attendance is needed for inspection or to witness tests, the project engineering organization might send one or more of its own engineers to assist the purchasing agent. To avoid expensive overseas travel when foreign suppliers are involved, it is sometimes possible to engage a local professional engineering company to undertake these inspection and expediting visits.

Alternative sourcing

When expediting appears to be failing, the design engineers might be able to suggest an alternative item that can be obtained more quickly. The solution might instead mean finding another source of supply. If the original order does have to be cancelled because the supplier failed to make the agreed delivery, there should be no penalty for the purchaser, because the supplier broke the contract by failing to fulfil the delivery conditions.

Purchase order status reports

The purchasing department or agent is usually required to keep the project manager informed of the progress status of all current purchase orders. This responsibility, in addition to the inspection and expediting reports already described, extends through all stages of the journey to the project site. This reporting can be done by means of regular order status reports, which list all purchase orders in progress, giving outline details of shipping and delivery dates, and highlighting any problems and corrective actions.

Efficient communications and dedicated, regular reporting by all the purchasing agents and freight forwarding agents involved on the project are essential if purchase order status reports are to be of any value. It is also important to arrange the reports so that potential or real problems are highlighted.

Priority allocation in manufacturing projects

Occasions will arise when work cannot be carried out by manufacturing departments in a sequence that suits every project. If the production control department is able to pick up all orders and load them sequentially, or according to their own machine and manpower schedules, no serious problem need arise. Sooner or later, however, an order is going to be placed that is wanted urgently. Then the production controller will have to take account of job priorities.

ABC priorities

Some organizations attempt to allocate order priorities by labelling factory jobs as priority A, B or C. A job carrying *Priority A* would be perceived as being more urgent than jobs at *Priority B*, leaving *Priority C* jobs as the poor relations. It is easy to understand why such systems break down. Delayed *Priority C* orders will eventually become wanted urgently. In the end, everyone will label their orders as *Priority A*, so that everything is wanted at once and nothing receives special attention.

Wanted-by dates

A preferable arrangement is to schedule manufacturing orders by 'wanted by' dates. The production controller can then attempt to schedule to meet these dates, and inform those who are likely to be disappointed. Scheduling software that can work to such target dates has long been available to production management. If any project item is expected to be delayed beyond its critical date, the possibility of subcontracting the work must be considered.

Resolving conflicting priorities in manufacturing

Special project work often has to take its place in the production organization alongside routine manufacturing work or jobs for other projects. Conflicts can arise between jobs with different priorities, and it is a brave person who attempts to intervene between two rival project managers who are fighting for the same production resources.

Critical path analysis and multiproject resource scheduling are the logical ways for deciding priorities but, unfortunately, some managers are not over impressed with logical reasoning when they see their project being delayed (taking advantage of available float) in favour of other work. If such problems cannot be resolved on the spot without bloodshed, a sensible approach is to refer the problem to a third party arbitrator - preferably a manager at a more senior level in the organization, capable of assessing the relative merits of the conflicting jobs.

Progress meetings

Regular progress meetings provide a suitable forum where essential face-to-face communication can take place between planners, managers, subcontractors and other project stakeholders. The main purpose of progress meetings is to review progress periodically, with the intention of identifying problems and discussing remedial actions needed to keep project activities on schedule. This important subject is discussed in Chapter 37, so will not be commented upon further here.

Progress reports

Internal progress reports to company management

Progress reports addressed to company management must set out the technical, fulfilment and financial status of the project and compare the company's performance in each of these respects with the scheduled requirements. For projects lasting more than a few months, such reports are usually issued at regular intervals, and they may be presented by the project manager during project review meetings. Discussion of a report might trigger important management decisions that could lead to changes in contract policy or project organization. For this and many other reasons it is important that data relevant to the condition and management of the project are presented succinctly and factually, supported where necessary by explanations and forecasts. Internal reports often contain detailed information of a proprietary nature. They might, therefore, have to be treated as confidential, with their distribution restricted.

Progress reports to the client or customer

The submission of formal progress reports to the client or customer could be a condition of contract. If the customer does expect regular reports then, quite obviously, these can be derived from the same source that compiled all the data and explanations for the internal management reports. Some of the more detailed technical information in the internal reports may not be of interest to the customer or relevant to their needs. Customer progress reports are therefore often edited versions of internal management reports.

Whether or not financial reports accompany customers' progress reports will depend on the main contractor's role in each case. Under some circumstances cost and profitability predictions will have to be regarded as proprietary information, not to be disclosed outside the company. In other cases, the project manager may have to submit cost summaries or more detailed breakdowns and forecasts as part of an advisory service to the customer.

Although customer reports may have to be edited in order to improve clarity and remove proprietary information, they must never be allowed intentionally to mislead. It is always important to keep the customer informed of the true progress position, especially when slippages have occurred that cannot be contained within the available float. Any attempt to put off the evil day by placating a customer with optimistic forecasts or unfounded promises must eventually lead to unwelcome repercussions. Nobody likes to discover that they have been taken for a ride, and customers are no exception to this rule.

References and further reading

Devaux, S.A. (1999), *Total Project Control: A Manager's Guide to Integrated Planning. Measuring and Tracking*, New York: Wiley.

Kerzner, H. (2009), *Project Management: A Systems Approach to Planning, Scheduling and Controlling*, 10th edn, Hoboken, NJ: Wiley.

Lock, D. (2013), *Project Management*, 10th edn, Farnham: Gower.

Marks, T. (2012), *20:20 Project Management, How to Deliver on Time, on Budget and on Spec*, London: Kogan Page.

Meredith, J.R., Mantel S.J., Jr. and Shafer, S.M. (2017), *Project Management: a Strategic Managerial Approach*, 10th edn, Hoboken, NJ: Wiley.

Turner, J.R. (ed.) (2014), *Gower Handbook of Project Management*, Farnham: Gower.

32
CONTROLLING CHANGES

Dennis Lock

One thing that can be guaranteed for any project lasting more than a few weeks is that the end result will be changed in some way from the original concept. As experience accrues through the design and fulfilment stages it is inevitable that minor corrections and changes will be needed to correct errors or to adapt to conditions and circumstances that were not foreseen when the project was authorized. People also tend to have second thoughts about designs and specifications and those thoughts can materialize into actual changes. When correctly managed, changes can ensure that the final as-built condition of the project will satisfy all the stakeholders. But those two words 'correctly managed' are the key to success - and what this chapter and the next are all about.

Change factors in relation to project control and performance

The first thing to consider when talking about change controls is to identify what actually constitutes a project change. Imagine a designer who has been working for a week on a drawing but realizes that there was a flaw in the calculations, meaning that the design work has to begin all over again. Well, to the individual designer that certainly means a change. It could also delay project progress by a week. But in this case the flawed design had not been released outside the design department. The necessary design rework will not change the original purpose and specification for the project. No materials or work will have to be scrapped.

So, early corrections of design errors, no matter how serious, are not really changes within the meaning of project controls. No drawing or specification has been issued to other departments, no purchase order or external contract has been placed, and nothing (except the flawed design) has to be scrapped. So it could be said that the definition of a project change (for the purposes of time, cost and quality control) is any event or decision that requires a reissue of manufacturing or construction drawings and specifications against which work has already been ordered, committed or started.

Now consider a different example. A hotel refurbishment project has been in progress for a few weeks when a visitor from senior management comments that all the bathroom taps and fittings would look far better if they were gold plated. Well, maybe that would be an improvement, but it is certain that the change would increase the project costs. Further, if the original stainless steel or chromium plated fittings had already been ordered (or even worse, delivered to site) the change would cause scrap or penalty costs and progress delays.

Every project is at risk from suggestions from well-meaning people that (if implemented) would increase project costs, cause delays and (in some circumstances) introduce unforeseen problems in manufacture or construction. I used to call this risk 'creeping improvement sickness', but now the universally accepted name is 'scope creep'.

One way of regarding change requests is to label each proposal as either essential or desirable. If a proposed change is essential for operational or health and safety reasons, then clearly it has to be made. But changes that are simply desirable generally have to be disallowed, unless the stakeholder making the request is in a position to accept the delay and pay for additional costs.

Design freeze

One important aspect of changes and their control is that changes made later in a project lifecycle will cost more than changes made near the project beginning. Changes made to correct errors in drawings and specifications during the initial project design phase might delay progress but should not be expensive. However, once physical activities have begun on a project, the costs of work and materials (the costs of work in progress) will begin to accrue. These accrued costs are sometimes also known as sunk costs. Clearly the accrued costs will grow as the project progresses through its lifecycle (as illustrated in Figure 32.1). So a proposed change near the beginning of a

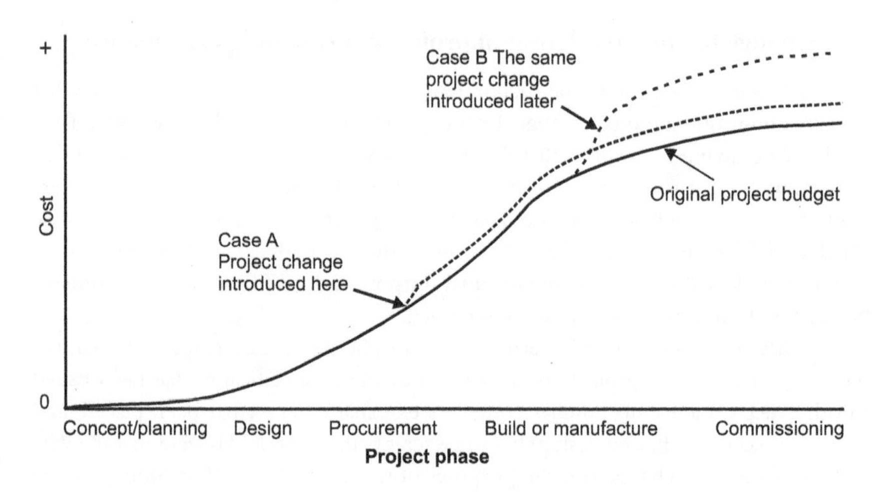

Figure 32.1 Notional cost of a project design or specification change in relation to the project lifecycle.

project might be more acceptable and cost far less than the same change considered later in the project.

There comes a point in every project where changes must be forbidden unless they are essential for health and safety or to correct some catastrophic design error. Some companies recognize this milestone by announcing that their projects have reached a state of 'stable design'. The more common name for this project stage is 'frozen design' or 'design freeze'.

Change requests and documentation

Requests for changes can come from many directions. Even a suggestions box, accessible by all members of staff, can eventually generate change requests. So it is clear that a filter must be in place to prevent a project from being disrupted, delayed and overspent as a result of unnecessary changes. That filter must have four procedural stages, which are as follows:

1 Evaluation and consideration of the proposed change.
2 Authorization or rejection.
3 Implementation.
4 Documentation.

Change board or change committee

Even the simplest change can sometimes affect the costs and progress of a project far more than the change originator envisaged. For that reason every change proposal needs to be considered by senior representatives from the key project departments. Thus in most companies that conduct projects on a regular basis a committee is established for the sole purpose of considering change requests. This committee might be known as a change board or change committee. A change committee might comprise the following people (with others co-opted as necessary), although this certainly cannot be described as typical owing to the many different kinds of projects and project organizations:

- Chief engineer or project engineer (the design authority).
- The project manager.
- A senior manager from the construction or manufacturing department.
- A purchasing manager or officer who can advise on purchase orders that could be affected.
- The relevant commercial manager or officer who has knowledge of the contract and the client.

The change committee members should be capable of assessing the need for any proposed change and for predicting the effect that the change would have on project progress, costs and profitability, reliability, quality, health and safety. They must be sufficiently senior to make executive decisions that could affect costs, progress, profitability and relationships with external stakeholders (particularly the customer or client).

Change requests

There is no reason why absolutely anyone (even the most junior person in the organization) should be denied the opportunity of suggesting or requesting a project change. A change might indeed arise as a result of an apprentice posting a note into a staff suggestions box. To quote a popular maxim, 'There's no harm in asking'. It's up to the change committee to decide how to deal with each change request. But it is clearly necessary to establish an efficient and standard discipline for dealing with all change requests. Office disciplines usually mean having to fill in forms, and change requests are no exception. It is usually found that two different procedures are needed, one for changes that arise within the organization and another for changes requested by a customer or client.

Internal change request forms

To ensure that every change request is fully detailed with all the information needed to help with the authorization decision it is usual to ask the change originator to record details of the proposed change on a standard change request form. An example of a change request form is shown in Figure 32.2. It is very important that each change request is fully supported by details of why the change is considered necessary. Other aspects of the request procedure are apparent from the example in Figure 32.2. If any information written on the request form is lacking or unclear, then the change committee will have to delay their decision until the originator has supplied the missing details, and that could delay the process by weeks.

External change request forms

External change requests usually come from a client or customer. A request might be verbal or made in writing. However it is received, each external change request must be recorded and progressed using a standard form. The document commonly used for this kind of change is a contract variation order, an example of which is given in Figure 32.3. Clearly if the request comes from a paying customer it will be given sympathetic consideration. In fact, there are companies that make little profit from the work in the original contract, but do very well by pricing contract variation work more generously.

Although contract variations can mean increased profitable work, they nonetheless often constitute a considerable nuisance. Also, it is possible that what appears to be a simple customer request could in some way affect the project adversely. There are many possible risks. For example, extra load might be placed on a structure, requiring a design modification not foreseen by the client, or work in progress might have to

Change request	Serial No. Request date:
Project title	Project number.
Change details (attach drawing or separate sheets if necessary):	
Reason for change:	
Change requested by:	
Emergency action requested (if any):	
Estimated cost:	
Effect on schedule and work in progress:	
Chargeable to client: No ☐ Yes ☐ (if YES give order details below):	
Change committee instructions. Change approved ☐ Request denied ☐	
Signed:	Date:

Figure 32.2 Typical elements of a change request form.

be scrapped and done again. So contract variations should be subject to the same change committee scrutiny as other change requests, although they will of course be approved wherever possible. The subject of contract variations was also discussed in Chapter 28.

Contract variation order	CVO number:
	Project number:
Project title:	Issue date:
Summary of change (use continuation sheets if necessary):	
Originator:	Date:
Effect on project schedule:	
Effect on costs and price:	Cost estimate ref:
Customer's authorization details:	Our authorization:
Distribution:	

Figure 32.3 A contract variation order

Change coordination

It is essential to nominate a person (a change coordinator or change control clerk) who can register and serially number all change requests and then ensure that every request is progressed reliably and promptly through all stages from consideration and full implementation or rejection. The logical home for this coordinator is the PMO. It is unlikely that the change coordinator will be fully occupied with changes, so this activity can usually be combined with one or more other duties. Every approved change request must be implemented not only on the physical project, but also in cost schedules, drawing registers, purchase schedules and other relevant project documentation. That follow up is part of the change coordinator's duties.

Change documentation will feature heavily should any contract disputes arise later in the project, when customers might deny liability to pay for changes that they have caused. There must be an audit trail for every change requested by the client or customer. The project company's position in respect of claims for additional pricing, or of allowing resulting delays in project delivery will be badly weakened if evidence of all the customer's change requests cannot be produced.

The change decision process

The stage has now been set for the decision process, when the change committee must decide whether or not to approve each change request or decide upon some other action. The change decision process is shown in Figure 32.4. This decision-tree diagram is self-explanatory and needs little amplification. One branch of the decision tree not shown in the diagram is concerned with what happens when the change committee needs further explanation to help with their decision, in which case the committee will return the request to the originator via the change coordinator, asking for more reasons or details.

It is important that everyone involved in the project should know how and where to place a change request. For internal changes, the originator needs to obtain a change request form from the change coordinator, fill it in with as much detail as possible, and then return it to the coordinator for passing on to the change committee. External customers and clients will have their regular contacts with the project organization, so they might ask for changes through sales engineers, senior project engineers, the project manager or even a more senior manager. However the request is received, it must be 'entered in the system' by passing it through the change control clerk or change coordinator, who will then ensure that the request is put in front of the next change committee meeting.

The change committee members will all be senior managers, whose time is precious. To avoid constantly interrupting their other work it is customary to submit change requests to them in batches, with the corresponding meetings held at regular but not too frequent intervals. A balance has to be struck between causing possible project delays while changes are in the process of being considered and disrupting the normal work of the committee members. Monthly intervals between change committee meetings are common.

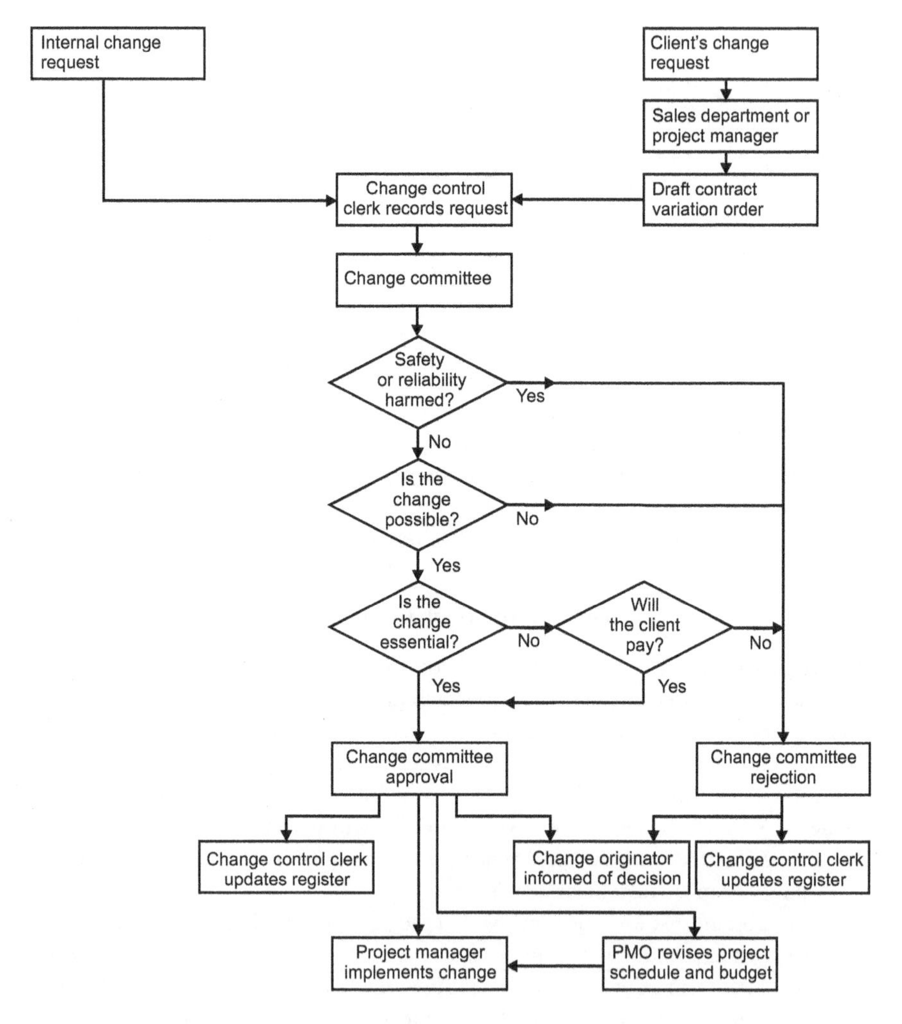

Figure 32.4 Lifecycle of a project change request.

Emergency procedures

It sometimes happens that the instructions contained in drawings or specifications cannot be followed exactly. A design error could be responsible, or perhaps the materials specified cannot be obtained. Maybe fixing holes on two components do not quite line up, and the assembly workers want permission to drill the holes out larger and use bigger bolts or rivets. Another possibility is that an inspection or test report will reveal results that are just a little outside the specified tolerances. It is inevitable that problems such as these will crop up on projects where the designs are new and unproven in practice. The remedies needed are often not really design changes, but are minor adjustments to allow progress to continue. However, any change from the design instructions must be approved by a qualified designer or engineer, otherwise allowing departures could lead to all kinds of problems.

Where a change is allowed, the original instructions, drawings or specifications must be updated so that the records are correct. Then if the same design is used in the future the problem should not recur. As described below, there is more than one way of coping with these problems.

Concessions and production permits

Requests for concessions or production permits are made when, for some reason, the manufacturing or construction people on the project are unable to comply exactly with the design instructions, but believe they can offer an alternative approach to get round the problem. So a formal documented procedure allows the request to be made via the change coordinator. Then the request document is subjected to the same change committee consideration as the change request procedure described above, but possible fast-tracked by the coordinator (see the following paragraph).

Fast-tracking a change request

When the need for change approval is very urgent, it is still essential that the documentation and consideration steps are all followed. However, when the urgency is really pressing the change coordinator can visit each member of the change committee in turn, and collect their comments or approval quickly so that the project can proceed.

Marked-up drawings

Marked-up drawings offer a method by which a senior engineer or the project manager can provide on-the-spot approval in a factory, assembly shop or production site when, for any reason, the work cannot be performed exactly as specified. Approval can only be justified in limited cases. Suppose, for example, that a decimal point has been misplaced or omitted from a drawing. The mistake might be obvious, but the quality inspector cannot sign off work that differs from the drawing.

Marked-up drawings are generally to be deplored and discouraged. However, when the solution is obvious, one way of surmounting the problem quickly is for the project manager, project engineer, or other authorized person to mark up the correction on two identical drawing prints and sign both. One signed marked-up print then accompanies the work through to completion and final inspection. Most importantly the twin print must immediately be passed to the change coordinator, who will ensure that the identical correction is made on the original design.

Documentation

Recording the 'as-built' project condition of a project is vital for a large number of reasons. In a mining project, for example, it is essential that plans of the shafts and galleries are kept up-to-date, both for safety and rescue reasons and to ensure that if workings should be closed and then reopened many years later, the engineers know what to expect. In industries such as aviation and automobiles, where products are sold in different versions, changes have to be recorded in such a way that the build state of every individual item sold can be accurately found.

It is an absolute rule that if a design change is made that renders a project component non-interchangeable with the same component used elsewhere on the project, the drawing number (and therefore the component identification number) must also be changed.

Actions after a project change has been requested

Someone has to take charge of every change request and follow it through from the time the request is submitted until the change has either been rejected or fully implemented. This job can be assigned to a clerk, who might be designated the change coordinator. The logical home for the change coordinator is in the PMO. In a small project this job will not occupy the coordinator full time, so other duties can usually be combined with the role.

The control process for ensuring the timely consideration and, where appropriate, the implementation of all approved change requests usually depends on having a change register.

Change register

Figure 32.5 shows one example of a change register page. This example has been reduced for the purposes of illustration here but in practice change registers might need to be formatted on A3 size pads. The initials ECR over the first column in this example refer to the change request, which might be termed an 'engineering change request'. All change requests should be serially numbered for identification and control purposes. Change registers serve to ensure that all change requests are dealt with

Change register

Project No: Project name: Sheet No:

ECR		Drawings/specs affected:	Brief description of change	Change committee		Drawing/spec reissue	
No.	Date:			Date:	Decision/PoE	Date:	New rev

Note: ECR is the abbreviation for engineering change request.

Figure 32.5 Essential features of a change register. Abbreviations are explained in the text.

promptly and placed before the change board or change committee at each of their meetings. They also record the subsequent actions relating to each request.

When the change committee has decided how to act on a change request they will issue instructions ranging from 'change request rejected', through 'more information required' to 'change approved'. In projects where more than one item could be affected by the change, such as when components or assemblies are being batch produced, the change committee might give instructions as to whether the change must be implemented retrospectively, or only in current manufacture and future batches. So in those cases the change committee will designate a PoE, which is the point of embodiment for the change among the repeating batches. Batch production is not common in projects but it can happen.

The change coordinator will scan the register daily to ensure that every change request is followed through, and that every authorized change does result in reissued specifications or drawings. The coordinator will also ensure that the originator of the change request is kept informed of progress. Note that if a change is introduced to the design of a component that renders it non-interchangeable with similar items made before the change, then the part number must be changed.

Build schedules and traceability

This section applies particularly to manufacturing projects where a design project results in the manufacture of items for sale or and use by other companies or individual members of the public.

When a motorist attempts to buy a replacement part for a vehicle, the first questions the store assistant will ask is 'What is the make and model of your car and the year of manufacture and, do you know the chassis number?' From those details the store assistant should be able to find the parts list version that applies to your particular car and, from that, the spare part identification number should be found. Now bear in mind all the other vehicles sold by that car manufacturer over many years. The only possible way to identify the correct replacement part for your car depends on the availability of complete parts lists for each batch of vehicles made, or even sometimes for your own particular vehicle if it was custom built.

In manufacturing and production management terms, each complete parts list that pertains to a specific product is known as a build schedule. A build schedule is a completely detailed family tree, similar to a WBS. It will include not only all the part identification numbers for the product as it was built, but should also show the modification details for any after-sale recalls. Your vehicle is identifiable uniquely by its chassis number. From those details it will even be possible to order retouch paint that matches the original colour of your car.

Now consider the following story. It was a grey, damp and cold morning at the airport and passengers boarding Flight xyz to the sunny Bahamas were looking forward eagerly to the sun and sands that would replace the English winter. The Lox Airways Skyliner 604 soared gracefully into the winter sky and soon, above the clouds, the aircraft was bathed in glorious sunlight. But tragically, an hour or so into the flight, all contact with the aircraft was lost. Subsequently wreckage was found scattered over a remote part of Southern Ireland. There were no survivors. But among the wreckage and pathetic remnants and remains of the men, women and little children the black box recorder was found and recovered intact.

Subsequent analysis of the black box data revealed that the aircrew appeared to have lost consciousness. Further investigation found that a faulty component in the enclosed flight deck had led to the crew breathing in noxious but (to them) undetectable gases. Further probing found that a faulty servo motor was responsible. The question now arose 'Could this happen again in another Skyliner 604?' So all similar planes were immediately grounded. The next question was 'How many other aircraft have been fitted with servo motors of the same design and from the same production batch?'

Finding the answer to such questions depends on procedures that are more specific to manufacturing and assembly routines than to one-off projects, so I shall only mention them briefly here. One essential requirement is the serial numbering of every significant component so that it can be uniquely identified and its modification status recorded. Even for small components, such as nuts, bolts and rivets, the records should show from which batch these were used so that, should one fail, every item can be traced back to its supplier and actual production batch.

I shall leave this subject here because we are entering the world of production management rather than project management. For those who are interested there is more detailed explanation of these procedures in Flouris and Lock (2007).

When project changes are welcome

Consider a large petrochemical project that is nearing completion and handover to a client. The proud project manager is showing senior representatives from the client over the project site. It is a beautiful summer day, and the sun is glinting off the various storage tanks, pipework and stacks, all of which have been painted matt medium grey. One senior client representative remarks 'Those stacks, tanks and pipes would look far better if they were painted emerald green, and that also happens to be our corporate colour. Is it too late ...?' Am I entering the realms of fantasy here? Well, actually, no I am not. Something like this actually happened on a large project from my experience when our client asked the project manager if the colour of some huge coal storage bunkers could be changed before we handed the project over.

How would you react if you were that project manager? I would answer 'It's never too late' and ask the client to put the request in writing (in the form of a contract variation order). Now consider how you would price that variation order. When your company originally bid for the contract you were pricing keenly, with a low mark-up and high risk, because you were up against fierce competition. But now all the competitors have left the scene. You are in control. You can, within reason, charge what you like. Father Christmas has arrived early this year. It's your birthday! So, some project changes are to be welcomed. In fact it has been said that there are contractors who expect to make most, if not all, their profits from variation orders.

Effect of changes on earned value analysis

Once EVA, prediction and reporting have been implemented in a project, any engineering change or contract variation will probably affect the factors used in earned value calculations and predictions. That's common sense and hardly needs amplification here. However, what does need to be said is that someone in the project organization should be made responsible for ensuring that the earned value implications of any change are taken into account. And, that 'someone' will most likely reside in a PMO.

References and further reading

Flouris, T.G. and Lock, D. (2009), *Managing Aviation Projects from Concept to Completion*, Farnham: Ashgate.

Lock, D. (2013), *Project Management*, 10th edn, Farnham: Gower.

33

PERFORMANCE MEASUREMENT AND ANALYSIS

Tony Marks

This chapter examines best practice methods for measuring performance and, just as importantly, for analysing what that means in terms of delivering the project scope on time and within budget.

Introduction

The primary purpose of setting a project specification, schedule and budget is to provide a reference that enables measuring and controlling the project parameters of cost, time and quality. The four main steps required in the control process are the plan-do-check-act cycle, as described in Chapter 4 and shown in Figure 4.1. A slightly different and more detailed interpretation of this control cycle is shown here as Figure 33.1, in which the four principal steps are named as:

1 Set a measure.
2 Record progress.
3 Assess the impact.
4 Take corrective action.

Measuring earned value

Most project practitioners regularly progress their plans and generate status reports indicating the events and milestones that have been achieved. They also predict events that will occur in the future and when those events are expected to happen. Another process is to record the amount of money spent and then estimate how much more will have to be spent to fulfill the project. However, in most cases the physical progress or achievement is measured and reported separately from the financial results and forecasts. Few project people measure their efficiency of working or the value of work that has been achieved. In other words, performance measurement tends to

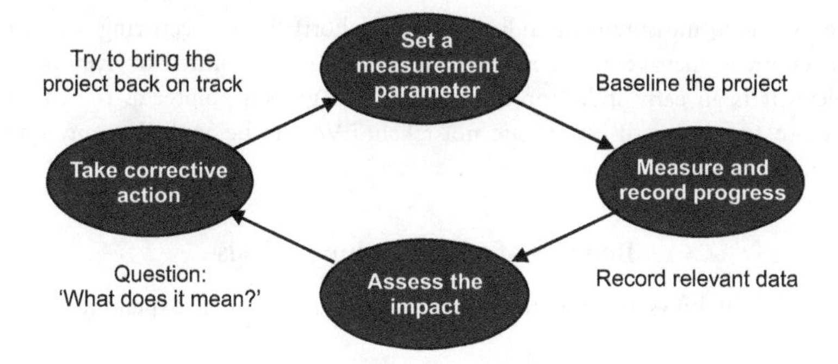

Figure 33.1 A control cycle. This interpretation of control by measurement and active feedback is sometimes known as a cybernetic control loop because of a similar process used to correct distortion in electronic circuits (such as high fidelity audio amplifiers).

be neglected. So now it is time to introduce and explain the concept of earned value measurement.

The most definitive measures of achievement are ideally obtained in terms of recognizable events such as milestones. But the earned value concept aims to measure progress more precisely and to attach a monetary value to the measured (or assessed) amount of work that has been achieved. Although this value is usually expressed in monetary terms it can also be stated in other appropriate units such as man-hours. In other words, the value earned when a specific milestone or deliverable has been achieved is based on the planned (budgeted) cost of achieving that milestone or deliverable.

Consider a project with a total estimated cost of £1m. Suppose that the actual cost of the work done at the point of measurement is £90,000 when the budget for that same work was only £80,000. That's a 12.5 per cent overspend. It clearly indicates that the planned earned value has not been achieved. If this same level of performance were to be perpetuated for the remainder of the project the final project cost would be 12.5 per cent over the original £1m estimate, which is £1,125,000. This kind of analysis and prediction depends on three factors, which are:

- A budget sub-divided into measurable packages of work.
- A method for measuring achievement or progress.
- A method for collecting the costs directly related to the measurable packages of work.

EVA (also known as earned value measurement) is a discipline that facilitates sound management and gives an early indication of the state of project health (principally from a cost and schedule perspective). However, all measurements of performance must be treated with caution. They are indicators of the efficiency or performance of a project. They should be used as one of the criteria upon which project managers base their decisions, after having interpreted the information and placed it in context.

Performance measurement indicates where shortfalls are occurring and where extra resources, management action or other support is required to overcome the problem. It is an early indicator of problems and gives a pointer as to what may happen to the project if actions are not taken. EVA can be applied to any type of project, irrespective of size.

Benefits of earned value analysis

Figure 33.2(a) shows the traditional way of depicting actual expenditure against budget over time.

This graph shows how much has been spent to date and compares it to the budget that was planned to be used by the report date. However, it does not show:

- If the project is ahead of schedule or running late.
- If the project is truly over or under spent.
- If the project is producing value for money.
- If money has been spent on the right things.
- If the problems are over or have only just begun.

The graph in Figure 33.2(b) is similar to the graph in Figure 33.2(a) except that now a measure of performance (or status value) has been included. The extra line represents the earned value or, in other words, the proportion of the budgeted work that has actually been achieved. The following additional information can be gleaned from this graph:

- The project is behind schedule because the amount of work completed is less than that scheduled.
- The project is overspent because the cost of the work completed is greater that the budgeted cost of that work.
- The project is spending money incorrectly and is costing more to achieve than the budget.

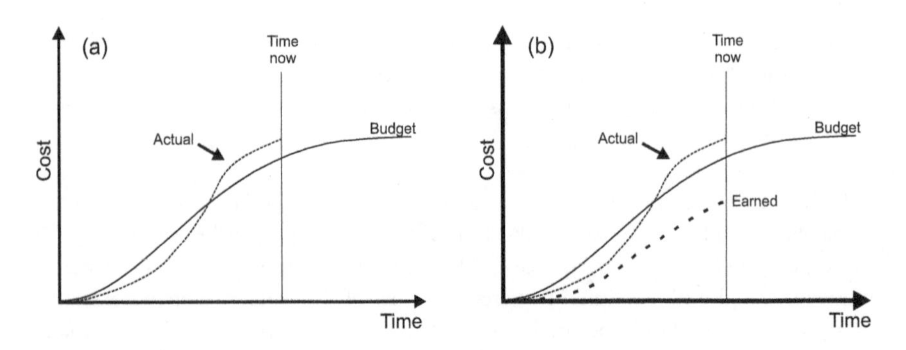

Figure 33.2 Two different ways of indicating project expenditure against time.

- The problems do not appear to be contained, since the slope of the actual cost line is greater than the budgeted cost line. This indicates that the amount overspent is going to increase, even though the earned value line does look as if it will intersect the budget cost line at some time in the future.

Key performance indicators used in earned value analysis

EVA has its own specific terminology and relies upon the following key performance indicators:

- Performance measurement baseline (PMB). This is the time-phased budget plan against which project performance is measured. It is formed by the budgets assigned to scheduled cost accounts and undistributed budgets. It equals the total allocated budget minus management reserve.
- ACWP. This is the total cost incurred and recorded for the work performed in a given period and which is being subjected to EVA.
- The planned value (PV) or budgeted cost of work scheduled (BCWS). This is the planned cost of work scheduled to be done during the period under earned value review. Specifically, BCWS is the sum of all budgets for all packages of work and, in total, forms the PMB at the time of measurement.
- Earned value, or the BCWP. This is the sum of the budgeted costs for completed work packages and completed portions of open work packages at the time of the earned value review.
- Cost variance (CV). This is the difference between the BCWP and ACWP. Thus:

  CV = earned value (BCWP) – actual cost (ACWP)

A negative CV value indicates an overspent budget and a positive value indicates spending below budget.

- Schedule variance (SV). SV can be considered either in terms of cost or (more usually) time.

For cost, the SV is the difference between the BCWP and BCWS. A negative value indicates that less work has been done than planned and a positive value indicates more work has been done than planned.

SV (cost) = BCWP – BCWS

For time the SV is the difference between original planned duration (OD) and the actual time expended on the work to date (ATE).

SV (time) = OD – ATE

An earned value case example

The following example is taken (with permission) from Lock (2014). It shows how to calculate the earned value parameters for a project to build a brick wall.

Imagine a bricklayer and a labourer engaged in building a new boundary wall enclosing a small country estate. If the amount of progress made had to be assessed at any time, the work achieved could be measured by the area of wall built or more simply (as here) by the length of wall finished.

The scope, budget and schedule for this project have been defined by the following data:

- Total length of wall to be built = 1000m.
- Estimated total project cost = £40,000.
- Planned duration for the project is 10 weeks (with five working days in each week) = 50 days.
- The rate of progress is expected to be steady (linear).

When the above data are considered, the following additional facts emerge:

- Budget cost per day = £800.
- Planned building rate = 20m of wall per day.
- Budget cost for each metre of wall finished = £40.

At the end of day 20 the project manager has decided to carry out an EVA. The data at the end of day 20 are as follows:

- Work performed: 360m of wall completed.
- ACWP: £18,000. This is the total cost of the task at the end of day 20.
- BCWS: 20 days at £800 per day = £16,000.
- BCWP: the 360m of wall actually completed should have cost £40 per metre = £14,400.

Cost implications

The cost implications of these data can be analysed using EVA as follows:

$$\text{The cost perfomance index (CPI)} = \frac{\text{BCWP}}{\text{ACWP}} = \frac{14400}{18000}$$

The implication of this for the final project cost can be viewed in two ways:

1 We could divide the original estimate of £40,000 by the cost performance index and say that the predicted total project cost has risen to £50,000, which gives a projected CV of £10,000 for the project.
2 Alternatively, we can say that £18,000 has been spent to date and then work out the likely remaining cost. The 360m of wall built so far have actually cost £18,000, which is a rate of £50 per metre. The amount of work remaining is 640m of wall and, if this should also cost £50 per metre that will mean a further £32,000 predicted cost remaining to completion. So, adding the costs measured to date and the remaining costs predicted to completion again gives an estimated total cost at project completion of £50,000.

Schedule implications

Earned value data can also be used to predict the completion date for an activity or an entire project. The first step is to calculate the schedule performance index (SPI). For the wall project at day 20 the SPI is found by:

$$SPI = \frac{BCWP}{BCWS} = \frac{14400}{16000}$$

The original estimate for the duration of this project was 50 working days. Dividing by the SPI gives a revised total project duration of about 56 days.

Methods for assessing the progress on a task

Most EVA must be performed not just on one activity (as in the brick wall project just described), but on many project activities. At any time in a large project, three stages of progress can apply to a measured activity. These stages are as follows:

1 Activity not started. Earned value is therefore zero.
2 Activity completed. Earned value is therefore equal to the activity's cost budget.
3 Activity is in progress or has been interrupted. For tasks in construction projects, the earned value can often be assessed by measuring the quantity of work done (as in the brick wall example). For other tasks it is necessary to estimate the proportion or percentage of work done, and then take the same proportion of the current authorized cost estimate as the actual value of work performed.

Earned value analysis reliability

Early predictions of final costs can be unreliable for at least four reasons:

1 Estimates of progress, or of work remaining to completion are subjective judgements, and people usually err on the side of optimism.
2 During the first few weeks or even months of a large project, the sample of work analysed in earned value calculations is too small to produce valid indications of later trends.
3 There is no guarantee that the performance levels early in a project will remain at those same levels throughout the remainder of the project.
4 Additional work might be needed on activities previously reported as complete. For example, drawings and specifications often have to be re-issued with corrections and modifications when they are put to use in manufacturing or construction. Unless due allowance has been made elsewhere for this 'after issue' work, the project budget might eventually be exceeded in spite of earlier favourable predictions from EVA.

What if the prediction is bad?

Suppose the actual hours recorded for a project during its progress greatly exceed the earned value expected for that time. The resulting prediction will indicate final costs well in excess of the budget. The first thing to note is that the project manager should be grateful for the early warning. Escape might be possible even from an apparently hopeless situation, provided that suitable action can be taken in time.

Strict control of changes or a design freeze can curb unnecessary expenditure and conserve budgets (although changes requested and paid for by the customer might be welcome because they will add to the budget). Unfunded changes should only be allowed if they are essential for project safety, reliability and performance. See Chapter 32 for more about controlling project changes.

In the face of vanishing budgets, the demands made on individuals will have to be more stringent. But this can be helped through good communications, by letting all the participants know what the position is, what is expected of them and why. It is important to gain their full cooperation. The project manager will find this easiest to achieve within a project team organization. If the organization is a matrix, the project manager must work through all the departmental managers involved to achieve the essential good communications and motivation.

People's performance can often be improved by setting short- and medium-term objectives, but these must always be fair and feasible. Targets should be expressed in measurable quantities (for example time or money) so that assessment can be fair and objective. Individuals should be encouraged to monitor their own performance. These personal objectives should equate with the project objectives of time, cost and performance through the WBS. When work is done on time, it is far more likely that the cost objective will be met.

If, in spite of all efforts, a serious overspend still threatens, there remains the possibility of replenishing the project coffers from their original source — the customer or client. This feat can sometimes be accomplished by re-opening a fixed-price negotiation whenever a suitable opportunity presents itself. An excuse to renegotiate may be provided, for example, if the customer should ask for a substantial scope change, or as a result of external economic factors that are beyond the contractor's control. Failing these steps, smaller modifications or project spares can be priced generously to help offset the areas of loss or low profitability. Care must also be taken to invoice the customer for every item that the contract allows to be charged as a project expense.

Remember that, without EVA, forewarning of possible overspending might not be received in time to allow any corrective action at all. The project manager must always be examining cost trends, rather than historical cost reports.

When the predictions are bad, despair is the wrong philosophy. It is far better to reappraise the remaining project activities and explore all possible avenues that could restore the original project targets.

References and further reading

Fleming, Q.W. (2010), *Earned Value Project Management*, Newtown Square, PA: Project Management Institute.

Lock, D. (2014), *The Essentials of Project Management*, 4th edn, Farnham: Gower.

34
FORECASTS AND CORRECTIVE ACTIONS

Dennis Lock and Shane Forth

There are times in the lives of all project managers when our projects hit snags or are at risk of failure. This chapter contains tips and advice about recognizing the warning signs and taking effective corrective action. Some possible remedies are well-known. Others require more ingenuity and courage.

Forecasting

Of course, the first forecasts for task durations and costs (leading to total project predictions) are made before the project begins, using estimating and planning methods such as those described in earlier chapters. But forecasting does not end there - it has to be a continuous process right through the project lifecycle to ensure that the completed project satisfies its design parameters and stays within the authorized schedule and budget. Schedule forecasting and control is particularly important because as a general rule any task or project that exceeds its planned timescale will also exceed its budgeted cost.

Quantities and parameters

The baseline for control must be the original project schedule and budget breakdown, as subsequently amended by authorized changes. The fundamental question is 'Will all work in progress and future tasks finish on time and within their budgets?' Continuous monitoring and forecasting during project execution should examine one (or possibly more) of the following parameters, bearing in mind that task durations are usually the most important factors, not only for project completion time, but also for the total project costs:

- the remaining effort (man-hours or cost) needed to finish each unfinished task;
- the expected start and finish dates for future tasks (especially critical tasks) compared with their original scheduled dates;

- any discovered schedule error that, when corrected, would change logical relationships, task criticality and the expected project completion date;
- any authorized, imminent or expected change in project scope;
- the actual or reputational performance of external suppliers and subcontractors.

If any failure to meet the planned schedule for remaining in-house tasks is detected, the project manager must also consider whether planned arrangements with suppliers or contractors might also be affected, in which case they must be consulted and kept informed.

Effective forecasting

It must be recognized that a forecast is not a fact. It can only be an estimate of what is to come. All forecasts should ideally be made in the first instance by those who are going to supervise the work. However, the PMO or project manager will challenge forecasts, especially where the estimators are known to be unduly pessimistic, optimistic, or otherwise unreliable.

The project manager and the PMO will want to challenge and seek remedies for any forecast or estimate that threatens the established project budget or schedule. Also, an organizational culture of continuously revising targets and then failing to achieve them is never acceptable. Remedies, however, must always be acceptable, practicable and achievable.

Warning signs

If a project manager becomes aware that a project is failing when it is too late to do anything about it, then (with the exception of unpredictable events such as natural disasters) it is usually safe to assume that the fault lies principally with the project manager. In normal circumstances if a project is correctly planned, sensibly scheduled and diligently monitored, then any indication of possible failure should become apparent early, while there is still time to take some form of corrective action.

One very common threat to projects is scope creep. Various people are usually prone to make comments such as 'wouldn't it be nice if ...' and ask for changes that could delay the project, cause work done to be scrapped, introduce unforeseen risks and create budget overspends. An effective project manager, with backing from higher management, will ensure that no unauthorized change is allowed to happen. Change control procedures were described in Chapter 32.

Many projects are heavily dependent upon external companies, such as suppliers and subcontractors, over whom the project manager has no direct control. So a diligent project manager will keep an eye on the business news, company reports and so forth to gain early warning of any potential financial or operating difficulty facing suppliers and subcontractors upon which the project relies to any great extent. One essential and simple remedy for reducing such risks is for the project manager to consult the purchasing manager and/or the CFO before committing to any large contract, because one important duty of purchasing and financial managers is to watch the financial news and be aware of the status and performance of relevant suppliers, contractors and competitors. However, there have been cases (for example in the construction industry) where sloppy auditing and over-optimistic (thus untruthful) company reports have hidden such risks until they become apparent too late.

Methods for accelerating progress

Concurrent engineering

Concurrent engineering is a fairly rare but ultimately rewarding process that relies on close cooperation between a project client and the project contractor. It can also apply to cooperation between two companies within a group, where one company is conducting an engineering project for another. The process is best described by the following case example.

A company that designed and manufactured heavy special purpose machines for the automotive industry would discuss the proposed machining methods in great detail with its clients. Suppose that a project required the machine manufacturer to design and build a single special purpose machine for the accurate machining of aluminium gearbox cases. Such projects always give rise to many problems, such as how to clamp the car components in the machine to facilitate accurate machining.

Much time and money was saved for both companies by allowing their design engineers to cooperate in the early project design stages. So the automotive engineers would collaborate with the machine designers and ensure that the automobile components did not present terrible and expensive problems for the machine designers. Thus some potentially serious future manufacturing problems were foreseen and prevented in advance.

Difficulties with suppliers

Clearly the first port of call for the project manager when difficulties with suppliers arise is to the purchasing manager or materials controller. Routine (and recommended) purchasing and expediting measures should usually give early warning of any potential supply problems. However, there are sometimes rare occasions when routine purchasing and expediting methods appear not to be satisfying urgent project needs. Although unconventional (and possibly causing annoyance to the purchasing manager) a direct approach from the project manager to the supplier can occasionally produce the desired results. In normal times it is best for the project manager not to interfere, but the following case is an example of an exception to this rule.

A case example of overcoming supply difficulties

This case happened some years ago. It concerned the medical division of a company whose product range included heart-lung machines - precision-made mechanical devices used in operating theatres during heart surgery to take over the heart function. These machines were manufactured to order, usually in single units, so they were treated as special projects. Response to an urgent order from a hospital for one of these machines was delayed because the manufacturer had no motors or rev-counter cables in stock. So these components were placed on order from the suppliers.

The two electric motors had to be flameproof because of operating theatre regulations. The manufacturer quoted a long delivery time for these motors. Two rev-counter cables (similar to automotive tachometer cables in all respects except for their short length) were even more of a problem, with a quoted delivery of 16 weeks. The purchasing manager informed the project manager that he could do no better. The project and the customer would, apparently, just have to wait. But the project manager had the customer and the hospital patients firmly in mind. So he wrote two personal letters, one to the CEO of each supplier, explaining the urgent need for these components.

As a result the cable manufacturer very obligingly interrupted a long production run for the automobile industry and produced the two special cables in less than two weeks. The delivery time for the motors was halved. So the company was able to manufacture and ship the heart-lung machines to the hospital far more quickly than if the project manager had simply followed routine procedures and done nothing.

Successful project managers are reluctant to accept defeat. When things get tough they will examine all possible ways of keeping their projects on track and satisfying the stakeholders.

Increasing available resources

When the order books are full it sometimes happens that some project tasks have to be put on hold simply because people are busy with other work or a particular production facility is out of action or not free. Here is another challenge for the project manager who has urgent tasks to fulfill. Alternative resources will have to be found.

Overtime working

It is a general rule in project management that routine project tasks should never be planned on the assumption that people will be asked to work outside normal working hours. However, if project activities do run late, the project manager will be very grateful when critical tasks can be accelerated by asking people to work occasionally for longer hours or during weekends. It is recognized that this approach will soon result in diminishing returns if used too often, because people work faster and better when they are not tired.

Priorities between different projects and clients

Now we are entering dangerous territory. When a company is engaged in a number of projects, it can be the case that the project manager who shouts loudest gets the prize. The driving force can, of course, be the project clients, some of whom can shout louder than others, even possibly threatening to place all future work elsewhere. These situations demand tact and diplomacy, and the project managers involved must usually consult the senior executives and seek their help in mediating and dealing with the clients.

However, projects for external clients are programmed to suit the programmes of those clients and can often be components of the clients' larger capital projects. Just as in-house projects are at risk of running late, so the clients' capital projects can also be delayed. Suppose, for example, that a project to build a new machine for installation in a client's new factory is running late. It might be the case that the new factory will not be ready, so that late delivery of the machine would cause no difficulty and, indeed might be more convenient for the client. So, this is one possible escape route that the project manager might, with discretion, explore.

Subcontracting

Many project tasks (even including fundamental design) can be subcontracted to external agencies when the home organization is in danger of becoming overloaded. It is well known that the demand for project resources within any company can fluctuate widely, dependent on the number of current projects and their lifecycle stages.

Special care is needed to maintain the organization's reputation for quality and reliability when entrusting project work to subcontractors. Prudent companies will develop and maintain good relationships with potential subcontractors, even to the extent of occasionally outplacing work that could be done in-house (just to keep the wheels oiled). A satisfactory situation is reached when the project manager can call not only upon resources within the project organization, but has fall-back arrangements with potential subcontractors that will enable work to be subcontracted at short notice when required.

One danger is that the external companies will ask the project manager to guarantee them a minimum amount of future work. Such requests have to be resisted because they require commitment, defeating the purpose of subcontracting (which is to provide a flexible supply of resources without commitment). The solution for the project manager is to keep in friendly contact with more than one potential subcontracting agency.

Fast-tracking the schedule

Conventional project plans can often be accelerated by taking a few small risks and overlapping tasks. The network fragments in Figure 34.1 illustrate the process. In the upper part of this illustration the conventional and familiar sequence of design-check-order materials-wait-make is shown. Fast tracking, as shown in the lower part of the figure, aims to cut out unnecessary delays as far as possible. In the lower half of Figure 34.1 the plan has been changed so that materials are placed on order before the design drawings have been checked. This will accelerate the task but incur a small risk that incorrect materials could be obtained as a result of unchecked design errors.

By the simple process of ordering materials as soon as the design is available, and then sending a vehicle to collect from the supplier, 17 days have been shaved off this sequence of tasks. If these tasks were critical that could possibly mean 17 days earlier

Normal scheduling - total planned duration = 43 days.

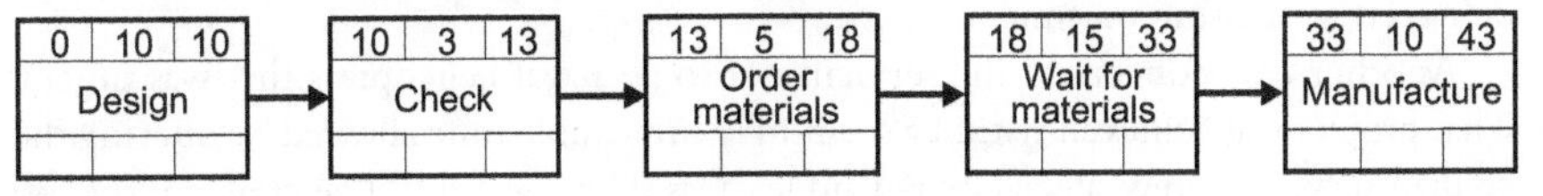

Fast-track scheduling - total planned duration = 26 days.

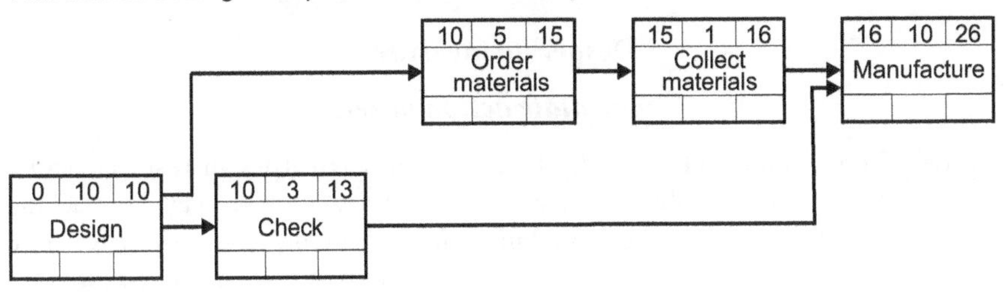

Figure 34.1 Fast-tracking to gain time.

completion for the entire project. If, on the other hand, the schedule contained other critical (or near-critical) paths the project manager should look to see if those other tasks could also be crashed. When taken to extremes, it is sometimes possible to crash many project tasks, so that most tasks in the schedule become critical. Then there will be multiple critical paths through the network.

Some time-saving methods can increase the costs of individual tasks. Then the prudent project manager will ensure that first priority for crashing is given to the tasks that are cheapest to crash. Also, crash considerations will be given first where the least risk is added to the project.

Unconventional rescue methods

Bribery (but never corruption)

Two examples follow that show how bribery – sorry, perhaps that should read incentives – can sometimes claw back project delays.

The first of these examples concerned the shipment of custom-built operating theatre patient monitoring equipment to an overseas hospital. The project had been severely delayed by many months owing to a total company reorganization that involved many management changes. The manager who found himself responsible for taking over this hospital project was determined to avoid further delay. The shipping department was asked to book space on a vessel. That set a fixed target. During the final assembly stages (which involved very difficult electrical assembly and cabling) the project manager placed a small bet with the small final assembly crew that they could not finish the work in time for this project to be transported to the ship. The task seemed impossible, but the assembly crew rose to the challenge. They and the project manager remained in the factory, working continuously for three days and nights without sleep until the job was done. The assembly crew were paid their tiny winnings (but they also received triple time wages). The project consignment was delivered to the ship on time.

Another case concerned the servicing of some naval helicopters that was urgent. That project was achieved within a seemingly impossible time allowed by offering the civilian servicing crew special extra paid leave as their incentive. The crew got the job done on time and were then able to take leave.

Desperate measures

Immediate action orders

Please refer once more to Figure 34.1. In that example a total design-to-manufacture time was shown to be capable of reduction from 43 days to 26 days by fast-tracking the schedule. That process involved taking some risk by ordering materials before the design drawings had been checked but would not have caused any disruption to other work. If more time had to be saved on this task, maybe the estimated manufacturing time of 10 days could be reduced by giving this task extreme priority. Under traditional circumstances special jobs in a factory have to undergo waiting and transit periods as they are moved from one process to another.

One extreme process that can remove all waiting times and greatly speed total task duration involves the use of an immediate action order procedure. Immediate action orders can be guaranteed to cause disruption, displeasure and annoyance to many in the organization but they can be extremely effective in getting the almost impossible done in a very short time. The immediate action order process works as follows (for a paper-dependent system).

The project manager (or other applicant) must first persuade the CEO or other nominated senior executive that the project task (or job sequence) is of vital importance and that without special measures it would run late and cause untold and unacceptable difficulties for everyone concerned, including important project stakeholders. The immediate action order (IAO) will have to stand out from the ordinary, to be recognized by all who see it as something that is exceptional and demanding immediate attention. With a paper system that is achieved by printing the form with vivid diagonal fluorescent stripes. A flashing icon on the screen would be the computer equivalent.

All the required tasks are detailed on the special IAO job routing form, with their start and finish times left blank to be filled in according to the results achieved. Then the form is given to an eager and athletic progress clerk, who is also provided with a stopwatch. The following example was a real case.

An immediate action order case example

A very small 2400 hertz, 3000 volt transformer was needed as part of a prototype guided weapon project. The combination of high voltage and extreme operating environment meant that this transformer had to be totally encapsulated in epoxy resin. That required a specially designed and manufactured demountable aluminium mould.

A prototype transformer was wound and assembled, moulded in resin and then tested. The entire process took six weeks, partly because of normal routing delays between manufacturing stages, but also because of the intricate nature of the transformer clamps and the mould. During testing the transformer failed and one winding burned out. Of course, the transformer, contained in its hard resin block, was now scrap. Under normal circumstances at least another six weeks would have been needed before the design fault could be identified and another transformer produced. The project schedule did not allow anything like six weeks. Something extraordinary had to be done.

The project manager persuaded the company's general manager to authorize an IAO. The IAO document was soon made ready, signed by the general manager and then handed to a progress clerk. The first job station on the order was the electrical design department, and so the clerk arrived there without delay and presented the dismayed design manager with the IAO, time-stamping it as he did so. The work of an electrical design engineer was interrupted so that he could review the design of the transformer and its mould. He identified the problem with the first design, namely that the mould required more holes to prevent the trapping of air bubbles during the resin encapsulation under vacuum. Revised mould drawings were printed in the design office and handed to the progress clerk, who stamped the exit time on the IAO and resumed his travels.

The next port of call for the progress clerk was to the transformer department, where the production of a new transformer was immediately put in hand. Meanwhile the progress clerk went to the machine shop to demand the manufacture of a modified casting mould. That meant removing a job in mid-process from a milling machine and asking the machine operator to make the new aluminium mould parts instead.

And so the clerk, the IAO and the transformer components took their express route through all the relevant departments until the new transformer was made, cast in resin and (fortunately) tested successfully. The first, failed transformer had taken six weeks to make. Its successful replacement was ready in only three days.

The IAO process must not be abused, or it could greatly disrupt a company's routine processes. So the following special conditions must always be strictly observed:

1 Only one IAO can be allowed in the organization at the same time. It might, indeed, be wise to specify a minimum interval period of (say) one month between any two successive IAOs. IAOs must not be allowed to become commonplace.
2 The IAO must bear the signature of a nominated very senior executive. That gives the IAO organizational clout.
3 The task cost budget restrictions are removed. For example, if necessary someone can be sent by air travel to pick up vital materials.
4 Once authorized, an IAO takes precedence over all routine work, in all departments.

Creating a project task force

A project can sometimes be accelerated dramatically by changing the organization structure. Chapter 6 described the difference between team and matrix organizations. In a matrix organization the project manager is particularly reliant on cooperation from the managers of other departments, over whom the project manager usually has no direct authority. With a team, the project manager has direct command of the project staff and is therefore able to exert greater authority and potentially get faster results.

If a project is facing disaster it is sometimes possible to restructure the organization to give the project manager more immediate authority. A team organization is the clear choice, but even that can be improved upon by setting up a special project task force. In this arrangement everyone working on the project is seconded to the task force and reports directly to the project manager until the project aims are realized. Now there are no competing projects and no dissenting departmental managers. The project manager has supreme authority and is able to say what has to be done, when and by whom.

Generally a task force can be disruptive and will cause staff resettlement difficulties when the project ends. But desperate times call for desperate measures and creating a task force is occasionally the clear and only choice.

Calling in the doctor

When problems continue to arise that are beyond the scope and comprehension of the organization's general and project management, an independent external view can sometimes be helpful. When your problem is toothache your recourse is to a dentist. For health problems it's the doctor. For business problems it could be a management consultant. That approach might be too late for an individual project, but could be beneficial for future projects. I have to confess that I have occasionally acted as a consultant, and so have to declare an interest here.

Management consultants operate either as individuals or in companies. They have generally earned and deserved a reputation for being very expensive, ultimately producing glossy reports with costly recommendations that might (or might not) be favourably received by the client. Then, while the client is still attempting to understand the report, substantial claims for fees, travel, meals and other expenses follow. However, a competent consultant who investigates the organizational structure, workings and objectives of a company can occasionally yield positive benefits and improve future project performance (always provided that there is a future for the organization after the project failure that the consultant was too late to cure).

Some business schools offer consultancy services, which can give mutual benefits to the client, the business school and advanced students. Accreditation or certification by the Chartered Association for Project Management, the Project Management Institute or other relevant professional organization can provide some guarantee of a consultant's competence.

Keeping the stakeholders informed

All clients and other stakeholders of large capital projects expect (and deserve) to be kept informed of progress. This is usually achieved by means of written and illustrated reports that are issued at monthly or quarterly intervals (depending on the size and duration of the project). There is always a natural tendency to give the best possible impression in such reports. Occasionally, when difficulties arise, the project contractor might wish to hide these from the client with the firm expectation (or desperate hope) that the difficulties will disappear with time. With the occasional minor glitch this might not be a problem. But when the glitch is not minor the truth will eventually emerge and, to put it mildly, the contractor-client relationship will undergo a change for the worse.

So, while progress reports do not have to list every little problem, they must not be written with the intention of deceiving the client. Early and honest disclosure of impending difficulties might elicit some sympathy – even helpful suggestions – from the client. But dishonesty will eventually reap awful repercussions.

References and further reading

Buttrick, R. (2005), *The Project Workout*, 3rd edn, London: Prentice-Hall.
Kerzner, H. (2000), *Applied Project Management: Best Practices on Implementation*, New York: Wiley.
Lock, D. (2013), *Project Management*, 10th edn, Farnham: Gower.

PART VIII

Information and communications

35
COMMUNICATIONS AND DOCUMENTATION

Dennis Lock

Projects depend to a very great extent on effective communications between all the stakeholders and also upon keeping that information so that it can be retrieved when needed. Some information may be confidential or even secret, and that places additional responsibilities on the project manager. Much of the information accrued during the life of a project will still have importance long after the project has been finished.

Communicating work instructions

For the routine issue of work instructions to a manufacturing department or construction group (those who perform the 'hard' tasks) every established company will have its own routine methods and these are outside the scope of this handbook. The project manager and the PMO will, however, also have to ensure that instructions for 'soft' project tasks (such as producing or approving designs, drawings and specifications) are issued to the relevant departmental managers. An efficient way of achieving this begins with the preparation of a detailed schedule of all project tasks, preferably derived from a critical path network. Ideally all current tasks for all projects should be scheduled together within the known available resources, with all scheduled dates driven by the critical path data.

Every departmental manager should receive his or her specific schedule of fresh project tasks from the PMO on a regular basis. Good practice is to issue such schedules periodically (monthly or more frequently), with each fresh issue updated to take account of progress reported on existing tasks. Reliable distribution, ensuring that managers get schedules specific to their departments, can be achieved by allocating a unique sort code to each department (or its manager) and then filtering and sorting all project schedule report data accordingly. Thus each manager should receive regular work schedules giving the following data for each project task that is due to be started (or is already in progress) within a particular department and the report period:

- task ID code (for reporting and use on timesheets);
- task name or description;
- scheduled start date;
- scheduled finish date;
- earliest possible start date;
- latest permissible finish date;
- task float or slack;
- amount and type of resources required (for example, 1DE might signify one design engineer).

Achieving that demands a competent PMO, a detailed network diagram, sensible coding systems (see Chapter 5) and reliable project management software. The techniques needed to produce such schedules have been available for very many years. However, in complex organizations it is quite possible for some managers responsible for project tasks to be missed out when routine work instructions are issued, as the following case example illustrates.

The push-pull nature of project communications

Managers cannot rely on receiving all the communications they need. Communication is often a push-pull process – the communicator pushes the data out but the recipients must always be looking out for and pulling in data that they need. The following case example illustrates what can go wrong when an element of the push-pull process is missing.

The project in question was for the manufacture of special-purpose transfer line machinery for machining automobile engine cylinder heads. The scheduled engineering design tasks included the preparation of detailed drawings for all the machines along the transfer line, the transfer line itself and the jigs and clamps for positioning the cylinder heads at the various machining stations.

I made the decision to keep the network diagram and its resulting schedules as simple as possible by not including the design of the lubrication, hydraulics and cutting oil systems, because those tasks followed on naturally from the machine component and assembly design and did not need to be scheduled separately. That assumption was valid. If the design and drawing schedules were sequenced correctly for the machine components, anyone reading those schedules would know when work on the lubrication and hydraulic systems would be needed. However, I made the terrible mistake of not including the manager responsible for lubrication and hydraulics (let's call him David) on the monthly work schedule distribution. The following short dialogue occurred in a tense and unpleasant disciplinary meeting with the engineering director (ED).

ED to David: 'Why didn't you produce the lube and hydraulics drawings?'
David: 'Because Dennis didn't give me the schedule' (which was perfectly true).
I now expected to be fired. It was that sort of company. But, wait ...
ED to David: 'But it was your job to find out.'

I could not believe my luck. It *was* my fault. But it was David's office that became vacant at the end of that week. That was a lesson learned. This had been a double communications failure. I had not pushed the schedule information to David, but (crucially) he had not pulled it from me.

There are many instances in the lives of projects where information has to be pulled in with determination from external sources. Here are a few examples:

- receiving design approval from a project client for layout drawings;
- receiving reports from quantity surveyors;
- receiving operating and maintenance instructions for equipment sourced from external suppliers;
- receiving progress reports from external contractors.

Following up progress within the organization

Issuing project task requirements by the means of schedules is, of course, only half of the communication process. The other half requires active follow-up to ensure that the scheduled instructions are being carried out.

One system works by having all the departmental managers mark up and return their schedules at the end of each reporting period to show whether or not each scheduled task has been carried out. That has the great advantage of simplicity. All those managers need only to tick jobs that have been done according to plan. Of course, for jobs not started or running late more explanation will be needed. If the procedure is made too complicated, feedback will be late or missing altogether.

A very useful occasional check, which might require some courage, is for the project manager to visit a department without prior warning and ask its manager to point out the people who are actually working on the project at that instant. Then heads have to be counted. Sometimes the results can be revealing and very disappointing. The work schedule might show, for example, that 10 designers should be hard at work on the project for which you are responsible, but the head count might reveal only five – because one or two are off sick and others are working on projects whose managers have been progressing their projects more aggressively.

Controlling work subcontracted to external design offices

Some project tasks can be dependent upon outside offices or manufacturers. Some companies even subcontract important design tasks to independent external companies. That arrangement can work very well to relieve resource overloads. Establishing good rapport is essential. For critical design work where it is especially important for the designs to comply with company traditions, standards and quality, it is good practice to arrange for a senior engineer from the external company to be brought in-house to work under the supervision of the project design manager for a month or so. Then, when that engineer returns to the external office he or she will be imbued with the desired approach to quality and design practice and can confidently be expected to supervise the external work.

One way of following up external work is to appoint a subcontract liaison engineer whose main job is to tour the external facilities to issue and authorize new project tasks, follow up progress on work already issued and answer occasional technical questions. The liaison engineer can be employed within the PMO and report to the PM0 manager.

Project communications standards

On any project for an external client or customer a document should be prepared that gives the names and contact details of all the principal parties (external and internal) involved in the project. In some organizations this information will exist in the form of a secretaries' manual.

For large projects some companies produce a 'project handbook' (also known by various other names such as 'project control manual'). That document should be kept up to date and issued to all departmental managers. The manual will usually specify various technical and project scheduling procedures that must be followed, but will also describe preferred communication channels (preferably enhanced by the inclusion of one or more organization charts). A project handbook might also include such data as the levels of authority owned by individuals, for example giving financial limits for approving expenditure.

It is important that everyone working on a project is aware of what they may, or may not be able to authorize according to their position and seniority in the organization. One particular danger that must always be kept in mind and guarded against is to ensure that unauthorized instructions are not given to issue or amend instructions to external subcontractors and suppliers. It is a well-known and common danger that a design engineer (for example) might make technical inquiries from a potential supplier, or inquire about an existing purchase order and then agree verbally to a new order commitment or an order amendment without the due process that should at least involve the purchasing department and possibly the organization's established change control procedures.

Escalating issues

The term 'escalating issues' is an unfortunate example of 'management speak'. It does, however, describe an essential process for attempting to resolve project difficulties that are beyond the scope of the project manager. An 'issue' can mean any problem from a slight mistake in a drawing to something far more serious. Escalating issues is the process of referring problems up the chain of command in the organization so that senior management are made aware of the risk and, where possible, can take or authorize remedial action.

When any problem occurs in a project, the project manager will obviously attempt to take corrective action. However, there will be occasions when the problem is outside the control of the project manager. For example, it might be that a departmental manager is failing to carry out project tasks according to the schedule, perhaps giving priority to other work. The project manager has no direct line authority over the recalcitrant departmental manager, but has a clear duty to take some kind of action. That means appealing to a more senior executive, who does have the required organizational power and can, if deemed appropriate, come down like the proverbial ton of bricks on the reluctant departmental manager.

Another way of describing this process is called exception reporting. The term 'exception' in this context can apply to any task or cost item that is in danger of failing to meet the planned project needs or falling outside its planned time or budget limits. The remedy might be to increase a budget allowance, authorize special measures or allocate more staff (all of which approaches would usually be outside the limits of authority enjoyed by the project manager and require the approval and intervention of senior management).

Exception reporting also means that senior managers are not faced with regular and detailed time-wasting reports that simply list the many tasks and processes that are on plan, within budget, and need no senior management attention.

Document ownership and retention

Every project of significant size will result in the accumulation of drawings, specifications, correspondence and so forth, some of which will remain valuable and will need to be kept safely not only during the project lifecycle but also for some years after the project has been finished. Project document archives are discussed in Chapter 36, but some thought has to be given as to who actually owns the designs and specifications generated during the active project.

I take the view (probably old-fashioned) that whoever owns the documents after project delivery; the project contractor has a duty of care to the client in retaining copies of all documents that might need to be accessed even some years after the project has been handed over. Then expert advice can be given if the project owner runs into operating or maintenance difficulties.

Now suppose that the project outcome happened to be a bridge carrying road and rail over a deep gorge and that a few years after being put into service the bridge and all those travelling upon it at the time finished up in the river below. Then the inevitable public enquiry would need evidence of designs, and that means having access to the original design calculation documents and the resulting drawings. And that, in turn, means knowing where to look not only for the design documents, but also for records showing which companies constructed the various steel or concrete bridge components.

This is straying into the territory of the following chapter, but it highlights the need to determine who will own the project design documents after completion. The time for asking that question is before the project begins and (in projects for external clients) ownership of drawings and specifications should be specified as part of the contract terms and conditions.

Registers and schedules

Registers generally split into those that pertain to the organization as a whole, and those that refer only to the specific project. These two states are not mutually exclusive. Registers are important for several reasons, and their value often extends well beyond the active project lifecycle. One or two of the following descriptions also appear in other chapters, but are included here to give the whole picture.

Project register

Project registers were described in Chapter 3 but are included here because they are a very important starting point when attempting to retrieve information about a finished or abandoned project. A project register is used to allocate all new project numbers and record details such as:

- project ID number;
- project title;
- name of the client;
- name of the project manager;
- date of project authorization.

If often happens that records of previous projects need to be accessed, for example when the original designs need to be retrieved for possible inclusion in new projects, or when legal issues arise. The project register is usually a good starting point for the search.

Drawings schedules

At the beginning of a project it is useful to list all the significant drawings that will have to be drafted and issued throughout the project lifecycle. Drawing schedules can be a valuable tool for helping to plan design activities, and very useful long after a project ends for information retrieval. Essential elements of a drawings schedule are shown in Figure 35.1. In a large project the drawings schedule can extend to many pages, which might be subdivided into groups for the various design disciplines.

Purchase control schedules

In many projects it is customary for the project team to list all the foreseen major component purchases in a purchase enquiry schedule, which will be identified by the project number and broken down into parts for the various departments engaged on

Drawings schedule					
Project: Shane Villas Estate, Luton Project number: SF 0456		Page 1 of pages Issue date:			
Drawing number	Title	Client appro (if requ'd)		Issues	
		Requested	Received	Date	Rev No.

Figure 35.1 Typical headings for the first page of a project drawings schedule.

Purchase control schedule								
Project: Shane Villas Estate, Luton Project number: SF 0456					Page 1 of pages Issue date:			
Spec. No.	Rev No.	Description	Qty	Supplier	Order No.	Amdt No.	Date	On-site cost

Figure 35.2 Typical headings for the first page of a purchase control schedule.

the project (such as fluids, mechanical, electrical and so on). Enquiry schedules are essential controls when reaching out to potential suppliers for quotations. Enquiry schedules morph into purchase control schedules once the suppliers become known and orders result. So purchase control schedules are another useful source when trying to retrieve information on purchased goods and components from current and past projects. Figure 35.2 illustrates typical page headings for a purchase control schedule.

Build schedules and the modification status of manufactured products

WBS have been described elsewhere in this handbook but when they are taken down to the detail of all included components and parts of a manufactured project they become parts lists. Now suppose that an initial project design results in a parts list for a consumer project such as an automobile. As each product comes of the production line it will have a manufacturer's serial number. Over the course of time, modifications will result in changes to the original parts list. That could eventually lead to chaos and customer dissatisfaction when trying to order replacement parts.

The solution is to ensure that each production batch can be related directly and unequivocally to the relevant issue of the total detailed parts list. In some industries those specific parts lists become known as build schedules. When the project end user requires a replacement component, the product's serial number will identify the relevant build schedule, from which the part and modification number of the correct replacement part can be found.

Risk register or risk log

Risk registers (or risk logs) for a project list all the perceived project risks and should be compiled as part of the risk management process. They can help when beginning to identify possible risks for a new project.

Change register

A project change register records all changes made during the design and manufacture (or construction) of a project. It should show the correct issue of every drawing and specification used on the finished project. That information can clearly be important when retrieving drawings should the project need rectification after delivery.

Contract variations and dayworks orders

It is most important, especially on large projects, to keep careful records of all approved contract variation orders and dayworks sheets. Difficulties in agreeing the final payments can arise if all parties fail to maintain records and keep a financial tally throughout the project lifecycle. Even on relatively small construction projects it can be difficult for the parties to agree the final amount if there have been a large number of variation orders and (especially) the more casual dayworks orders.

References and further reading

Lock, D. (2013), *Project Management*, 10th edn, Farnham: Gower.

36
CONTROLLING ARCHIVES

Dennis Lock

The success of every project depends to a very large extent on communications and information flows. During the project lifecycle much of that information will accumulate in files of drawings, specifications, purchase orders, change records, letters, emails, contracts, minutes of meetings and other documents. When the project ends, many of those files will remain as a valuable resource that must be stored - and stored in such a way that data can be found and retrieved years into the future.

Data handling and storage policy

Every business organization should have a well thought-out policy for how it handles and stores data. For data held and communicated about people there are important legal requirements set out in the Data Protection Act (2018). This legislation requires that a data protection officer (DPO) is appointed who reports to senior management. This is not usually a very time-consuming role and the DPO will often be a manager or other suitably reliable member of staff who has other duties. It is unlikely that this aspect of handling personal data will have to be considered in the context of project controls, so it will not be discussed further in this handbook.

The most important decisions about storing project data have to be made when the project is nearing its conclusion. Just as every project should begin with a formal authorization document, so it is important that the same degree of thought is given by senior management when a project ends. Every formal project closure document should include a section that lists the various types of documents (contract letters, drawings, technical specifications, claims for payment, contract variation orders, change notices and so on) and specifies a retention policy for each document type.

Reasons for retaining project documents

There are many reasons why documents should (and in some cases must) be stored. One fairly obvious reason is to keep designs and engineering calculations for future reference. In some cases documents have to be kept for legal reasons. A contractor's duty of care to clients can also require the storage of documents for several years after projects have been finished and handed over.

Retained engineering

Retained engineering is a collective name sometimes used to describe documents that enshrine all the design work, calculations and specifications resulting from project design engineering. Another frequent description is 'lessons learned', but that clearly also has wider implications. The idea behind retained engineering is that some designs and calculations used on one project might be suitable for repetition on future projects, with obvious savings in time and design costs. However, as technology advances, so the value of retained engineering diminishes and today's engineering becomes tomorrow's outdated ideas. Nonetheless retained engineering is not to be despised. It can save time and money and reduce the risk of technical mistakes.

Legal and moral obligations

Documents pertaining to company accounts and taxation have to be retained for several years. The legal obligation for retaining most financial and taxation documents has a time limit, which, according to the Statute of Limitations, is usually six years. However, this is dangerous territory for non-legal people and project practitioners. Company policy on the retention of these documents is best left to the decision of the company's legal advisors.

Design calculations and drawings can be needed as evidence when structures fail, especially where personal injuries result. Here no time limits apply and it is essential that all relevant data are retained and also indexed to facilitate identification and recovery.

Every company that designs and builds operating plant (such as that required for a petro-chemical or mining installation) has a moral obligation to support the plant operator with technical advice long after any warranty period has expired. That obligation applies not only to items designed, manufactured and installed by the project contractor, but also in some cases to plant and equipment purchased from subcontractors and external suppliers. Some subcontractors and suppliers will inevitably merge or go out of business after a project ends (or more worryingly even during the active project lifecycle).

Although it might not be general practice, the company responsible for a design and build contract would be well advised to keep not only its own design drawings, specifications, operating and maintenance instructions, but also copies of the maintenance and operating instructions provided by the external suppliers of the major installed equipment items. If your company develops a good reputation for after-care service it will be more likely to attract future work.

Coding data for filing and retrieval

Anyone who has ever borrowed a book from a library will be aware that books are catalogued and arranged on the shelves according to a comprehensive system of subject coding known as the Dewey Decimal System. The nearest equivalent of that system for project managers is the project WBS, which (unlike the universal Dewey Decimal System) will always be unique to each company (and very probably also specific to each project).

Searching the archives to retrieve data

To search a database for a particular piece of project information, the starting code sequence should be project ID number/WBS code. That depends on the existence of a suitable document control and coding system in the company, which is not always the case. An extreme example of failure in this respect came to my notice during the course of a consultancy assignment, when I discovered that my client (a highly competent engineering organization) had no fewer than 10 different drawing numbering systems operating simultaneously within the same organization, one system for each of 10 different departments. At the other extreme, it was once necessary for me to establish norms for several aspects of engineering and manufacturing (such as labour times and the average shipping weights for large machine components) and the coding and accuracy of that US company's records were so exemplary, going back over many years, that my search task was made very easy.

Where coding systems are badly designed, or otherwise fail to turn up the desired information, the alternative when trying to find a piece of historical information is to search the database using keywords. That, of course, assumes a suitable database has been set up and maintained. If the project organization is small, and the amounts of data involved are also small, searching should not be too difficult. For a large organization handling more project work, searching will be more difficult. However, the larger organization should be able to afford the employment of a full-time professional project librarian or documents controller.

An important resource when searching for past project details is the project register. Thus all pages of the project register should be kept, regardless of age. Then any keyword such as the client's name, project title, project date and name of the project manager should help to find the entry in the project register. Once the project register entry has been found, the project ID number will be revealed and that should lead to the WBS. So, it is clear that project register pages should never be removed from the archives. Other important documents that can help in future searches are the drawings register and the purchase control schedules.

Storage media

Storage media can be classified within three groups, which are:

1 Original hard copy (on paper or translucent film).
2 Microfilm.
3 Digital.

Hard copy storage

Although the digital age has been with us for a few years, much project information is still stored as hard copy. Companies that conducted projects many years ago might still have drawings and schedules on file as original documents. Such files can occupy large amounts of storage space and it is good practice to consider having historical documents scanned and placed on digital media, such as compact discs. However, quite often the hard copy files fade over time, or become creased, torn, stained or otherwise impaired and that can mean that information will be lost or corrupted when attempts are made to scan or film them.

There are occasions when a project has to be closed before completion, but with the option of a restart at a later date. One example was a design-and-build project where the client was a copper mining company in Zambia that was proposing to increase its refining capacity. Some way into the project design the market price of copper fell. The Zambian company had to ask for the project to be put on temporary hold. So the London company that was designing the new installation had to close project design operations and put all the completed drawings, specifications and design calculations into stored files that could easily and safely be recovered when the price of copper recovered.

Equipment, such as plan chests, for storing drawings up to AO size, is still readily available. So it seems that hard copy storage is not yet a completely dead option. If storage space does become a problem (which it usually will) there are companies that provide storage for documents in secure premises on a rental basis. However, it is very easy for such stored files to become forgotten as time passes, but the rentals will continue, thus adding to overhead costs. So, if your company does use external file storage facilities, the usefulness of the files and the need to retain them should be kept under periodic review.

Microfilm

Various microfilming options have been available for document storage over the last 50 years or so and many organizations will have project archives stored on microfilm. Although this section is written in the past tense, microfilm reading and printing equipment is not yet completely obsolete and can still be purchased readily. Microfilming generally relied on two different formats:

- Photographing small documents onto film sheets in a grid pattern. Each sheet was known as a microfiche. These sheets were popular for filming original documents up to A4 size. This format was not used widely for project engineering documents.
- 35mm roll film, a popular choice because of its use and availability in the motion picture industry. Thus film stock was readily available. One AO sized drawing could be filmed on each 35mm frame.

The 35mm format was particularly used for project drawings. Reel-to-reel copies could easily be made, using either the same silver halide based material as the originals, or a cheaper and faster method using diazo film. The life of silver microfilm would depend on the storage conditions but was claimed to be as long as 100 years. Diazo microfilm was not quite so durable, but still had a shelf life measured in several decades.

Both silver and diazo 35mm film could be cut into individual document frames for mounting in aperture cards. These cards owed their format to the old Hollerith 80-column punched cards that were once used to enter data into early mainframe computers via card readers. Each card had a rectangular cut-out section with adhesive edges, sized to accommodate one 35mm frame. Microfilmed drawings mounted in aperture cards were easy to store and could easily be duplicated. Viewers and printers enabled the images to be viewed and retrieved by engineers. The areas of these cards not sacrificed to the microfilm aperture could still be punched according to the old Hollerith code system. The resulting punched aperture cards could be sorted quickly and automatically using a card sorter machine. Aperture cards could also be bar coded.

Original sized prints on plain paper or translucent paper could be made from microfilm using the Xerox process. These prints could be of very high quality but occasional errors could arise (for example) when decimal points on the original documents were faint, so that they would not appear at all on the prints. The Xerox process has always been a very convenient means of reproduction from microfilm.

Digital storage

CD-ROMs are a safe and compact medium for data storage, especially if they are duplicated to guard against loss or accidental damage. One good thing about files on CD-ROMs is that they are out of bounds to cyber criminals – provided they do not already contain viruses. The discs are durable and take up little space. All this method needs is a good librarian. As an example, all the files resulting from the creation of this handbook are stored on CD-ROMs.

Every company large enough to employ at least one IT engineer will know that back-up off-line files of working documents should be made at the end of each day. Computer hard drives can fail and (in one spectacular case known to me) computers left on overnight can even catch fire. A strong case can also be made for duplicating stored data in two different places, to safeguard against fire or other loss or damage.

Original documents can be scanned for digital storage. Scanning equipment for large format documents can be expensive and occupy large areas of office space. So, as with bulk microfilming, the use of an external bureau can be a suitable option.

Care of documents in transit

When original documents are sent to an external bureau for microfilming or digital scanning they are temporarily out of your custody and you have to rely on the relevant bureau for safe collection, transit and return. If documents are lost or damaged and you have digital back-up files in house, then any loss or damage might not be serious because the data can easily be recovered.

In two cases from my experience original documents were lost as a result of being sent to an external bureau for microfilming. This happened in two different companies. In both cases the original drawings plus the resulting microfilm were returned safely by the bureaux. But these returns happened just as the offices were closing for the evening, and the original documents were left out on counters overnight – from where the office cleaners removed them as trash. One of these instances occurred in a very small town, and the distraught drawings registrar miraculously recovered them intact by driving to the local council rubbish tip and searching. The other case happened in Central London, and 350 original drawings belonging to a client were irretrievably lost, leaving the company with just one reel of 35mm film and, soon after, a replacement manager for the department responsible. The lesson learned here is that no matter how much care is taken for the security of documents held in house, special extra care has to be taken when valuable original files have to be transferred to or from an external location.

Disposal

When original documents are no longer required at all they should be disposed of in a way that prevents unauthorized people from reading them. We are no longer allowed to take them into the yard and burn them, so shredding is the usual option. If the documents are particularly sensitive, criss-cross shredding (chipping) is better than simple strip shredding. There are commercial contractors who will undertake this work for you. But be sure you choose a company with a good reputation before you trust them with disposal of confidential files. And also be certain to double-check that the files sent for disposal do not contain documents that you intended to keep and which are not backed up digitally.

Acknowledgment

I am indebted to Angela Pammenter for her advice and for reviewing this chapter.

37
MEETINGS

Dennis Lock

Successful project delivery depends to a very large extent on having efficient communications and good human relationships, both within and outside the company. Unresolved disputes, personal quarrels, late responses to questions, delayed approvals for designs or expenditure can all spell disaster for projects that might otherwise have been successful. Most day-to-day project communications are informal but occasionally meetings are needed to build relationships, motivate people, resolve problems, reach decisions by consensus and generally put communications on a formal basis.

General

Most of this chapter is written assuming that the meetings described will take place in one location with all or most of the participants physically present. Virtual meetings, where people are brought together using communications technology are mentioned later in this chapter.

The meeting organizer should ensure that, as far as possible, an environment is provided that will allow discussion in reasonable comfort and without distraction or interruption from external sources. Here is a short checklist:

- Will the proposed venue and time be suitable for all the delegates?
- Has a suitable room been reserved?
- Have the delegates been given the meeting agenda in advance, so that they can come prepared and know what to expect?
- Are the audio-visual aids in place and working?
- Has everything possible been done to prevent unnecessary incoming telephone calls to the room (which includes asking delegates to switch off their mobile phones)?

- If the meeting is expected to be a long one, have refreshments been arranged? (However, a cynical meeting chairman might argue that a little discomfort can be an incentive for delegates to accelerate the proceedings and get things over quickly).

Emergency meetings or informal meetings sometimes have to be called at short notice where there might not be time to trawl through a checklist of all the above steps. On such occasions the chairman will be very grateful for the existence of a meeting room specially reserved for the project. For some small urgent meetings a project war room (described below) will be invaluable.

Meetings should not be timed to start too early in the day if that would cause inconvenience to visiting delegates (especially those from a potential or an actual project client). However, some people argue that meetings should be scheduled fairly late in the day so that there is an incentive to get proceedings over and done with as quickly as possible. On the other hand, people are more alert early in the day, when more creative and collective brainpower is likely to be available. So, perhaps it would not be a good idea to begin a brainstorming session for the start of a critical path network sketch for a large project at 4.30pm on a hot summer afternoon.

Too many meetings?

All readers will at times have felt aggrieved when asked to attend a meeting, when they feel they could have been somewhere else, doing something more productive. 'Meetings, bloody meetings!' was the general complaint.

I once undertook a project management consultancy assignment for the highly competent engineering division of a successful company connected with the automotive industry. When the meetings room was needed it was frequently unavailable. I remarked to the engineering director that they always seemed to be having meetings. The engineering director was unrepentant. 'When we have a problem we kill it with a meeting' was his answer. He ran a very efficient department and it would have been wrong to quarrel with him.

So it might be that an organization has rather too many meetings, but that is far better than having too few meetings if that would mean allowing difficulties and delays to continue without resolution. Meetings are a vital form of communication that can resolve conflicts, provide solutions to problems and help to promote a team spirit.

Project meetings room

Wherever available office space allows, it is a very good plan to set aside one room for the exclusive use of members of the project team (or, in a functional matrix organization where no actual team exists, for the use of staff assigned to work on the project). That creates a project 'war room'. The war room is the project playpen. Project staff can have immediate access to discuss progress and problems, lay out drawings and schedules, display scale models and so forth. A war room provides a focus and physical home for a project.

For projects where commercial or national security is particularly important, the project war room must be kept locked outside working hours (which can disrupt the work of office cleaners). It will probably also be necessary to screen the interior from view by using obscured glass in some windows.

A war room provides an ideal venue for immediate impromptu meetings. It is especially useful when people working on project design are dispersed over several offices in a large building complex and need to meet frequently for ad hoc discussions. Critical path schedules can be sketched and reviewed in a war room, where the inconvenience of having to put drawings away outside working hours and then bring them out again each day is avoided.

A project manager enjoys the best of all worlds when the PMO, project manager's office and the project war room are all located close to each other.

Conduct of formal meetings

If a meeting were to be considered as a micro-project, then the published agenda could be regarded as the project schedule. Keeping to that schedule as far as possible and not allowing the meeting to veer off course requires strong chairmanship. It sometimes happens that meetings are allowed to depart from the published agenda, with most of the delegates becoming increasingly disconcerted while two or more delegates discuss technical issues that should be resolved elsewhere. A competent chairman will limit such distractions to a minimum and keep the meeting on the track set out in the agenda.

Arguments in meetings can provoke creative discussion and may not be undesirable, but they must be resolved before the meeting ends so that people do not leave feeling resentment or dissatisfaction. The chairman has a duty to uncover the facts that underlie any inter-departmental or inter-functional problems, not so much to apportion blame as to resolve difficulties and assure the continued progress of the project.

The chairman's role does not end when the discussions are over and all decisions have been reached. Good practice is for the chairman to run back through the events of the meeting, summarize the discussions that have taken place, and ensure that everyone leaves with the same interpretation of what has been discussed and agreed. It is important that delegates understand and agree to all the actions asked of them and accept them as their personal responsibilities.

No ambiguity must be allowed after any statement as to who is directly responsible for taking action. Every person listed for taking action must receive a copy of the minutes (an obvious point that is sometimes overlooked, especially when the action is required by someone who did not attend the meeting). Times must be stated with precision. Expressions such as 'by the end of next week' or 'towards the end of the month' should be avoided in favour of actual times and dates.

Publication and distribution of the minutes, including directions for actions to be taken by named individuals, must be rapid. There is no point in issuing minutes a week after the meeting when the time for some urgent actions have passed. Trevor Bentley (1976) had a very good solution to this problem. In his entertaining lectures he described a combined minutes and actions control sheet (one A4 page) that could be annotated with decisions and actions required during the meeting, with photocopies given to all the delegates as they left. So distribution of the minutes, complete with a list of all the agreed actions, was achieved with no delay whatsoever.

Bentley's proposal went further. He recommended that one copy of the combined minutes and action plan form should be sent to a senior manager after each meeting. If the minutes showed that a meeting had reached no conclusion, with no resulting actions, the senior manager would be expected to ask what purpose the meeting had served and why had it been called in the first place?

When any meeting breaks up, it will have been successful only if all the members feel that they have achieved some purpose and that actions have been agreed that will benefit the project. Meetings can be very productive when they are efficiently managed, but they can waste expensive time and damage morale if the delegates depart feeling that nothing of use has been achieved. The last thing one needs to hear from departing delegates is 'Well, that was a complete waste of time!'

Kick-off meetings for new projects

Every significant new project should be launched with an inaugural meeting, but the atmosphere and conduct of such meetings will depend very much on the kind of project that is being kicked-off. Most people, when asked to describe a project kick-off meeting, would envisage a new project, ordered by an external customer with all the opportunities and risks attached to such a new venture. So I shall begin with that vision. However, the opening of a project for internal change requires a very different approach, and that will be discussed later in this chapter.

A graph of project costs against time, or of progress against time, almost always has an 'S' shape and is therefore known as an S-curve. As time passes through the project lifecycle, such curves rise slowly from the first few days or weeks, then steepen for the most intensive part of the project, and finally flatten out horizontally to become asymptotic with the final result. The start-up period can be very slow, while people are being organized, information is gathered and a project spirit is developed. A common way in which to speed proceedings and steepen the S-curve over those first few weeks is to begin a project with an inspirational 'kick-off' meeting.

The ideal kick-off meeting will be attended by people who are to become key members of the project workforce. The best kick-off meetings are supported and attended by a member of the organization's senior management, who can announce the project, summarize its objectives, say a few good words about the customer or client and then introduce the person chosen to lead the project as project manager. A kick-off meeting should be a joyous occasion. Every company likes to see its order book filled. Staff can be given a positive view about the future, learn the basic parameters of the new project and generally be motivated positively.

Initial meetings for in-house management change projects

Management change projects often provoke fear and apprehension. People generally dislike any change to their familiar work environment and routine, or any suggestion of such change. Changing the familiar organization structure or the procedures to which people have long become accustomed can provoke resentment, fear or even hostility. The time leading up to the announcement and implementation of change, when rumours abound, can disrupt daily progress and actually cost money. I often warn against the danger of creating unfounded rumours by telling the tale of what happens when people in grey suits are seen touring the premises using a tape measure and making notes on clip boards. They might only be measuring for replacement floor coverings but their appearance suggests change (and the fear of change) in the eyes of their beholders. People generally prefer the continuum and order of their working existence.

Studies leading to any significant management change project usually have to be conducted initially with some secrecy to avoid spreading unfounded rumour. Mergers and acquisitions can be particularly difficult times. Downsizing can be even worse. But when the time comes to implement any project that will in some way change the organization or its working methods, staff will clearly have to be informed. That is probably best done in a general staff meeting, rather than by the more formal, impersonal (and possibly cowardly) way of posting notices or issuing personal letters.

Calling a general staff meeting gives senior management the opportunity to outline the impending change, point out any advantages and answer questions. A principal aim is to dispel fear and rumour. As with so many meetings, the test as to whether or not the meeting was successful is the general impression that all the attendees have when they depart. There is more advice on handling staff relations in management change projects in Fowler and Lock (2006).

Project progress meetings

On any project with a lifecycle exceeding a few months it is customary and sensible to hold progress review meetings at regular intervals. The number of people invited to attend such meetings will clearly depend on the size of the project and complexity of the project organization. Where external contractors and subcontractors are playing a significant role in the project it is usual to ask each of those external parties to send at least one representative.

Since progress is the main subject of discussion at these meetings, evidence of progress will need to be produced. This can take various forms, such as evidence of design progress (number of drawings released for production or construction, purchasing progress and so on). Milestones from the project schedule are usually clear indicators of progress. For construction projects reports or certificates from quantity surveyors and other evidence of progress will usually be needed. Recent photographs can often be useful.

In addition to reporting progress these meetings provide a good opportunity for all parties to outline any current or foreseen difficulties, and ask for help or information needed to keep the project on the move. Now is the time, for example, for the project manager to ask external delegates why drawings due from them for approval have not been received, or why the recent photograph of a project site looks exactly the same as the photograph of the same site taken a month earlier.

External delegates also have the opportunity to ask why their companies' invoices for stage payments have not been paid. It is not unknown for contractors to slow work or stop altogether when they are owed substantial amounts of money by the project owner. In one case from my own experience we refused to ship completed equipment worth hundreds of thousands of pounds until payment for our outstanding invoices was forthcoming.

As with all meetings, it is important that advantage is taken of the fact that principal project participants are gathered together and to ensure that all problems and difficulties affecting project progress are either dealt with immediately, or are referred to the relevant people for immediate attention as soon as the meeting ends. A progress meeting will have been unsuccessful if one or more delegates leave feeling that questions have been left unanswered or that disputes have not been resolved.

A case example of progress meetings becoming unnecessary

The above account of progress meetings adheres to the conventional view that progress meetings are an accepted way of project life. Here is some food for less conventional thought.

A heavy engineering company in the English Midlands had long been accustomed to holding project progress meetings. Depending on the particular project manager, these were held either at regular intervals or randomly whenever things looked like going badly adrift.

Meetings typically resulted in a set of excuses from participants as to why actions requested of them at previous meetings had been carried out late, ineffectually, or not at all. Each meeting would end with a new set of promises, ready to fuel a fresh collection of excuses at the next meeting. This is not to say that the company's overall record was particularly bad, but there was considerable room for improvement and too much time was being wasted at meetings. Too many promises were being made and broken, and too many activities were running late.

Senior company management supported a study which led to the establishment of a small PMO. The engineers in this PMO used critical path networks for all projects and a computer produced detailed work-to lists. The scheduling software provided resource smoothing within the stated capabilities of all departments. The work-to lists were filtered and sorted using WBS and OBS codes so that departmental managers received schedules that pertained only to the tasks within their own departments.

The task of allocating specific jobs to individuals was left to the departmental managers, but with the improved scheduling there was reasonable confidence that no department would be asked to do work for which it had insufficient resources.

The PMO had a progress coordination engineer, whose job was to follow up all the tasks scheduled, using detailed work-to lists. This engineer knew which jobs should be in progress at any time, the scheduled start and finish dates for those jobs, how many people should be working on each task, how many people should be working in total on each project at any time, and the amount of remaining float available for every task. The progress coordination engineer's remit extended to include external offices that provided subcontracted design services.

If a critical or near-critical task seemed to be in danger of running late, steps were taken to bring it back into time (by arranging overtime working during evenings and weekends if necessary). All the managers and external subcontractors cooperated well, and they all experienced a new sense of order in their working lives. After several months under this new system it dawned on all the functional managers that they were no longer being asked to attend progress meetings. Except for kick-off meetings at the introduction of new projects, progress meetings had become unnecessary.

Meetings to inform people about the proposed outcome of a project

It sometimes happens that a meeting has to be called to advise non-project members about a proposed project. Although this might be necessary for various occasions, we can identify two main categories of such meetings.

Sometimes, early in the life cycle of a large public project, one or more consultative meetings have to be arranged to which members of the general public are invited, along with civic dignitaries and all other kinds of interested parties. That might be part of a lengthy public enquiry into the project. Such meetings fall into two parts, which are first a presentation of the facts of the proposed project, followed by a question and answer session during which representatives of the project management or investment team listen to questions or statements from those present.

Another kind of presentation meeting is seen when a management change project is nearing its conclusion and the time comes to announce the changes to all the staff whose working lives will be affected by changes in procedures or the organization. Very often people become concerned for the security of their jobs, worry about their status under the changed regime, or have doubts about their ability to master the new challenges presented to them.

Although these two kinds of meetings take place at opposite ends of the project life cycle and are very different in scale, they have some similarities and have two requirements in common.

The first requirement is that the project management team must be fully prepared. They need to have 'done their homework' and have all the facts of the project available. A good presentation, showing the project in its most attractive aspects, makes a good start. The reasons for the project and the advantages for the community or the organization must be presented in the most attractive light.

For public projects models and plans should be placed on public display for perhaps a week before the meeting takes place, so that people can familiarize themselves with the proposals and come to the meeting prepared.

The second essential factor for both public meetings and in-house meetings is that the project management team must anticipate the questions that will be asked - and be ready to answer them as fully and truthfully as possible. The audience, whether public or in-house, must be given the feeling that they are (as far as is practicable) being consulted. They must feel that their questions and apprehensions are being taken seriously. They need reassurance that any fears they might have about the project are either groundless, or will be taken into account.

So, the overriding piece of advice here is 'be prepared'.

Virtual progress meetings

With current communications technology it is not always necessary for delegates to attend meetings in person, although face-to-face meetings are usually preferable. Virtual meetings can be held using telecommunication links, allowing groups of people to assemble in meeting rooms that are geographically separated by considerable distances. One particular application of this method is when a parent company needs to have meetings with a staff from one of its subsidiary companies that happens to be located in another continent.

Virtual meetings should imitate face-to-face meetings as closely as possible. Screens should not be too small (they should preferably be at least 50 inches). Ideally they will be mounted on a wall that is visible to everyone easily, without need to bend necks and sustain aches and pains. Cameras should be positioned so that everyone in each meeting room is clearly visible to all at each receiving end.

Virtual meetings save time and travel expenses, especially for international projects. However the meeting organizer must always bear in mind that people of different nationalities working across the world can have very different ideas about which days are working days and which are festival or Sabbath days.

There is also always the issue of international time zones to be considered. For example, I once needed to contact a person in Mumbai (let's call him Mr Kardomah) and was not quite sure of his telephone number. I rang the number listed in my records and a polite gentleman answered to tell me that there was no one at his address who answered to the name of Kardomah. I hung up my phone, checked my records and decided that I must have made a mistake when calling. So I tried again very carefully and received the curt answer 'There is no Mr Kardomah here'. Then I realized that it would be 2.30am in Mumbai and I had called this unfortunate man from his bed twice.

References and further reading

Bentley, T. (1976), *Information, Communications and the Paperwork Explosion*, Maidenhead: McGraw-Hill.

Fowler, A. and Lock, D. (2006), *Accelerating Business and IT Change*, Aldershot: Gower.

Roberts-Phelps, G. (2001), *50 Ways to Liven up your Meetings*, Aldershot: Gower.

Streibel, B.J. (2007), *Plan and Conduct Effective Meetings*, New York: McGraw-Hill.

38

GIVING CONTROLS HIGH VISIBILITY

Shane Forth and Dennis Lock

It is well said that a picture is worth a thousand words. This chapter explores some of the ways in which charts, diagrams, photographs and computer technology can help people who control projects.

Flip charts, whiteboards and blackboards

We do not need to dwell for very long on the topic of handwritten wall charts and flip charts because these are so familiar, and their uses are well known. For project control applications flipcharts are particularly valuable in brainstorming sessions because they provide an informal means for gathering initial ideas when attempting to establish concepts and objectives. When used for collecting suggestions, any flipchart entries are usually encouraged, no matter how impractical, irrelevant or ridiculous they might seem. Then the items listed can be edited to eliminate entries that are clearly irrelevant or inappropriate. Here are some of the many possible flipchart applications:

- initial listing of project tasks when attempting to define the full scope of a new project;
- identifying possible risks;
- identifying and evaluating the potential benefits of proposed internal projects;
- provisional collection of suggestions for project control improvements;
- collecting ideas for solving specific problems;
- listing various options as an aid for decision-making.

Bar charts and Gantt charts

So-called Gantt charts are familiar to all of us who participate in projects. They derive their name from Henry Gantt (1861–1919), who worked with the American industrial management scientist Fredrick Winslow Taylor. You can read about those people and their work in Kanigel (1997) and in Chapter 31 of Lock and Scott (2013). However, it is not clear that project bar charts as we know them today resemble the charts devised and commonly used by Gantt. His charts appear to have been based more on times and dates written or chalked within grids and were used principally for loading jobs on to machines in factories.

Bar charts are so familiar in project control that there is hardly any need to illustrate one here but there is an example (for instance) in Figure 15.2. Tasks are listed vertically down a column at the left-hand side. Columns ranged from left-to-right are allocated to successive periods or calendar dates. Then the relevant jobs or items are shown in the main body of the chart as horizontal bars, with their lengths scaled according their duration. A typical staff holiday chart affixed to an office notice board is a good bar chart example. The initial task displays on the screen for Microsoft Project opening are bar charts.

At one time bar charts were the only available method for displaying project schedules. Proprietary kits for wall charts could be bought, some relying on a steel back plate on to which coloured horizontal magnetic strips representing tasks could be positioned. Another popular method used Lego strips that could be plugged horizontally into a backboard drilled with holes on a grid pattern. The Lego strips were available in a range of colours to signify such things as the type of resource needed for the task and they could be cut to scale length according to the planned task duration.

Charts with bars that could be repositioned sideways (in the time dimension) were a useful early tool for smoothing resource usage, before computer applications became available that could perform such tasks better and more efficiently. Bar charts start to become unmanageable and lose much of their value when they attempt to show more than 100 project tasks.

Histograms

Histograms are also (confusingly and probably inaccurately) sometimes called bar charts. However, in histograms the bars are vertically orientated. Histograms are particularly valuable for showing daily resource requirements for a project, and they will indicate potential resource overloads or idle times. Figure 38.1 shows the idea. Some computer applications have very good graphics for this purpose but they need also to have the capability of calculating levelled resource schedules (Open Plan is one example).

S curves

S curves are line graphs that plot the aggregation of information gathered over time. They are particularly valuable for showing things such as costs, labour time or installed quantities against project time. When actual quantities are plotted on the same graph as planned or budget quantities they indicate variances and trends very clearly. There is an example in Figure 33.2.

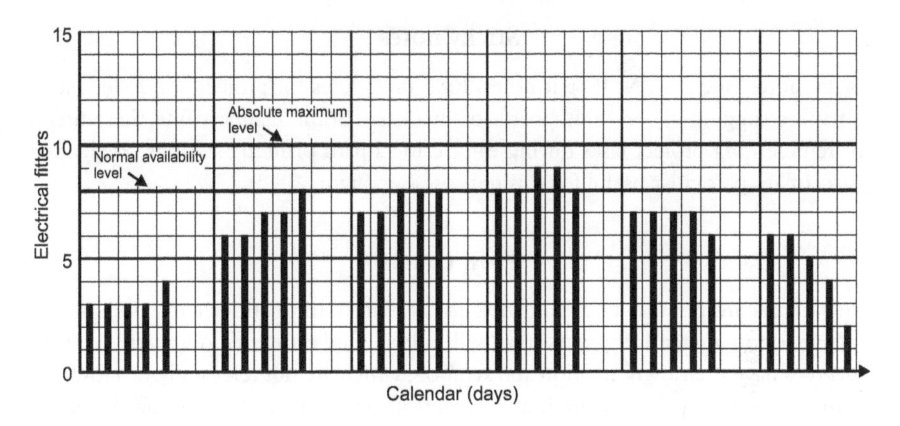

Figure 38.1 Histogram display of a project labour schedule after efficient levelling.

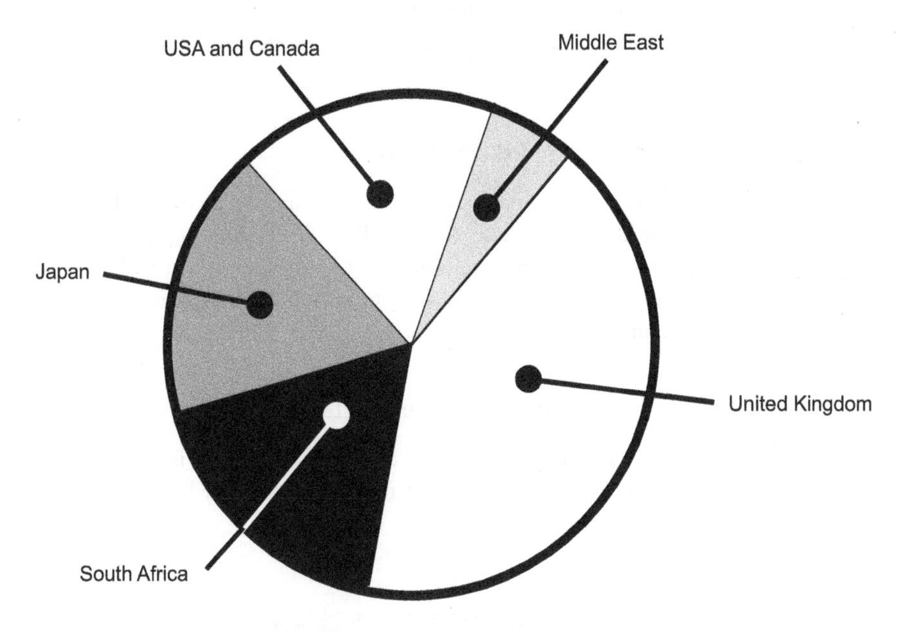

Figure 38.2 Fictional pie chart showing location of projects sold over past ten years.

Pie charts

Pie charts are so called because they resemble the plan view of a pie or cake that has been cut into slices. Their particular strength is in showing the relative proportions of things in proportion to the whole picture. These charts are not commonly used for project controls. However, should (for example) a company wish to show graphically how its projects have been sold to different areas of the world, a pie chart such as that shown in Figure 38.2 would do the trick.

Sticky notes

Sticky notes (such as 3M Post-it notes) have many uses for project controllers, such as pasting suggestions on whiteboards or flipcharts. Some planners find these useful for representing project tasks during early critical path planning sessions. Preprinted packs of are particular useful for this purpose (see Figure 38.3).

Flow charts

Flow charts are popular as design tools for chemical and petro-chemical engineers when designing chemical process plants, distilleries and refineries. All charts for these purposes have their own notation and symbols, which are not relevant here. However, in the context of project planning and control, critical path diagrams are based on flow chart principles, with all tasks running in time from left-to-right. Readers might be interested to know that we produced a flow chart in the form of a precedence diagram for the benefit of our contributors, showing all stages in the preparation of this handbook from contract initiation to publication. Unfortunately that chart is too big to be condensed legibly on to a page here.

Likert charts

Not common in project controls applications, but occasionally useful in IBRs and satisfaction surveys, the format of these charts will be familiar to any person who has received a questionnaire from a travel operator or other service provider. These charts are designed to show the opinions on topics such as the quality of service experienced. There is an example in Figure 30.1.

Line of balance charts

Line of balance charts exist in three quite different forms, each of which is intended to show particular aspects of project progress requirements related to a particular calendar date.

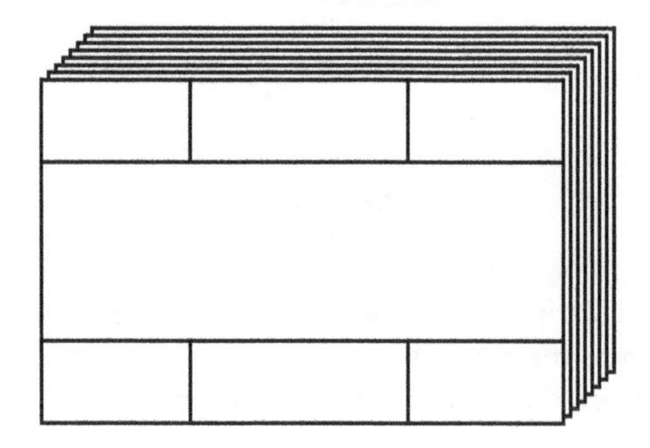

Figure 38.3 Use of preprinted sticky notes as an aid to drafting critical path networks.

Figures 22.4 and 22.5 show line of balance charts of two types that are applicable to construction projects for building houses.

Matrix charts

Matrix charts can be useful for decision-making and also for documenting and announcing the results of those decisions. Several matrix forms are useful in project controls. A matrix of stakeholders' interests is one useful tool (see Figure 1.2). Three applications, applicable to risk management, are shown in Figures 24.5, 24.6 and 26.7. Other forms include a responsibility matrix that shows which managers or other named people are responsible for particular tasks or decisions in a project organization (Figure 38.4) and a distribution matrix can be useful for helping project managers to decide and then inform everyone about who should get what in the distribution of project documents. That distribution might be by internal mail, by postal services or by electronic means (see Figure 38.5).

Task type: / Responsibility:	The client	Project manager	Project engineer	Purchasing mngr	Drawing office	Construction mngr	Planning engineer	Cost engineer	Project accountant		and so on
Make designs		✚		●							
Approve designs	●	■	✚								
Purchase enquiries		■	✚	●							
Purchase orders	■	■	✚	●							
Planning	■	■	✚	✚	✚	✚	●	✚			
Cost control		●		✚		✚		✚			
Progress reports		●	✚	✚	✚	✚	✚				
Cost reports		●		✚		✚		✚	✚		
and so on											

● Principal responsibility (only one per task)
✚ Secondary responsibility
■ Must be consulted

Figure 38.4 Linear responsibility matrix
A tool to help decide who does what and communicate the decision.
Source: Lock (2013).

Document type: / Recipient:	The customer	General manager	Project manager	Project engineer	Works manager	Production control	Buyer	Quality manager	Accountant		and so on
Bought-out parts lists				1		1	2	1			
Material specifications				1			1	1			
Purchase requisitions				1			1*				
Purchase orders							1		1		
Shortage lists			1			1	1				
Committed cost reports		1	1	1			1		1		
Drawing list			1	1	1						
Drawings for approval	1			1*							
and so on.											

1 Number of copies per recipient

* Retains original signed document

Figure 38.5 Document distribution matrix
A matrix tool to help decide who gets what and communicate the decision.
Source: Lock (2013).

Bubble diagrams

Bubble diagrams are used where choices have to be made between different project opportunities for their inclusion in project programmes and portfolios. They have little relevance at the control level of individual projects and are thus not illustrated here. However, for those who are interested in project portfolio management, there are several examples of bubble diagrams in Lock and Wagner (2019).

RAG indicators

RAG indicators take their name from the red, amber and green colours that are familiar in the traffic control signals designed to keep all road users safe. In the project controls context these colours can be used as text colours or highlights or as coloured symbols alongside items reported in lists or dashboards. Thus the attention of managers reading reports would be drawn to the items with red signals, indicating that these are running out of control and are in need of urgent corrective action (which might already be too late). Amber signals highlight those areas or items in a project that are critical or borderline (for example, activities that have zero float in a work schedule). Green signals signify items that are running well according to schedule and budget, and which need no higher management intervention.

Photography

Photography, particularly showing progress stages at open sites in building projects such as construction and shipbuilding, has long been a useful means for recording project progress at intervals. The results often prove useful for discussion at regular progress meetings. With the automatic features of modern digital cameras anyone can take pictures of reasonable quality. Site cameras are also valuable security tools, provided that lighting is adequate. Images can also be valuable to companies for publicity, perhaps even using fixed location cameras for time-lapse photography. In recent years the use of drones for photography has become prevalent, provided always of course that the project is not within the restricted area of an airport.

Dashboards

Dashboards take their name from the dashboards of automobiles, where the requirement is to show a combination of important factors such as speed, oil pressure, engine rpm, distance travelled, remaining fuel or battery charge and so forth. The factor common to all dashboard presentations is that they are designed to present operating parameters so that their scale and importance can be instantly recognized. In the project controls context a dashboard can display multiple objects, such as Gantt charts, histograms, S-curves and pie charts. Dashboards can also include objects that more closely resemble automobile gauges showing project performance and RAG indicators.

As with any tabulation or presentation of data, it is of course essential that the project control data displayed in dashboards are relevant, recent and accurate. Otherwise the information could give unwarranted cause for alarm or, conversely, a false indication of success.

Since the late 1990s the 'data warehouse' approach has matured rapidly, enabled by advancements in technology. The trend is for project controllers to move away from creating schedules, estimates, budgets and various reports (functions that were sometimes performed by people working independently and then distributing these documents as hard copy or email attachments). A few years ago one of us (SF) was consulted during a project control audit, where the auditor was surprised that he did not see the usual large pile of progress update papers. He thought that much information was missing. However, he became very impressed when the planning engineers showed how all the data were present, automated through the data warehouse and dashboards. The auditor was very impressed to see all the information present succinctly in one place, with high visibility and quality of information available directly to the project manager and the team. It was apparent that the project manager and team were enabled to improve task prioritization, assignment of resources and forecasting. The outcome was a good result for both the audit and the project.

The web-based, data centric and truly digital project controls environment is becoming a reality. High visibility dashboards are being enabled by a new breed of business intelligence software that can consolidate and integrate data from multiple project control software tools. All this provides easy and instant access directly to project team members at their desktops, in real time or aligned snapshots on demand. The information can be filtered and sorted according to the individual needs of people, so that they can understand the current status of work for which they are responsible and compare that against the baseline plans, budgets and KPIs for which they are responsible. Also possible, is drilling down by exception through project coding structures, such as the WBS. That provides 'actionable insight', which is the ability to examine the lower detail where needed. That allows root causes of variances to be identified so that corrective actions can be taken to keep the project on track.

A dashboard example

A typical dashboard will display the following information:

- schedule milestones;
- budget, earned value, actual and forecast costs;
- approved and pending changes;
- time-phased data;
- KPIs;
- 'narratives' giving written descriptions of progress, performance, scope of work and safety issues.

The example shown in Figure 38.6 displays following data:

- time-phased cost performance (budgeted, earned value and actual);
- CPI and SPI;
- variance analysis;
- a forecast for the entire project;
- key WBS elements of over- and under spend.

In an executive summary dashboard (not illustrated here) the same graphical time-based data are displayed as those shown (for example) in Figure 38.6 but with the following additions:

- time-phased previous, current, to-date and 'at completion' data (with RAG indicators for CPI);
- schedule milestone performance;
- a pie chart of high-level WBS budget elements.

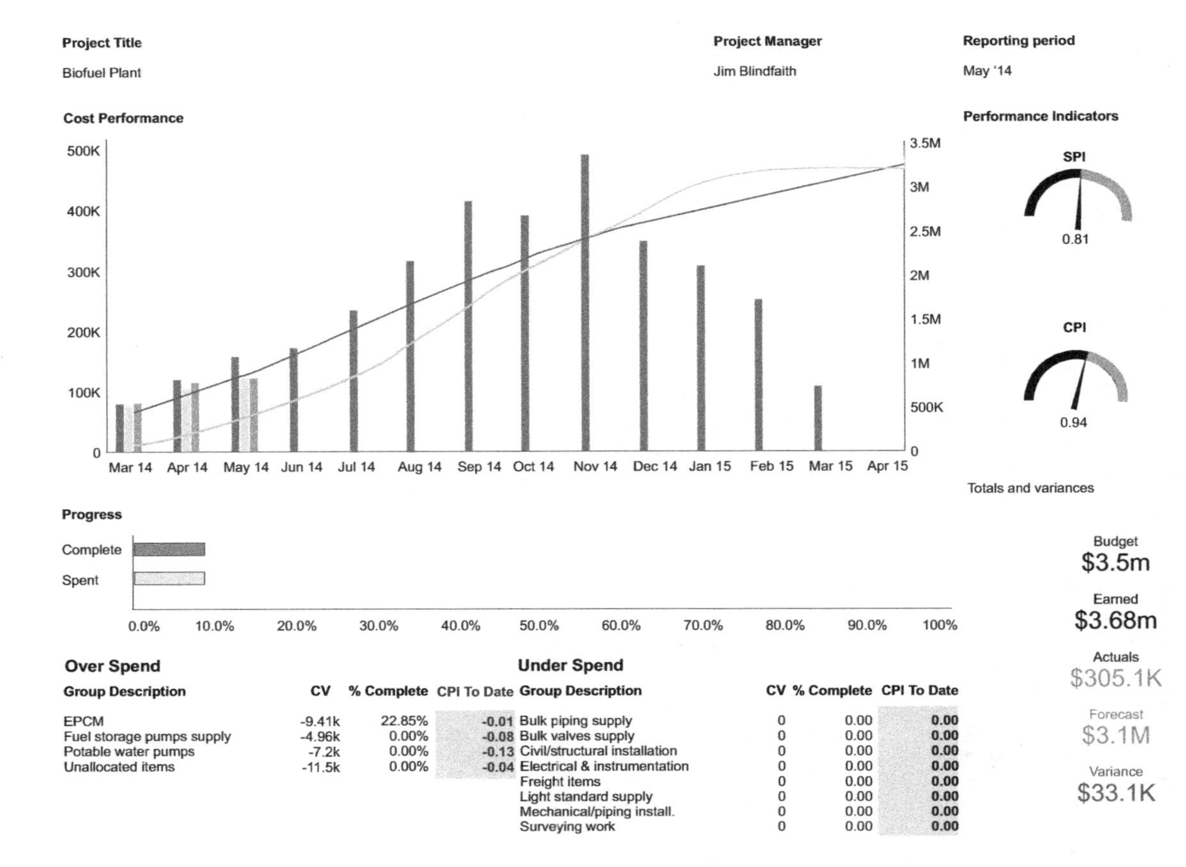

Figure 38.6 Format of a cost performance dashboard. This example has been redrawn and slightly modified to improve clarity. Clearly a small screen is unlikely to be suitable for viewing dashboards.

Source: Courtesy of ARES PRISM.

4D building information management

In the 1990s, Microsoft Windows project management software such as Microsoft Project and Primavera replaced the older mainframe and early text-based PC tools with a graphical user interface. The progression of engineering design from drawing boards (2D) to 3D CAD (computer aided design) was in its early days. Those technology advancements led to a new field of innovative research into the integration of planning and scheduling software with the 3D CAD model. Using this approach, the 3D CAD model components were linked to activities and dates in the project schedule to simulate the planned construction build sequence and progress. That could be done on a cumulative basis (monthly for instance) or to visualize actual construction progress.

All this change in working practice was somewhat daunting. The potential benefits of implementing the research and further integrating the 3D modelling approach with the project schedule were at this stage considered to be too expensive and too challenging. Today, however, the 3D CAD model is widely used in engineering design and has become known as 3D building information modelling (BIM). With BIM now adopted by the UK government on its projects, integration of the project schedule (which has become known as the fourth dimension (4D)) is increasing. 4D modelling allows the designer to 'walk through' the 3D model time-base representation of the physical construction and helps the planning engineer to visualize the project and prepare a more realistic project schedule.

Showing the construction build sequence to the rest of the project team so visibly gives them the opportunity to work together and establish the best construction methods. This can save both time and money. For example, the coordination between the different construction companies involved in the project is much improved and can promote improved productivity and safer methods of working.

References and further reading

Kanigel, R. (1997), *The One Best Way: Frederick Winslow Taylor and the Enigma of Efficiency*, London: Little, Brown and Co.

Lock, D. (2013), *Project Management*, 10th edn, Farnham: Gower.

Lock, D. and Scott, L. (Eds) (2013), *Gower Handbook of People in Project Management*, Farnham: Gower.

Lock, D., and Wagner, R. (Eds) (2019), *The Handbook of Portfolio Management*, Abingdon: Routledge.

PART IX

Assurance and governance

39

PROJECT REVIEWS AND AUDITS

Alison Lawman

The purpose of project reviews is to examine the current state of play – where the project is in terms of progress and whether or not it is on target regarding the budget, schedule and scope. Reviews provide a validated insight into the progress of the project for its stakeholders and in that respect they are an essential aspect of project governance. The use of reviews is part of organizational learning and process improvement, but they can also highlight gaps in knowledge or capability. They provide opportunities for facing reality, and while that might be a little daunting (and perhaps somewhat unwelcome) it is far better to be alerted to potential problems so that they can be dealt with.

Introduction

The nature of project reviews and audits can be confusing because the terms are often used interchangeably. However, whether conducting an internal review or a full-blown audit, we are ultimately looking for answers to some key questions, such as:

- Are the costs under control and is the budget being adhered to?
- Is the timescale for the project realistic and are key milestones being achieved?
- Is the project meeting its quality objectives and is the work being carried out to the required standard?
- Are the processes fit for purpose?

A review might consider whether or not the recruitment practices conform to the organizational policies and how well the team is motivated. It can also examine whether the project has been adequately planned and if appropriate control techniques are being used. Are the control measures sufficiently robust and do these effectively trigger action? This is not an exhaustive list, but essentially a review can focus on the processes being followed and/or the performance against them.

These areas are measured during both reviews and audits. However, audits tend to be process driven while reviews are more likely to be performance focused. For example, an audit will look at the accounting systems or processes when considering financial aspects, but a review will consider the return on investment or look for any CV. For quality, an audit will look at the procedures while a review is likely to focus on customer perceptions and satisfaction.

Purpose of reviews

The purpose of a project review is to examine the current state of play in terms of three essential control factors, namely:

1 Progress and whether we are on target regarding schedule.
2 Cost, in terms of expenditure against the authorized budget.
3 Scope (is the project satisfying the original specification, or has there been scope creep?)

Reviews provide a validated insight into the progress of the project for its stakeholders. In that respect they are an essential component of project governance. The use of reviews is part of organizational learning and process improvement, but reviews can also highlight gaps in knowledge or capability. They are an opportunity for facing reality, and while that might be a little daunting (or even somewhat unwelcome) it is far better to be alerted to any potential problems so that they can be dealt with.

Types of reviews

Reviews are important throughout the project lifecycle and the most common types of reviews are gate reviews, peer reviews and post-project reviews. Timing of reviews will depend on such things as project type, project size and conditions, but generally there should be a gate review at every key decision point (for example, a gate review is often undertaken at the end of the concept or initiation phase). Reviews are usually conducted during the main execution or development phase. For larger scale projects, it is recommended that multiple reviews are conducted to ensure that progress is being made against the plans. Peer reviews can be conducted at any point throughout the project lifecycle. A post-project review (or project post-mortem) is undertaken upon completion.

Gate reviews

Gate reviews are associated with 'stage gate project control'. They are designed to consider the work completed before moving through the 'gate' to the next stage. Stage gate reviews are particularly relevant to research projects, where the project funds are released in tranches. Those who provide the funds need to have regular progress reports to decide whether to continue funding the research or 'pull the plug' and end the project.

In business projects a concept gate review should be undertaken to focus on the business case, while a definition gate review should assess the project management plan. The exact focus of each review will depend on the decision point. It might be to review the delivery strategy undertaken before committing to any major procurement. Or it could be an investment decision review prior to the release of further funding tranches.

Gate reviews can also be made to check that nothing has changed in terms of objectives or in the project environment, and that the business case remains viable. In the event that circumstances have changed, such go/no go decisions act as a control measure to determine whether or not to proceed with the remainder of the project.

The benefits of gate reviews are that they provide assurance to stakeholders that the work is progressing to plan, but they can also highlight any areas that may require attention. A planned framework of gate reviews provides a formal approval mechanism for the release of funds, ensuring controlled investment of capital and protecting against wasteful use of finite funds. Gate reviews also provide a documented statement of the current state of the project because they assess performance in line with KPIs and other metrics. Equally important, is the delivery of a formal assessment of the effectiveness of risk management on the project.

Peer reviews

Peer reviews can be conducted at any point throughout the duration of the project, and multiple times if required. They are an opportunity to learn from peers who can bring a level of objectivity to the review process. Often those closest to the day-to-day management of a project can fail to see the wood for the trees, so a fresh set of eyes can be helpful.

The process of peer review should provide a positive experience, not directed in any kind of punitive manner. The person or persons chosen to undertake the review should be respected by both the project team and the senior management in order to provide credibility to the outcome, as the purpose of conducting such a review is to improve the likelihood of project success. The challenge for peer reviews is to maintain rigour. By definition peer reviews tend to be conducted by individuals from other project teams that are subject to similar time and resource constraints, and therefore the availability of peers might be limited.

Some key areas of focus for a peer review include consideration of the business case and whether or not it provides a suitably robust justification for the project, and indeed, whether it continues to be relevant and viable.

Determining the level of team morale can provide a useful indicator of the leadership of the project. A poorly motivated project team may stem from feelings of antipathy, frustration, or lack of recognition, although these conditions might not be the fault of the project manager but could indicate a wider problem. Other elements for consideration are whether the project scope is clearly defined and controlled; clarity as to the ownership for deliverables and are there appropriate metrics in place to track progress. The presence of metrics does not always equate to valuable measures.

A gate review should also examine the level of executive support for the project, which can either help or hinder the project manager's efforts. Without solid support, the project manager's job becomes very challenging indeed, or even impossible in some cases. Finally, a gate review should consider whether or not the delivery timeframe and resourcing levels are realistic.

Post-project review

A post-project review should be standard practice as part of the project close-out and handover phase. It is a key element of project governance and its rationale is simple – it is an opportunity to capture lessons while they are still fresh in everyone's minds. This review has to consider both what went wrong and what went well. Too often the focus is placed on the negative aspects. Adequate attention and consideration should also be given to behaviours and actions that went well and thus need to be repeated on future projects.

A post-project review also helps to identify correlations between estimations and actual results. It is particularly useful for determining the level of accuracy achieved in early project estimates, by comparing those estimates with the actual results. Thus a post-project review is part of the continuous improvement process for managing projects.

The process of a post-project review should also include recognition of individual and team performance. This is particularly important for personal morale and for career development, especially when team members have been seconded away from their regular roles.

The review document itself provides valuable data for knowledge management and can contribute to the organization's project maturity going forward. But a key proviso in this process is that it should be an open and honest reflection on the project performance, and not an exercise in recriminations.

The output from the post-project review must be documented and made available for use on future projects – not simply filed away within the archives. It is essentially a formal report on the project performance, which is important not just to the project team but also to the client (whether internal or external).

Audits

When we hear the term 'audit' we immediately think about finance. Although that might be covered, it is rarely the only focus in project audits. The thought of an audit tends to strike fear into those on the receiving end, but an audit should not be a problem or be viewed negatively. Unfortunately, while that is a common perception, a project audit should be seen as a valuable opportunity to reaffirm that the project is being managed correctly. An audit considers the current project status and examines whether or not we are making best use of all the appropriate control techniques available.

A project audit is an independent review that examines whether the work complies with the statement of work, as well as with any regulatory requirements such as the General Data Protection Regulation and others that might be industry specific (including health and safety).

As with a project review, an audit seeks to uncover potential problems before they become issues. An audit should also identify opportunities for improvement and for doing things differently to create efficiencies. All parties should be told why a project audit is being conducted.

Types of audit

There are many different types of audit. Some might be specific to a particular type of project but common types include standard project audits, which are typically conducted as part of the continuous monitoring and control project activities. Such an audit can be requested by senior management at any stage or if there are particular concerns. An audit could be one of the following:

- A general project audit (described below).
- A quality audit, usually conducted during the main execution phase, and part of the quality assurance of the project.
- A risk audit specifically focused on risk responses.
- A procurement audit to examine the process and activity of all procurement throughout the project. This would normally be conducted during the final project closeout phase and could consider such aspects as value for money from suppliers and billing activity.
- Post-project audit.

There are three core elements of any audit, which are economy, efficiency and effectiveness.

Project audit

A project audit is an independent examination that is concerned with how well the project is being managed. It seeks answers to the following questions:

- Will the project objectives be achieved?
- Have stipulated processes been adhered to and were they the right processes?
- Which techniques should be used again - and which should not?
- How effective have the contracting relationships been?
- Should any suppliers be changed?

A project audit can be done while the project is in progress (which is particularly important on a large longer-term project), but it is often done upon completion. However, sometimes it is wise to wait. If a project has encountered some difficult challenges these may still be uppermost in the client's mind immediately after handover. Satisfaction may be less evident until the time when benefits are eventually realized.

The auditors will check whether the right things have been addressed during each of the project phases, and whether or not they have been done correctly. This will involve a detailed inspection of the project methodology, techniques and procedures. Key documents will be examined, and these should be readily available. If they are not, that should signal a problem. Part of this audit is ensuring that reviews have been undertaken at key points throughout the project to demonstrate solid governance. The auditors will want to see evidence that the work has been done and that processes have been followed.

Of course, the audit itself will need to have a clear process that is confirmed to all parties. Part of the process is to capture supporting evidence for all aspects that are reviewed. It may be that a templated approach is taken here which allows for tracking.

The project audit process

An audit process should begin with a clear plan of how the audit will be carried out, who will be involved and the roles they will play. Clarification is also important at the outset, as to the specific focus of the proposed audit, the reporting expectations and any time constraints.

The next step is to undertake the audit by examining the required elements and capturing supporting evidence to demonstrate completion of tasks or adherence to procedure. Once all of the data have been collated, the audit team needs to analyse and summarize the findings. This step is particularly important. It requires an experienced auditor to ensure that the data are interpreted correctly. With clear findings, the results should then be presented in a report to the appropriate individuals, who might be senior managers within the organization and/or an external client.

It is common for project audit reports to identify any concerns or potential issues and assign severity ratings to those that convey their level of importance. It is not uncommon for a project audit to highlight areas for attention, in which case the report should also suggest actions. These suggestions might go as far as an action plan, but it is often the case that the project organization or project team is tasked with drawing up a suitably detailed action plan to address the weaknesses. It is important to ensure that report recommendations are followed up to ensure that all necessary actions take place.

A project audit provides an independent, objective review of the project, which should identify areas for improving performance. On certain types of projects specialist audits may be required and a third party firm can be engaged for that purpose.

Construction projects are particularly prone to contractor issues, such as over billing, materials storage charges and so forth. In those cases an audit provides a degree of protection for the client or the project owner, ensuring that the project has been conducted on a level playing field. That said, the auditing firm needs to take care not to damage existing relationships.

Among other aspects, the auditing firm will be looking to ensure fair billing to the project, checking that the contractor is qualified to undertake the work and that the necessary insurances are in place.

Challenges

It is impossible for an audit to review every single transaction or activity on a project. So a sampling approach has to be taken, although this will not always provide the necessary deep insight. Sometimes this sampling may be based on a judgemental selection or it might be truly random. However, whatever the sampling process might be, how do we know whether the sample is truly representative, or it includes red herrings? There could be a very reasonable explanation for the resultant negative data. It might be the case that we would not expect something to be completed by that point. If goods were delivered later than scheduled but there might be a good reason. For example, a replacement supplier might be providing better quality items, which will save money in the longer-term by reducing operational costs.

So, the auditor needs to examine the facts fully, make observations and draw appropriate conclusions. For this reason alone an experienced practitioner is a key requirement. Objectivity and independence of the auditor are also very important. There must be no conflict of interest in the context of the project and the individuals working on it, although, in reality, that might not always be achievable.

There could be resistance from the project team and a reluctance to cooperate with the audit team. For example, the project team might provide only the specific information requested, but with the knowledge that it doesn't necessarily tell the whole story. This risk can be exacerbated if the project manager is not supportive of the effort.

Another challenge when conducting an audit can be the metrics in use on the project. These might be poorly chosen or carefully tailored to measure only what the team wants to focus on rather than what is needed. If the metrics are skewed, that could give an unreliable indication as to whether or not the project is successful. Of course, another challenge is that metrics notoriously look backwards; they might not be a true indicator of future performance. As with any performance measurement it is always a case of how the data are interpreted and analysed.

One of the key challenges around project audits is that there is no specific audit standard in the context of project management. Instead many organizations adhere to more general standards relating to auditor ethics, independence, professional competence and performance of the fieldwork (such as ISO 19011).

References and further reading

APM (2019) *APM Body of Knowledge*, 7th edn, Princes Risborough: Association for Project Management.

Maylor, H. (2010) *Project Management*, 4th edn, Harlow: Pearson.

Nalewaik, A. and Mills, A. (2017) *Project Performance Review: Capturing the Value of Audit, Oversight and Compliance for Project Success*, Abingdon: Routledge.

Oakes, G. (2008) *Project Reviews, Assurance and Governance*, Aldershot: Gower.

40
GOVERNANCE OF CONTROLS

Tony Marks

Introduction

It's probably best to begin this chapter by defining what is meant by 'governance'. The APM defines governance in the context of project management as follows:

> Governance refers to the set of policies, regulations, functions, processes, procedures and responsibilities that define the establishment, management and control of projects, programmes and portfolios.

This is a very broad definition and project practitioners are often confused about the best way to ensure good governance across such a wide brief. The Project Management Institute (PMI) has a similar definition, which also emphasizes the different levels of governance at project, programme and portfolio management levels:

> Governance is the framework, functions and processes that guide activities in project, program and portfolio management. In organizational project management (OPM), governance provides guidance, decision making and oversight for the OPM strategic execution framework.

Maturity models have become a useful tool for assessing organizations' current governance capabilities and for helping them to improve these capabilities in a structured way. Each of these models is a hierarchy that describes the characteristics of effective governance processes.

The Portfolio, Programme and Project Management Maturity Model (P3M3®) is a good standard among maturity models. It provides a framework within which organizations can assess their current performance and put in place improvement plans with measurable outcomes based on best practice. It provides a different way of assessing governance at project, programme and portfolio management levels within any organization, so it is worth reviewing it in some detail.

The OGC and P3M3

The Office of Government Commerce (OGC), owner of P3M3, is an office of Her Majesty's Treasury within the UK government and is responsible for improving value for money by driving up standards and capability in public sector procurement. It strives to achieve this through policy and process guidance, helping organizations to improve their efficiency and to deliver successfully. As shown in Figure 40.1, P3M3 is a hierarchical model that has three levels:

- Portfolio Management Maturity Model.
- Programme Management Maturity Model.
- Project Management Maturity Model.

There are no interdependencies between these models that allow for assessment at the appropriate level of detail. In terms of project controls the Project Management Maturity Model is usually the most important level. However, it is important to see it in the overall framework. A lower level of maturity in some aspects of the programme or portfolio models can still have an adverse impact at project level. It is therefore worth gaining an appreciation of the wider P3M3 context, even if only the project management model is implemented.

Maturity levels

For each model within the P3M3 framework there is a five-level maturity framework. The five maturity levels are:

Level 1 – awareness of process.
Level 2 – repeatable process.
Level 3 – defined process.

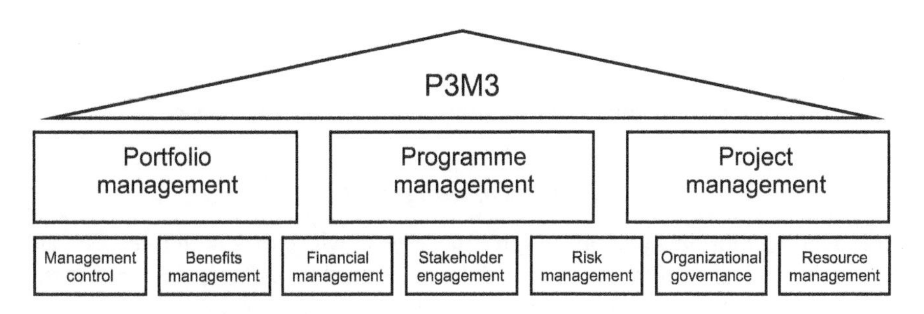

Figure 40.1 The three P3M3 levels.

Level 4 – managed process.
Level 5 – optimized process.

These levels comprise the structural components of P3M3 and are characterized as shown in Figure 40.2.

Process perspectives

P3M3 includes seven process perspectives. These exist in all three models and can be assessed at all five maturity levels. The process perspectives are as follows:

- Management control.
- Benefits management.
- Financial management.
- Stakeholder engagement.
- Risk management.
- Organizational governance.
- Resource management.

Because P3M3 has a modular structure it is possible to review the seven process perspectives across all individual models or review specific process perspectives across one or more models. This flexibility allows for the targeting of P3M3 in particular areas of concern such as the following:

- Organizational governance across all three models (project, programme and portfolio).
- Project management as a whole, across any number of process perspectives.

Attributes

At a lower level of detail, each process perspective has a number of attributes. These can vary depending on the process perspective in question. There are some generic attributes that are common to all process perspectives at a specific maturity level. These include planning, information management, and training and development.

When assessing an organization against specific attributes for process perspectives such as those shown in Figure 40.3, it is important to see the result as a benchmark against which improvements can be made. Some targeted initiatives can produce quick results, but overall improvement is usually a longer-term prospect.

Benefits of using P3M3

It is important for organizations to agree the levels at which they need to be. Not all organizations will be able to reach the highest levels and for some the level of investment to reach the highest levels will not provide a payback. However, in general most organizations will want to use P3M3 to provide a benchmark against which they can

Maturity level	*Portfolio management*	*Programme management*	*Project management*
Level 1 Awareness of process.	Does the organization's executive board recognize programmes and projects and maintain an informal list of its investments in programmes and projects? (There may be no formal tracking and documenting process).	Does the organization recognize programmes and run them differently from projects? (Programmes may be run informally with no standard process or tracking system).	Does the organization recognize projects and run them differently from its ongoing business? (Projects may be run informally with no standard process or tracking system).
Level 2 Repeatable process.	Does the organization ensure that each programme and/or project in its portfolio is run with its own processes and procedures to a minimum specified standard? (There may be limited consistency or coordination).	Does the organization ensure that each programme is run with its own processes and procedures to a minimum specified standard? (There may be limited consistency or coordination between programmes).	Does the organization ensure that each programme is run with its own processes and procedures to a minimum specified standard? (There may be limited consistency or coordination between programmes?
Level 3 Defined process.	Does the organization have its own centrally controlled programme and project processes and can individual programmes flex within these processes to suit particular programmes and/or projects? Does the organization have its own portfolio management process?	Does the organization have its own centrally controlled programme processes and can individual programmes flex within these processes to suit the particular programme?	Does the organization have its own centrally controlled project processes and can individual projects flex within these processes to suit the particular project?
Level 4 Managed process	Does the organization obtain and retain specific management metrics on its whole portfolio of programmes and projects as a means of predicting future performance? Does the organization assess its capacity to manage programmes and projects and prioritize them accordingly?	Does the organization obtain and retain specific measurements on its programme management performance and run a quality management application organization to better predict future performance?	Does the organization obtain and retain specific measurements on its project management performance and run a quality management organization to better predict future performance?
Level 5 Optimized process	Does the organization undertake continuous process improvement with proactive problem and technology management for the portfolio in order to improve its ability to depict performance over time and optimize processes?	Does the organization undertake continuous process improvement with proactive problem and technology management for programmes in order to improve its ability to depict performance over time and optimize processes?	Does the organization undertake continuous process improvement with proactive problem and technology management for projects in order to improve its ability to depict performance over time and optimize processes?

Figure 40.2 Checklists at the P3M3 maturity levels.

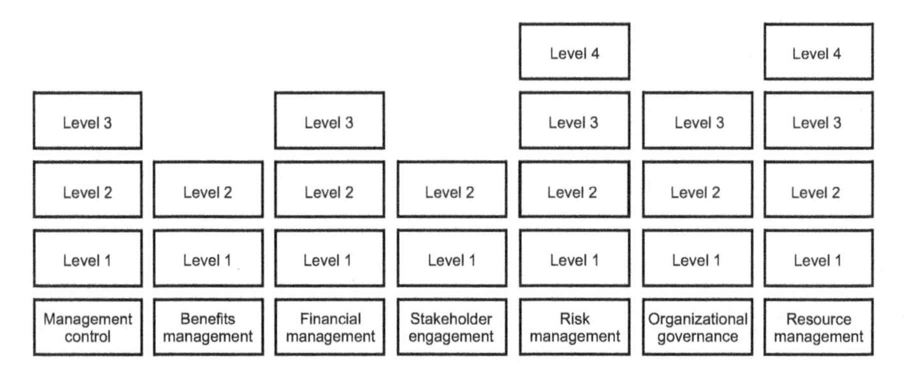

Figure 40.3 Example of the assessment of process perspectives.

improve as part of a long-term process. The benchmark results can be used to justify investment in portfolio, programme or project management improvements - and gain a better understanding of their strengths and weaknesses in order to enable improvement to happen.

P3M3 was originally created to provide a set of common standards and its owner, the former Office of Government and Commerce (OGC) made it available for all organizations to use. Organizations can now use the P3M3 model via materials available on the Axelos website on a self-assessment basis (see www.axelos.com/best-practice-solutions/p3m3). Axelos also hold a register of approved consultants who are accredited to help organizations apply P3M3.

When the OGC developed the original version of P3M3 they targeted some specific organizational benefits of using P3M3 compared with other maturity models. The claimed benefits of P3M3 are as follows:

- It acts as a health check of strengths and weaknesses judged against an objective standard, not just against other organizations.
- It helps organizations to decide the level of performance capability they need to achieve in order to meet their business needs.
- It focuses on the organization's maturity rather than on specific initiatives (good results are possible even with low levels of maturity, so are not in themselves a reliable indicator).
- It recognizes achievements from investment.
- It justifies investment in portfolio, programme and project management infrastructure.
- It provides a roadmap for continual progression and improvement.

Other, more tangible benefits include:

- increased productivity, with shorter cycle times;
- greater time and cost predictability;

- fewer defects, leading to higher-quality outcomes and a lower cost of quality;
- improved customer satisfaction;
- enhanced morale of employees.

Causes of failure

The OGC carried out extensive research into the common causes of programme and project failure and how those causes impact on outcomes. The causes of failure can be grouped into the following categories.

- Design and definition failures, where the scope of the change and the required outcomes and/or outputs are not clearly defined.
- Decision-making failures, where there are inadequate levels of sponsorship and commitment to the change – and where there is no person in authority able to resolve issues.
- Discipline failures, such as weak arrangements (or even no arrangements at all) for risk management and an inability to manage changes in project requirements.
- Supplier management failures, such as a lack of understanding of suppliers' commercial imperatives, poor management and inappropriate contractual set-ups.
- People failures, such as disconnection between the programme/project and stakeholders, lack of ownership, and cultural issues.

Maturity levels

The descriptions and characteristics of the five maturity levels apply equally to each of the three sub-models (portfolio, programme and project management). P3M3 recognizes that organizations may excel at project management without having embraced programme management (or vice versa). Similarly, an organization might be accomplished in portfolio management but immature in programme management. P3M3 therefore allows an organization to assess its effectiveness against any one or more of the sub-models independently.

The maturity levels enable organizations to identify a process improvement pathway as part of a long-term strategic commitment rather than as a quick fix. Although rapid short-term improvements can be targeted to achieve specific goals, the real benefits of P3M3 come through continual process improvement.

The OGC summarized the characteristics of each of the five maturity levels in the following paragraphs. Achievements at a given level must be maintained and improved upon in order to move up to the next level.

Level 1 – awareness of process

At this level processes are not usually documented. There are no, or only a few, process descriptions. They will generally be acknowledged, within managers possibly having some recognition of the necessary activities, but actual practice is determined by events or individual preferences and is highly subjective and variable. Processes are therefore undeveloped, although there might be a general commitment to process development in the future.

Undeveloped or incomplete processes mean that the necessary activities for better practice are either only partially performed or not performed at all. There will be little (if any) guidance or supporting documentation. Even terminology might not be standardized across the organization. For example business case, risk, issues, and so forth might not be interpreted in the same way by all managers and team members.

Level 1 organizations may have achieved a number of successful initiatives, but these are often based on key individuals' competencies rather than on organization-wide knowledge and capability. In addition, such 'successes' are often achieved with budget and/or schedule overruns and, owing to the lack of formality, Level 1 organizations often over-commit themselves, abandon processes during a crisis, and are unable to repeat past successes consistently. There is very little planning and executive 'buy-in', and process acceptance is limited.

Level 2 – repeatable process

At this level an organization will be able to demonstrate, by reference to particular initiatives, that basic management practices have been established (for example tracking expenditure and scheduling resources) and that processes are developing. There are key individuals who can demonstrate a successful track record and that, through them, the organization is capable of repeating earlier successes in the future.

Process discipline is unlikely to be rigorous, but where it does exist, initiatives are performed and managed according to their documented plans. For example, project status and delivery will be visible to management at defined points, such as on reaching major milestones. Top management will be taking the lead on a number of the initiatives but there may be inconsistency in the levels of engagement and performance. Basic generic training will probably have been delivered to key staff.

A significant risk remains that cost and time estimates will be exceeded. Key factors that may have preconditioned the organization to experience difficulties or failure include:

- inadequate measures of success;
- unclear responsibilities for achievement;
- ambiguity and inconsistency in business objectives;
- lack of fully integrated risk management;
- limited experience in change management;
- inadequacies in communications strategy.

Level 3 – defined process

At this level the management and technical processes necessary to achieve the organizational purpose will be documented, standardized and integrated to some extent with other business processes. There is likely to be process ownership and an established process group with responsibility for maintaining consistency and process improvements across the organization. Such improvements will be planned and controlled (perhaps based on assessments) with planned development and suitable resources being committed to ensure that they are coordinated across the organization. Top management is engaged consistently and provides active and informed support.

There is likely to be an established training programme to develop the skills and knowledge of individuals so that they can perform their designated roles more readily. A key aspect of quality management will be the widespread use of peer reviews of identified products, to understand better how processes can be improved and thereby eliminate possible weaknesses.

A key distinction between levels 2 and 3 is the scope of standards, process descriptions and procedures (stated purposes, inputs, activities, roles, verification steps, outputs and acceptance criteria). At level 3 processes can be managed more proactively using an understanding of the interrelationships and measures of the process and products. These standard processes can be tailored to suit specific circumstances, in accordance with guidelines.

Level 4 – managed process

Level 4 is characterized by mature behaviour and processes that are quantitatively managed – controlled using metrics and quantitative techniques. There will be evidence of quantitative objectives for quality and process performance, with these used as criteria in managing processes. The measurement data collected will contribute towards the organization's overall performance measurement framework and will be imperative in analysing the portfolio and ascertaining the current capacity and capability constraints. Top management will be committed, engaged and proactively seeking out innovative ways to achieve goals.

Using process metrics, management can effectively control processes and identify ways to adjust and adapt these to particular initiatives without loss of quality. Organizations will also benefit through improved predictability of process performance.

Level 5 – optimized process

At this level the organization will focus on optimization of its quantitatively managed processes to take into account changing business needs and external factors. It will anticipate future capacity demands and capability requirements to meet delivery challenges (for example through portfolio analysis). Top managers are seen as exemplars, reinforcing the need and potential for capability and performance improvement.

This will be a learning organization, propagating the lessons learned from past reviews. The organization's ability to respond rapidly to changes and opportunities will be enhanced by identifying ways in which to accelerate and share learning.

The knowledge gained by the organization from its process and product metrics will enable it to understand causes of variation and therefore optimize its performance. The organization will be able to show that continuous process improvement is being enabled by quantitative feedback from its embedded processes and from validating innovative ideas and technologies.

There will be a robust framework addressing issues of governance, organizational controls and performance management. The organization will be able to demonstrate strong alignment of organizational objectives with business plans, and this will be cascaded down through scoping, sponsorship, commitment, planning, resource allocation, risk management and benefits realization.

Process perspectives

There are seven 'process perspectives' within P3M3, defining the key characteristics of a mature organization. These apply across all three models and at all maturity levels. Each perspective describes the processes and practices that should be deployed at a given level of maturity. As organizations move up through the maturity levels, the quality and effectiveness of the processes and practices increase correspondingly. This incremental nature of process improvement is a key feature of P3M3.

P3M3 is flexible, enabling organizations to assess their maturity with respect to any one or more of the three sub-models. Organizations may also choose to assess their effectiveness against any particular process perspective(s) across all three sub-models – for example to measure risk management maturity in portfolio, programme and project management.

The OGC outlined seven process perspectives and gave a summary of their characteristics as follows:

Management control

This covers the internal controls of the initiative and how its direction of travel is maintained throughout its life cycle, with appropriate break points to enable it to be stopped or redirected by a controlling body if necessary.

Management control is characterized by clear evidence of leadership and direction, scope, stages, tranches and review processes during the course of the initiative. There will be regular checkpoints and clearly defined decision-making processes. There will be full and clear objectives and descriptions of what the initiative will deliver. Initiatives should have clearly described outputs, a programme may have a blueprint (or target operating model) with defined outcomes, and a portfolio may have an organizational blueprint (or target operating model).

Internal structures will be aligned to achieve these characteristics and the focus of control will be on achieving them within the tolerance and boundaries set by the controlling body and based on the broader organizational requirements. Issues will be identified and evaluated, and decisions on how to deal with them will be made using a structured process with appropriate impact assessments.

Benefits management

Benefits management is the process that ensures that the desired business change outcomes have been clearly defined, are measurable and are ultimately realized through a structured approach and with full organizational ownership. Benefits should be assessed and approved by the organizational areas that will deliver them. Benefit dependencies and other requirements will be clearly defined and understanding gained on how the outputs of the initiative will meet those requirements.

There should be evidence of suitable classification of benefits and a holistic view of the implications being considered. All benefits should be owned, have realization plans and be actively managed to ensure that they are achieved. There will be a focus on operational transition, coupled with follow-up activities to ensure that benefits are being owned and realized by the organization.

There will be evidence of continual improvement being embedded in the way the organization functions. This process will identify opportunities that can be delivered by initiatives and also take ownership of the exploitation of capabilities delivered by programmes and projects. Change management (and the complexities that this brings) will also be built into the organization's approach.

Financial management

Money is an essential resource that should clearly be a key focus for initiating and controlling initiatives. Good financial management ensures that the likely costs of the initiative are captured and evaluated within a formal business case and that costs are categorized and managed over the investment life cycle.

There should be evidence of the appropriate involvement of the organization's financial functions, with approvals being embedded in the broader organizational hierarchy. The business case, or equivalent, should define the value of the initiative to the business and contain a financial appraisal of the possible options. The business case will be at the core of decision-making during the initiative's life cycle, and may be linked to formal review stages and evaluation of the cost and benefits associated with alternative actions. Financial management will schedule the availability of funds to support the investment decisions.

Stakeholder engagement

Stakeholders are key to the success of any initiative. Stakeholders at different levels, within and outside the organization, will need to be analysed and engaged with effectively in order to achieve objectives in terms of support and engagement. Stakeholder engagement includes communications planning, the effective identification and use of different communications channels, and techniques to enable objectives to be achieved. Stakeholder engagement should be seen as a continuous process across all initiatives and a function that is inherently linked to the initiative's life cycle and governance controls.

Risk management

Risk management is about the way in which the organization manages threats (and also opportunities) presented by the initiative. Effective risk management maintains a balance of focus on threats and opportunities, with appropriate management actions to eliminate or minimize the likelihood of any identified threat occurring (or to minimize impact should a risk materialize) and to maximize opportunities. It will look at a variety of risk types that could affect the initiative, both internal and external, and will focus on tracking the triggers that create risks.

Responses to risk will be innovative and proactive, using a number of options to minimize threats and maximize opportunities. Risk reviews will be embedded within the initiative's lifecycle. There will be a supporting process and structures in place to ensure that the appropriate levels of rigour are being applied, along with evidence of interventions and changes made to manage risks.

Organizational governance

Organizational governance examines how the delivery of initiatives is aligned to the strategic direction of the organization. It considers how start-up and closure controls are applied to initiatives and how alignment is maintained during a project's lifecycle. This differs from management control (which views how control of initiatives is maintained internally). Instead, this perspective looks at how external factors that impact on initiatives are controlled (where possible, or mitigated if not) and used to maximize the final result. This should be enabled by effective sponsorship.

Organizational governance also looks at how a range of other organizational controls is deployed and at how standards are achieved (including legislative and regulatory frameworks). It also considers the levels of analysis of stakeholder engagement and how their requirements are factored into the design and delivery of outputs and outcomes.

Resource management

Resource management covers management of all types of resources required for delivery. These include human resources, buildings, equipment, supplies, information, tools and supporting teams. A key element of resource management is the process for acquiring resources and how supply chains are utilized to maximize the effective use of resources. There will be evidence of capacity planning and prioritization to enable effective resource management. This will also include performance management and exploitation of opportunities for greater utilization. Resource capacity considerations will be extended to the capacity of the operational groups to resource the implications of change.

Attributes

Within each process perspective there are a number of attributes. These are indicators of process and behavioural maturity. *Specific attributes* relate only to a particular process perspective, while *generic attributes* apply equally to all process perspectives at each of the five maturity levels.

Inevitably, organizations will perform well against some process perspectives and not so well against others. Attributes describe the intended profile of each process perspective at each maturity level, and the topics, processes and practices covered will change and build as the maturity level changes.

Conclusion

The structure of P3M3 enables organizations to benchmark where they are currently with respect to any of the process perspectives in all or any of their portfolio, programme and project management capabilities. This, along with knowledge of where the organization needs or wants to be in the future, provides the basis for an improvement plan to be devised and for progress towards the target to be tracked.

References and further reading

www.axelos.com/best-practice-solutions/p3m3.

PART X

People

41

MANAGING PROJECT PEOPLE

Nigel Hibberd

Project control means nothing unless the project manager understands people and knows how to communicate with them, motivate them and meld them into an enthusiastic team.

Communication

Project control is heavily dependent upon communication. Most people connect this subject with the passing of information such as work instructions and reporting data. But equally important are the interpersonal communications and relationships between members of the project team and between all the other stakeholders. Often individual people have an agenda that can get in the way of efficient and accurate communication, such as not wanting to convey bad news up the organizational tree, or perhaps wishing to suppress detailed information relating to poor performance that could be career-limiting for them.

In addition to language, communication includes such simple things as facial expression, body attitude (aggressive, friendly, passive and so on). The project manager needs to ensure good rapport with everyone at the top and lowest levels of the organization. That means being open and approachable and always giving encouragement. It is not a good idea to attempt project management by hiding one's self in an office and simply relying upon communication by telephone or the written word. The most successful managers adopt the practice of 'managing by walking about' (an expression that needs no explanation). The opposite of management by walking about is 'management by surprise' (which refers to the manager who fails to communicate and is then surprised when things go wrong).

Sometimes communication is difficult, especially when the message contains bad news. I feel that being a project controller is something like being an acupuncturist – we stick the needles in the project and people find that irritating but they know it is for the good of the project. Project controllers must have the strength of character to convey the truth, however good or bad.

Motivation

Motivation is a strange thing and is not as clear-cut as many people would imagine. Much work has been done by management theorists on this subject. I want to mention just two examples.

Elton Mayo and the Hawthorne experiments

Elton Mayo was an American management scientist who studied people's behaviour at work. He conducted a series of experiments at the Hawthorne works of the Western Electric Company (near Chicago) around 1930. His work included observations of a factory department that manufactured electrical relays. Various changes were made to the operatives' working conditions during the investigations and each time it was found that an improvement in output resulted. For example, lighting in the relay assembly room was improved and the production rate increased. But it was also found that the work rate improved again even when some of those changes were reversed. The conclusion (and lesson learned) was that the workpeople were motivated to perform better simply because someone was showing an interest in what they did (Mayo, 1933). So there's a lesson here for any manager, which is to show a genuine interest in what people do, give praise where appropriate and do not take their work for granted.

Maslow and his hierarchy of needs

Maslow (1954) understood that the motivation of people is a complex subject that goes well beyond simply giving people a salary raise. He identified a hierarchy of needs that can be represented in the form of a pyramid, with the most fundamental needs at the pyramid base. Figure 41.1 shows a simplified version of this pyramid.

The most basic human needs are physiological, which are the requirements for survival (such as air, water, food and shelter). Moving one step up the pyramid, the next level of motivational needs is associated with security and safety. This means not only physical safety, but also things such as job security and being able to look forward to a retirement pension in healthy old age. Another step up the pyramid brings us to the belonging and love needs. In the project management context this includes the need for friendship and collaboration within a team. We can all think of a few people from our work experience who seem to lack these needs. Moving nearer to the pyramid top are the esteem needs, which drive our ambition to perform well and succeed. At the top of Maslow's pyramid are needs that are more esoteric and difficult to define. Maslow called these the 'self-actualisation' needs. They can be quite different from one person to another, but can be summarized as wanting to attain the highest possible level of fulfilment, performance or moral conduct.

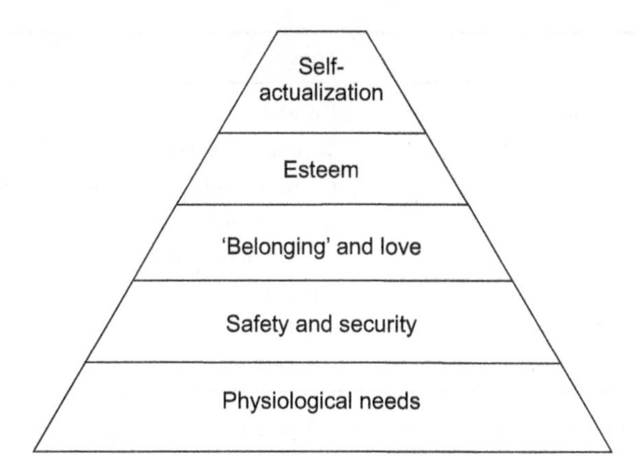

Figure 41.1 Maslow's hierarchy of human needs.

Maslow went far beyond the simple summaries presented here and the subject is complex. The balance of these needs varies from one individual to another and can change throughout a person's life. To add a further dimension and more complexity, the needs described here can also be applied to groups of people who share common circumstances.

Herzberg's motivational theories

Frederick Herzberg was an American psychologist who specialized in human motivational factors. He was responsible for the dual structure theory, with the factors divided into 'motivator factors' and 'hygiene factors' (see Figure 41.2). He also gave the hygiene factors the alternative names 'maintenance factors' or 'environmental factors'. Herzberg reached his conclusions after studies involving 200 engineers and accountants.

Herzberg considered both sets of factors from several different viewpoints. He found that the motivational factors could affect an individual's satisfaction in varying degrees. At the lowest level, with no motivating factors at all, there was no satisfaction, but neither was there dissatisfaction. Conversely, the absence of a hygiene factor could cause dissatisfaction, although the hygiene factors are not themselves motivators and do not cause satisfaction. One very interesting point is that pay is included among the hygiene factors and is not by itself a motivator.

The project team

So far I have discussed aspects of management concerned with people as individuals. However, in projects successful completion depends to a large extent on being able to create a team spirit, so that everyone can work together, all pulling in the same direction, with the combined aim of achieving the project objectives. Here some project managers will face problems.

Motivating factors	*Hygiene factors*
Sense of achievement	Pay and benefits
Recognition	Technical supervision
The work itself	Company policy and administration
Responsibility	Interpersonal relationships with subordinates, peers and supervisors
Status	Job security
Promotion opportunities	Working conditions
Personal growth	Personal life

Figure 41.2 Motivational and job satisfaction factors according to Herzberg.

Building a project team in a functional matrix organization

One potentially difficult problem faces project managers who work in a company that operates on the basis of a functional matrix (which was described in Chapter 6). The project workforce is split into functional groups, with each group reporting to a specialist manager. Thus there could be a structural design group, an electrical design group, a piping and fluids design group, an IT group and so forth. The people nominated to work on a project are scattered over these functional groups, with each person reporting directly to his or her specialist chief engineer or manager. When the manager of a new project wishes to meet the new project team for the first time there is no readily identifiable team of any sort. So the project manager has to create a virtual team, and that will need the full support and cooperation of all the functional group managers.

Every individual in that virtual project team will have two managers; namely the functional head and the project manager. This violates a very important law of management, which is the principle of unity of command. In other words, no individual should be expected to work at the same time for two different managers. In one such case the project manager was divorced from the sister of one of the functional managers. You can imagine the difficulties that caused for individuals working under those two different managers. The project manager in a functional matrix has to take particular care to communicate well with all the functional group managers and win their support.

When a project is first launched it is very good (if not essential) practice to gather as many people as possible who will be working on the project and enthuse them in a project kick-off meeting. That meeting should have strong and visible management support, with at least one member of the C-suite being present to support the project manager if possible. The kick-off meeting might well be the first occasion at which members of the new project team have the opportunity to meet each other. That should give the project manager a great chance to kick the project off in the right direction. That's an opportunity that should not be missed or messed up.

Individual team members

People are all different. We all have our own personalities, interests, agendas, distracting problems and so forth. Every team member could have his or her own:

- role and associated interests;
- perception of other people's roles;
- agenda;
- behavioural traits;
- partner with slightly different cultures.

It is possible that in some projects team members will be on temporary secondment from other departments, to remain located with the team for as long as they are needed. For example, a fairly common practice is to take one or more purchasing officers from the purchasing department and locate them physically with the project team for the duration of the project. In these cases the individuals involved will have dual reporting loyalties, which have to be recognized.

Belbin on team development

Belbin's work on team development is very well known and highly regarded. He identified the following phases through which a newly-formed team must pass before becoming mature and fully effective (Belbin, 2004).

1. Forming:
   - the team comes together;
   - they break the ice;
   - they size other members up;
   - many are unknown quantities.
2. Storming:
   - the team begins working together;
   - the members try to jostle for position;
   - there is a lot of energy expended that is not always productive.
3. Norming:
   - people in the team begin to realise each other's strengths and weaknesses;
   - they understand what each person 'brings to the party';
   - rapport and commitment build.
4. Performing:
   - the team is 'on song' and focused on delivering;
   - members will support a team member who is struggling, for the good of the team;
   - commitment is everything.
5. Retiring:
   - the team has almost delivered and is on rundown;
   - the team breaks up gradually and goes on to new work, possibly in new teams.

Much of the above is self-explanatory and will be familiar to experienced managers. However, the retiring stage can be very demotivating for team members, some of whom might be wondering not only about their next project assignment but whether or not there will be any future work for them at all. During this stage much of the original enthusiasm felt when the project began will have disappeared. There might even be an incentive to delay progress, so that the end of the project and day of reckoning is put back. The project manager must strive to maintain the level of commitment from everyone to continue working efficiently and accurately right up to the final day of project delivery.

I have found that the impact of this 'end of project blues' can be minimized by keeping as many of the team as possible together and loading them up with new work. Or, alternatively, transfer part of the team into new projects, where they will carry forward the benefits of established rapport and commitment. Then that will become a 'win-win' for the business and its people.

People affected by management change projects

Projects that change the working systems, IT or organizational structure clearly can affect all the people in an organization, well beyond those who were engaged in designing the new systems. Now the project manager has to team up with senior members of the company management to ensure that all the company staff are kept informed of any change which might affect their way of working or (even in some cases their continued employment with the company).

The leader of a management change project needs to have a sympathetic understanding of what change will mean to the organization's people. When a manager announces a change to the staff they will respond according to one of the following two classical models.

1. The individual believes that they can influence and control the change.
2. The individual believes the change to be inevitable and out of his or her control (an example from everyday life would be learning that the family home is about to be subjected to compulsory purchase to make room for a new road).

People who are going to be affected as a result of a management change project are most likely to fall into the second of these two categories. Many will probably go through different stages of response. The following are just a few of the possible responses:

- Disbelief – 'Can it really be true? Surely it will not affect me?'
- Realization – 'It really is true and it seems that I am going to be affected – but how?'
- Denial – 'Maybe it will not be so bad. Let's wait for a few days see develops'.
- Rage – 'I've worked so hard and am good at my job; how can they do this to me?'
- Reluctant acceptance – rebellious behaviour can occur at this stage if people are not properly supported.

- Acceptance of the new ways of working. The changed rules and processes become embedded and confidence is gained.
- Confidence – 'I can do this and feel better about my ability. It could mean new opportunities for me. Let's give this a try'.

Everyone is different and people will react to change in different ways. The important thing for the project manager is to remember that people are sentient beings, not machines, and they have to be treated with consideration and respect. The manager who listens to people and tries to accommodate their wishes will gain more respect and, in the end, better results. Always remember: managers are only as good as their teams. They can't do it all by themselves. My experience is that by establishing effective rules and guidance, and supporting people effectively with key users, the level of demoralization people experience from change can be reduced significantly. Typically the prototype system testing and commissioning team will become the key users around the business.

There is advice on dealing with management change projects and the people affected in Fowler and Lock (2006).

References and further reading

Adair, J. (2002), *Effective Leadership*, London: Pan MacMillan.

Adair, J. (2009), *Effective Team Building*, 2nd edn, London: Pan MacMillan.

Belbin, R.M. (1993), *Team Roles at Work*, Oxford: Butterworth-Heinemann.

Belbin, R.M. (2004), *Management Teams: Why They Succeed or Fail*, 2nd edn, Oxford: Butterworth Heinemann.

Drucker, P.F. (1954), *The Practice of Management*, New York: Harper & Row.

Fowler, A. and Lock, D. (2006), *Accelerating Business and IT Change: Transforming Project Delivery*, Aldershot: Gower.

Kanigel, R. (1997), *The One Best Way: Frederick Winslow Taylor and the Enigma of Efficiency*, London: Little, Brown and Co.

Kerzner, H. (2000), *Applied Project Management: Best Practices on Implementation*, New York: Wiley.

Lock, D. and Scott, L. (eds) (2013), *Gower Handbook of People in Project Management*, Farnham: Gower.

Maslow, A. (1954), *Motivation and Personality*, New York: Harper & Row.

Mayo, G.E. (1933), *The Human Problems of an Industrial Civilization*, Boston, MA: Harvard Business School.

42
PERFORMANCE IMPROVEMENT METHODS

Dennis Lock

In this chapter three approaches are described that can help organizations and their people to improve their performance and project control capabilities.

Communities of practice

A long time ago, in the 1960s, the profession of project management as we know it today was in its infancy. A few of us were exploring the application and possibilities of critical path network analysis and the use of computers to process them and produce effective reports for project control. We were a small, select band comprising project planners and managers from various industries plus systems engineers from software companies. We met and communicated regularly to discuss the latest techniques and software possibilities. The project people improved their skills as a result, and the software companies were able to understand what we wanted and develop their applications accordingly.

At that time I was a manager with project management responsibilities at a heavy engineering company in the East Midlands. The people with whom I exchanged experiences and the application of innovative techniques included the following:

- Herbert Walton of the British Oxygen Company (a highly respected manager who later rose to eminence in the APM.
- A cost and planning engineer from Massey Ferguson in Coventry.
- The manager of a project planning department in a Warwick construction company who, even in those early days, was using advanced critical path methods with modular templates.

- Systems engineers from two software companies who were developing project management applications and needed to know what we needed their products to do.

All of us met or discussed our various needs and experiences frequently, and most of us attended the second Congress of the International Project Management Association in Amsterdam.

All of us were supported in these endeavours by our employers and we all advanced our capabilities to our personal advantage and the benefit of our employing organizations. Although we did not know it at the time, we had formed an informal partnership, which today would be called a *community of practice*.

When we send delegates to a seminar and ask them afterwards whether or not they had gained anything useful in return for the time and money spent, it frequently happens that they report that their learning experiences from the formal presentations and lectures were somewhat disappointing but they had made several useful contacts with other delegates, with whom they intended keeping in touch to share experiences and professional knowledge. Again, this has resulted in a community of practice from which everyone involved can share their experiences, discuss mistakes to be avoided in the future, and generally advance the knowledge and expertise of our profession.

The APM has a number of special interest groups who meet regularly and that is a more conventional and structured way in which those involved in project management can benefit from their shared experiences and form communities of practice. Although these communities of practice came about almost by accident the value of CoPs (as they are now known) is a widely recognized form of professional and management development and improvement.

Action learning

There was a time when I was working as an electronics engineer on aeronautical test equipment. One day I was hauled away from my desk and comfort zone and told that I was being transferred immediately for several weeks to the accounts department, where I would be reporting to the chief accountant. Total shock and dismay hit me and I felt, to say the least, discomforted and humiliated. The purpose of this temporary transfer was for me to learn the company's cost accounting procedures and suggest ways in which they might be improved or adapted so that our engineering project costs could be more conveniently be collected, analysed, reported and controlled. My immediate reaction was very negative and I saw the process as a sudden demotion.

However, I soon began to learn quite a lot about cost accounting procedures. I learned about overhead recovery and under recovery; about the difference between direct and indirect costs, about direct bookings by indirect workers and indirect bookings by direct workers. I became familiar with bought and sales ledgers. As an unexpected benefit I had access to the payroll system and thus learned the salaries of absolutely everyone in the company. In fact I came to enjoy the experience. The detailed cost accounting knowledge that I gained during that time has been of considerable benefit to me in my subsequent management career.

When my stint in the accounts department was over, new project management accounting procedures were put in place and everyone declared themselves satisfied. I was plucked from the accounts department and transferred to a new role in a different company division where, for the first time in my life, the word 'manager' appeared in my job description. So it was win, win for everyone.

Now, with hindsight, I know that this was an example of a management development process called *action learning*. The formal process of action learning was the brainchild of Professor Revans (Revans, 2011). It is a process where managers are taken out of their usual jobs and comfort zones and asked to work in completely different roles for a period that might be (say) six months. A professional observer keeps an eye upon the participants and reviews progress. So a bank manager might find herself working as an engineering manager in a mining company, and the former mining manager would be placed in charge of a section of the bank. The arrangements can even involve three-way swaps, with three participants being moved across three different organizations.

All that might sound like a recipe for complete disaster for the people and the companies involved. However, the experiment was tried in a range of companies and most (but not quite all) of the participants and their supporting organizations reported benefits. Action learning is an extreme example of management development. It clearly requires courage and dedicated commitment from all the individuals and companies concerned. However, organizations should always ensure that their managers (including those responsible for project controls) are given the best possible opportunities to increase their practical knowledge and enhance their performance.

Benchmarking

Benchmarking is not in itself a process that develops or improves personal performance or management controls but it can highlight performance deficiencies (after which it becomes the responsibility of company management or project managers to find and rectify faults in their systems). Benchmarking requires that a company compares its performance with other companies operating in their same business sector. Where other companies appear to be performing better, that is a clear indication that the organization under review needs to take action and improve. All this raises two obvious questions:

1. What parameters can we choose for measuring performance?
2. How can we gain the knowledge from other companies about how well they are performing?

Parameters for measuring the performance of a commercial organization

Ask a lay person about how one might measure the performance of a company and you would probably get the answer that performance or success can be measured by the annual profit reported in the company accounts. More specifically, and a more accurate answer might be that performance could be measured by the percentage profit made in comparison with the company turnover. That result is one of a number of management ratios that can be used to measure and judge various aspects of a company's performance. The following management ratios and other measures are a short selection from ratios that can be used to compare the performances of different companies operating in the same business sector:

- Profitability ratio. This is the ratio of gross profit made to the sales value achieved in a year.
- Liquidity ratio. One way of measuring this is to divide a company's current assets by its current liabilities. A company that has insufficient liquid cash might not be able to pay its bills. That, in terms of project controls, means that suppliers could withhold deliveries of goods and services until they get paid. In extreme cases the company could fail.
- Inventory turn ratio. This can be stated as the number of times within an accounting year that the value of stored goods and work in progress (the total inventory value) is turned over. Progress delays and work held up because of problems will depress this ratio. So good project control has a direct influence on this ratio.
- Earnings per employee. This ratio, found by dividing annual turnover by the total number of employees, is a direct measure of whether or not the company is managing its workforce efficiently.

Performance ratio results that might be considered acceptable in one industrial sector could be regarded as awful in other. So comparisons have to be made within the same business sector to be of any use. One significant question remains: How can we obtain the data? One can hardly write to one's competitors and ask 'How are you doing?' to get the raw data needed for benchmarking. But every company has to produce and publish an annual report, including the audited accounts.

This account of benchmarking has given only an outline of the concept. For a more professional and experienced view see Sylvia Codling's excellent book on the subject (Codling, 1998).

References and further reading

Camp, R.C. (2006), *Benchmarking: The Search for Industry Best Practices That Lead to Superior Performance*, Milwaukee WI: ASQC Quality Press.

Codling, S. (1998), *Benchmarking*, Aldershot: Gower.

Hughes, J., Jewson, N. and Unwin, L. (eds) (2007), *Communities of Practice: A Critical Perspection*, New York: Routledge.

Lock, D. and Scott, L. (eds) (2013), *The Gower Handbook of People in Project Management*, Farnham: Gower.
Revans, R. (2011), *The Art of Action Learning*, Farnham: Gower.
Wenger, E. (1998), *Communities of Practice: Learning, Meaning and Identity*, Cambridge: Cambridge University Press.
Wenger, E., McDermott, R. and Snyder, W. (2002), *Cultivating Communities of Practice*, Boston, MA: Harvard Business School Press.

43
MANAGING YOURSELF AND YOUR CAREER

Lindsay Scott

There was a time when working in project controls was just transitional - a starting place from which to become a project manager or some other senior role. Times have changed and, for practitioners in project controls, they continue to change as projects become increasingly complex and 'mega' in their scope. The work of today's project controls practitioners has become increasingly mature in nature, with established approaches well embedded in organizations. Advances in best practices and technologies have seen parallel activity in the quest to recognize project controls as a profession and to acknowledge those who work within it as true 'professionals'.

Introduction

Within this chapter I shall discuss the career of a typical professional project controls practitioner and the elements which that includes. I shall provide food for thought on the different approaches to take and ultimately help you to put your career firmly in your own hands. But first I must take a moment to explain what it means to be professional and to be part of a profession. The Professional Standards Councils (PSC) in Australia states that a professional:

> is a member of a profession. Professionals are governed by codes of ethics, and profess commitment to competence, integrity and morality, altruism, and the promotion of the public good within their expert domain. Professionals are accountable to those served and to society.

That council describes a profession in the following terms:

> A profession is a disciplined group of individuals who adhere to ethical standards. This group positions itself as possessing special knowledge and skills in a widely recognised body of learning derived from research,

> education and training at a high level, and is recognised by the public as such. A profession is also prepared to apply this knowledge and exercise these skills in the interest of others.

Bassot (2013) helps to breakdown the term 'professional' and provides the following summary of what it means to be a professional:

- You have a body of knowledge in relation to your particular profession.
- You work in a relatively autonomous way, without need close supervision of every aspect of your work.
- You are expected to show some initiative in your work.
- You understand the boundaries of your role and have a clear grasp of when you need to refer to someone else.
- You have the relevant skills to carry out your role well.
- Your attitudes are in line with the profession to which you belong.
- You adhere to the code or practice or ethical standards relevant to your profession.
- Your work is not straightforward. It will involve making professional judgements where you encounter situations in which there are no clear right or wrong answers.
- You aim to improve your practice all the time, reflecting on what you are doing and engaging in continuous professional development.
- You aim to keep up to date with new knowledge and skills in relation to your practice.

To be professional project controls practitioners and to take charge of our careers through professional development means being a part of the profession, in all the forms which that takes.

With Bassot as my guide, I shall look first at the body of knowledge and examine how that drives our careers in project controls.

Bodies of knowledge: professional associations

There are a few professional bodies and associations nationally and internationally that focus on project controls plus project management associations (including those described in Chapter 1). Additionally there are many professional associations and institutions worldwide that cater for related professions, such as civil engineering, cost accounting and control, construction, electrical engineering, mechanical engineering, surveying and so forth. With each of these comes the opportunity to learn, understand and practice the standards advocated by that association.

Becoming a member of a professional body demonstrates not only your own commitment to your chosen profession but is regarded as a requirement by organizations seeking to attract people with the appropriate talents for their needs. Many professionals tend to become members of two or more associations. For example, a British project manager might choose to become both an APM member and a member of Chartered Management Institute. Membership (say) of the Association of Cost Engineers will reflect their specific discipline within project controls.

Certification by one of these bodies will greatly enhance the professional status of the individual. If the professional association is a chartered body, then chartered membership will also be a big personal advantage.

The ability for a professional person to combine and blend different bodies of knowledge and best practice from a variety of sources to provide the best possible solution or ways of working is becoming increasingly common. It is here that the link to professional judgement comes sharply into focus. As our programmes and projects become increasingly complex and complicated, new boundaries are being pushed and high-profile decisions are being made. The commitment made on becoming a member of an association also helps us to continue our professional development. Many associations and people advocate 'continuing professional development' (CPD) in order to maintain a competent level of skill and professionalism.

Continuing professional development

Improving your practice throughout your career is a process that has been coined as, 'having agency'. That means it's up to you to drive your lifelong learning, to make plans and carry out actions, to take time for reflection and so on. True professionals know that their careers are ultimately their own responsibility. That includes making sure they continue to be fit to practice and have the right skills for the marketplace. They take the long view that no job is for life and they should also be prepared for potential career-limiting situations.

Several formal technical learning options exist within the project controls profession. These focus on specific subjects such as planning, scheduling, cost control, risk management and so on. Coupled with the range of different tool options available we could be forgiven for thinking that a project controls role is purely technical.

During the earlier part of a project control practitioner's career the development focus tends to be on the formal accreditations and certifications that are aligned with the job title. Then it's the technical aspects of the job to which we devote most of our development time. As we advance in our careers and look for promotion and ladder-climbing opportunities, the focus moves from purely technical aspects to what the Project Management Institute refers to as the two other sides of the talent triangle. These two sides are strategic/ business management and leadership skills. These competencies include the following:

1. Strategic and business management competencies:
   - strategic planning;
   - benefits management;
   - financial management;
   - commercial and legal issues;
   - customer relationship;
   - portfolio management;
   - business models and frameworks;
   - analysis and modelling;
   - business planning and forecasting;
   - industry standards.

2 Leadership skills:
  - coaching and mentoring;
  - different leadership styles;
  - emotional intelligence;
  - conflict management;
  - influencing;
  - listening;
  - negotiating;
  - problem-solving;
  - team-building;
  - communicating.

In addition to formal training, there are many options for attending industry conferences, seminars, webinars and 'meet-ups'. Books, podcasts, white papers and journals are also readily available. The options are there to be found. With the membership of a professional association, some meeting options can be recorded as proof of attendance, although that can provide no proof of understanding or usefulness!

One fact that is often overlooked is the origin of learning with CPD, in other words from where does the learning come? The reality is that the principal learning source is the job itself – working on different assignments and projects, pushing to work on tasks and activities that take you outside your comfort zone. The most valuable learning, and career enhancing thing you can do is to volunteer to undertake something that genuinely terrifies and excites you in equal measures.

The 70-20-10 Model (Lombardo et al., 1996) is a formula often used in learning and development and the training profession. It describes where sources of learning predominantly come from with professionals, as follows:

- 70 per cent of learning comes from on-the-job experiences;
- 20 per cent results from interactions with others and the feedback we receive;
- only 10 per cent results from formal education or training.

Unfortunately, most of our development tends to focus on that final 10 per cent. The remaining 90 per cent relies heavily on our organization's support to create those opportunities for us on the job or to provide the environment or mechanisms for us to be in a position to learn effectively from interactions with others.

One possibility for harnessing the 20 per cent is the growth of communities of practice within our own organizations (the CoP concept was explained in Chapter 42). Internal CoPs (which can take the form of lunch-and-learns, technical forums or away-days) are often initiated and driven by the PMO or project controls function. This is a case of 'Why wait for our organization to provide opportunities when it is within our own remit as a supporting function to help drive capability improvements not just for ourselves, but also for the wider delivery organization?'

Taking time to reflect

An approach to managing your own career, which has become increasingly common in a project professional's career is reflective practice. Reflective practice has long been an approach to continuous learning in professions such as healthcare and education. Reflective practice is concerned with practice-based professional learning, in which we learn from our own experiences. There are many models that can help professionals develop a way to reflect on the work they do (not just thinking about how they have carried out a particular task) and also for looking at the impact of our behaviours, emotions, choices and actions.

Schon (1987) gives a compelling insight into why we should consider reflective practice as part of our career development, declaring that 'Professional knowledge involves both "technical rationality" (rules) and "professional artistry" (reflection in action)'. Part of the 'crisis' for professionals arises from the fact that very often the 'theory' or rules espoused by practitioners is quite different from the 'theory' or assumptions embedded in the actual practices of professionals.'

Fook (2007) tells us that reflective practice 'involves the ability to be aware of the "theory" or assumptions involved in professional practice, with the purpose of closing the gap between what is espoused and what is enacted, in an effort to improve both. A process of reflective practice, in this sense, also serves to help improve practice, by helping to articulate and develop practice theory'.

Getting started with reflective practice can be as simple as starting a journal. Moon (2004) suggests that a reflective journal can help to us to:

- focus our thoughts and develop our ideas;
- experiment with ideas and ask questions;
- organize our thinking through exploring and mapping complex issues;
- develop our conceptual and analytical skills;
- reflect upon and make sense of experiences and the processes behind them.

Journaling has become a way for project professionals to drive their own development through self-analysis, recognizing that the only limit to learning is ourselves. Basic questions such as the following can start the process:

- What am I learning?
- If I had to do this piece of work again, what would I do differently?
- How could I improve my stakeholder relationships?
- What do I enjoy most about this project?
- What didn't go so well today?
- What did someone else do today that I admired or liked?

Relevant skills

You need a wide range of skills if you are to achieve success in your career. These will, of course, depend on the career that you have chosen to follow. Within the project controls profession there has notoriously been a lack of published information on best practices or standards, in choosing which path to take, the different roles available and the ability to assess one's self against industry standards. The APM sums this up well in their paper 'Considering a Career in Project Controls' as follows:

'With multiple routes into the profession and a multitude of tasks making up the project controller's day job, there's no 'one-size-fits-all' career path through project controls'. To complicate matters further, the profession is changing. The Guild of Project Controls (a community-led online association) has created a career framework, which aims to plug the gap for practitioners in project controls. It focuses on the role profiles commonly found across industries; across the disciplines (planning and estimating, cost, forensic and project controls); and levels of experience from apprentice through to expert. Alongside this are self-assessments and bodies of knowledge, created in response to the largest online project controls community.

Creation of formal career paths and career development support for project control practitioners has fallen to the individual organization's own approaches to careers in project controls. Figure 43.1 shows the common paths from entry-level roles (such as apprentice) through to the senior executive level. Much of the earlier career progression is related to specialisms in certain areas of project controls (such as from project planner to planning manager) or in relation to the size and complexity of the programme and projects. These are not the only routes to progress in project controls. There is also the 'integrated' project controls practitioner who combines specialisms. There is also the opportunity to make a sideways shift mid-career into a role that sees a move away from a supporting function to an actual delivery role (such as project or programme manager). Relevant skills can also be picked up via the wider bodies of knowledge and associations in project and programme management.

There is a further career progress route to explore. This sits at the enterprise, regional or departmental level. A role at these levels is responsible for the entire practice of project controls (and often project management) across the organization. This senior-level role is responsible for portfolio management and providing a mature ecosystem in which programmes and projects can succeed. Here we see advanced skills in leadership and business management making a big difference in the effectiveness of these roles.

Work-based behaviours

The remaining areas of Bassot's guide, describing what constitutes being a professional, comprises items that can collectively be called 'work-based behaviours'. These include the following:

- Working in an autonomous way without needing close supervision in every aspect of your work.
- Showing initiative in your work.

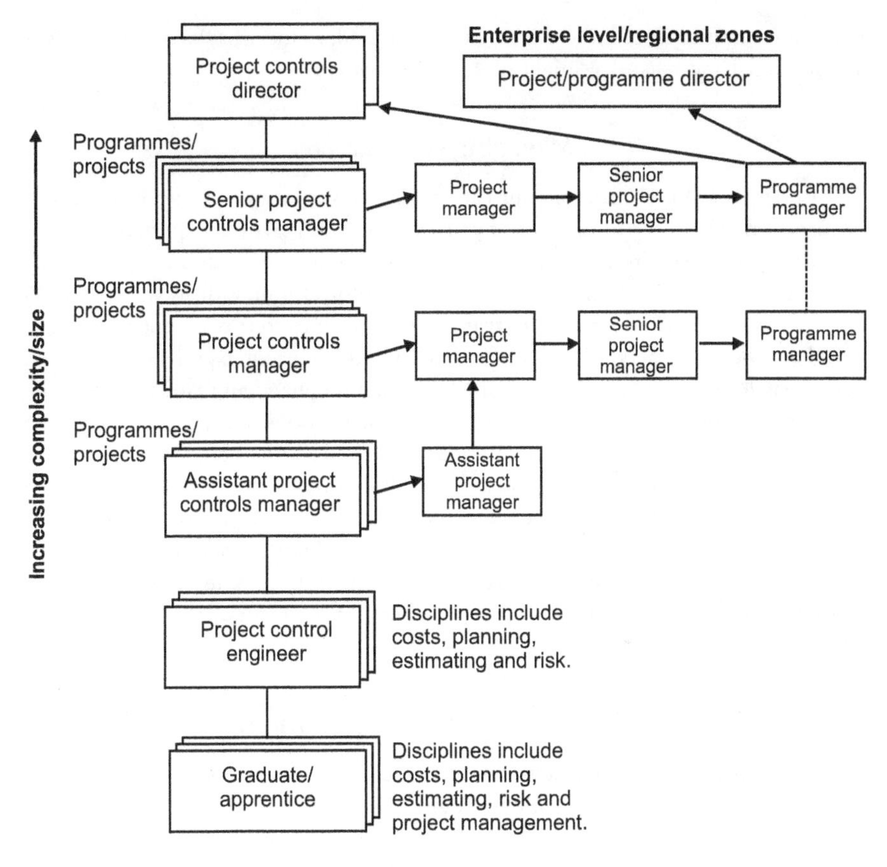

Figure 43.1 Project control roles in a group organization context.

- Having attitudes that are in-line with your profession.
- Adherence to a code of practice or acceptable ethical standards.
- Understanding the boundaries and knowing when to refer to others.
- Your work is not straightforward and involves using professional judgement.

Behaviours are what we see someone doing – how they choose to act or conduct themselves. We expect to see certain types of behaviours that conform to the level of professionalism expected. Our behaviours often make a difference when it comes to being singled out for promotion – or gaining the unwanted spotlight when mistakes are made. It is the work-based behaviours that help us to navigate uncomfortable conversations with customers or be a well-connected individual in a highly political workplace.

When we combine our technical ability, our people and leadership skills; our understanding about the business and the wider industry sector – and then wrap our behaviours around all of those – then we start to understand what we must possess as a professional and what that might mean for our careers.

Managing your career

What does it mean to 'manage your career?' At the very least this means being able to carry out your work to the best of your abilities and meet the expectations whoever pays your wages. Yet, if we look back to 2008 when a recession first hit, simply meeting these criteria did not help the scores of project professionals who suddenly faced redundancy. There was panic that training certifications were not up to date; that the professional network was not in place and that current experiences, skills and competencies were not a 'good fit' in the changing employment market. Since those times intelligent project professionals have learned lessons, knowing they have to manage their careers actively, always mindful that the bad times could return.

Managing your career actively includes all those things covered in the chapter so far plus having a portfolio that includes a decent resumé; an active on-line presence; a profile (both internally in your organization and within the wider external project community). It also includes choosing new assignments that are in line with a career plan and keeping your work/life balance in check. Subjects for a career portfolio are tabled in Figure 43.2.

Taking time to reflect on your career

I have already mentioned reflective practice in relation to our development; the same approach can also be used in relation to our career.

In the *Handbook of People in Project Management* (2013) Dennis Lock gave several questions to enable project professionals to reflect on their careers and these should be ideally revisited at least annually:

- Are you being suitably rewarded?
- Is your employer prosperous or do you work for a company that is in any kind of financial stress that could put your future at risk?
- Have you been unfairly overlooked for promotion more than once?
- Were you given inadequate recognition for a major project success?
- Are you continuing to develop in your career and are you given the management support essential to managing your work?
- Are you respected and placed in a position of trust with freedom to make important decisions?
- Do you like your boss and does he/she inspire you with ideas or relate experiences that enrich your own fund of knowledge?
- Are you happy in your job?

One of the weakest areas of career management for many professionals is the ability to make plans for their career, often leaving things to chance or waiting for opportunities to land on their laps.

Main subject	*Earlier career*	*Later career*
Curriculum vitae	• Focus is on the project control specialism and the core competencies of the role's technical aspects. • Clearly shows where you fit within the organization, department and the programme/project team. • Gives insight into the size and complexity of the programme or project worked upon.	• Clear leadership competencies are demonstrated. • Highlights of key programmes and projects that you have worked on, including span of control, client engagement and stakeholders. • Show achievements that are quantitative and business-focused.
Online presence	• Be connected to the project controls community and industry related groups. • Share insights and lessons from your own work experiences. • Share your work passions through articles or blogs.	• Share knowledge and experiences through articles and interviews. • Help others who are less experienced with insights to their questions. • Get involved in community activities, give something back.
Your profile	• Attend external events such as conferences, meet-ups, best practice forums and technology led seminars. • Volunteer to organize internal forums or gatherings. • Work on company initiatives or improvement activities outside the project environment.	• Put yourself forward in external events to share your experiences. • Get involved in cross-company initiatives and future of the industry groups. • Champion others within the organization. • Take time to coach and mentor.
New assignments	• It's not what you know: it's who you know. Make a connection with the resource scheduler! • 'Manage up' and work with your boss to make sure that he/she is making recommendations to others about you. • Make sure your career has SMART goals to make alignment and opportunity-spotting easier.	• Learn how to play the corporate politics game. • Create the new position or assignment yourself. • Be well connected and a 'go-to' in the organization for problem-solving and troubleshooting. • Become the futurist - scanning the horizon for opportunity and links with business strategy. • Use reciprocity to your advantage in your network.
Work/life balance	• Focus on self-development such as confidence and speaking up. • Learn how to manage your workload - get organized. • Learn to say 'No'.	• Self-care: recognize the signs of stress and overwork. • Provide pastoral care to your teams. • Learn to delegate. • Adopt different leadership styles. • Celebrate the successes.

Figure 43.2 Suggested contents of a career portfolio.

But what happens when an industry and profession are potentially facing huge changes that can cause a threat to our livelihoods?

Future careers in project controls

There is change coming down the line. A change that will impact project management and most definitely the project controls element of the project. The advances in technologies, which have collectively become known as Artificial Intelligence, are difficult to ignore. Where there are data, there is the possibility of automation and machine learning. The benefits of shifting to automation are numerous, 'higher productivity (and with productivity, economic growth), increased efficiencies, safety, and convenience' McKinsey (2017). Already the construction, manufacturing, rail and transportation industries are utilizing the skills of Robotic Process Automation (RPA) technicians and data scientists to automate repetitive work and gain deeper insights to projects which enable better decision-making.

The project controls aspects of projects are already being automated which has started to drive a focus on 'what does a project controller do with the additional time?'. This has led to conversations about upskilling, with project control practitioners being able to take a more consultative and coaching approach to their role whilst managing the exceptions in automation that the technology just can't handle, yet.

It has also led to the upskilling around project data analytics and how the project controls function can move from the role of data gatherer to data facilitator – improving the collection and cleansing of project data with a view to using technologies to provide accurate predictive analytics. The shift to providing stunning data visualization and data storytelling to prompt senior executives to act – not only utilizes technology but brings in skills around complex problem solving; emotional intelligence; creativity; critical thinking; negotiation and collaboration.

The project controls industry has nothing to fear in terms of job losses due to these technological advances, after all, project controls is the source of the data and job creation is more likely to increase due to the expanding remit of the role.

Right now, the industry has a lot of work to do. Artificial intelligence technologies are not there yet; unstructured data is still a problem; standardization of platforms and tools doesn't exist, yet the promise of improved project success rates will drive an organization's response.

For a project control professional today it is all about understanding what the future could look like and being in a position to drive the innovation. These technologies are not another generation away and they're driving the biggest shake up to the project control practitioner's career we've seen in decades.

References and further reading

APM Considering a Career in Project Controls www.apm.org.uk%2Fmedia%2F13168%2Fconsidering-a-career-in-project-controls.pdf Accessed: Jan 2020.

Bassot, B. (2013), *The Reflective Journal: Capturing Your Learning for Personal and Professional Development*, Basingstoke: Palgrave Macmillan.

PMI, Talent Triangle www.pmi.org/learning/training-development/talent-triangle Accessed: Jan 2020.

PSC (2019), *What is a Profession?* Professional Standards Council, www.psc.gov.au/ accessed 6 Nov 2019.

Guild of Project Controls www.planningplanet.com/guild Accessed: Jan 2020.

Lock, D. and Scott, L. (Eds) (2013), *The Gower Handbook of People in Project Management*, Farnham: Gower.

Lombardo, M. and Eichinger, R.W. (1996), *The Career Architect Development Planner*, Minneapolis: Lominger.

Fook, J. (2007), *Handbook for Practice Learning in Social Work and Social Care*, 2nd edn: Knowledge and Theory, ed. J. Lishman. Jessica Kingsley, pp. 363–75.

Moon, J. (2004), *A Handbook of Reflective and Experiential Learning: Theory and Practice*. London: RoutledgeFalmer.

McKinsey Global Institute (2017), *Technology, Jobs, and the Future of Work*. www.mckinsey.com/featured-insights/employment-and-growth/technology-jobs-and-the-future-of-work, Accessed Jan 2020.

Schon, D. (1987), *Educating the Reflective Practitioner*, San Francisco, CA: Jossey-Bass.

PART XI

Case examples

Case A
THE HERBERT-INGERSOLL TRAGEDY

Dennis Lock

Early in 1968 Herbert-Ingersoll Limited began its operations at their brand new premises on the Royal Oak estate in Daventry, Northamptonshire. Modern offices and a state-of-the-art manufacturing plant had been constructed on this 50-acre greenfield site. This project was the result of a joint investment by Alfred Herbert machine tools, based in Coventry, and Ingersoll Milling Inc, whose main operations were based at their plant in Rockford, Illinois.

The owners

Alfred Herbert Ltd occupied an enormous area of works and offices in Coventry. Herbert machine tools were renowned for their ruggedness and precision, although the factory was somewhat dated and in need of modernization. Ingersoll Milling Inc had a reputation for producing very high quality machines. Their principal range comprised plano-mills or scalpers (which in their terminology were known as adjustable rail milling machines or ARMMs), transfer line machines for the automotive industry and smaller one-off special purpose machines.

Ingersoll technology was at the forefront in the new Daventry plant. Ingersoll always put quality, accuracy and reliability first and it could have been argued that their products exceeded customer's requirements. They were proud to relate the story of an American arms manufacturer who had great difficulty in machining armour plate. That arms manufacturer approached Ingersoll in Rockford to ask if they could design and manufacture a special heavy duty ARMM for the purpose. 'Bring us in a sample' said Ingersoll's managers. They then astounded the arms manufacturer by machining a clean, heavy cut on the armour plate sample with no difficulty, using a standard Ingersoll ARMM in their factory.

Alfred Herbert owned 52 per cent of Herbert-Ingersoll and the remaining 48 per cent was owned by Ingersoll Milling Inc. All the technological expertise was supplied from Rockford.

The start-up plan

When new engineering offices and a brand new manufacturing plant are opened simultaneously, the problem arises of how to supply the factory with profitable work during the first months of start-up. In other words, how could the delay between opening and staffing the manufacturing plant, and the issue of working drawings from the design engineers be bridged? This difficulty had been foreseen and a brilliant start-up plan was in place. The new plant would manufacture transfer line machinery to fulfil an existing order from a British car maker, using drawings already specially produced at Ingersoll's engineering offices in Rockford. So, while the new Daventry design team was working on orders for future delivery, the Daventry plant would have work to keep its manufacturing staff and machines gainfully occupied.

It was during this time, in the spring of 1968, that I was recruited as Manager, Engineering Administrative Services, reporting to the engineering director in Daventry.

Work on anglicizing the drawings from Rockford began on time, but the senior engineers in Daventry disagreed with some of the design principles. Accordingly, instead of simply anglicizing the Rockford designs, the British engineers embarked upon a complete redesign of the transfer line machinery. The cost of that redesign approached £2m (which was approximately the total order value). Worse, there was a delay of several months before the main manufacturing operations could begin.

From the customer's point of view the project delay did not matter. Their new car development programme and the provision of space for the new transfer line machinery were also running late. So when the new transfer line was eventually installed and commissioned, the customer suffered no inconvenience. But Herbert-Ingersoll had now incurred a considerable financial loss.

The downward spiral to disaster

It was unfortunate that the Herbert-Ingersoll joint venture began operations at a time when the British and international economies were experiencing a downturn. At that time it was generally reckoned that the economy moved from peak, through minimum, back to peak demand in seven-year cycles. So potential customers were not investing in new machinery and the order books at Alfred Herbert and Herbert-Ingersoll were not healthy. With no customer receipts (no cash inflows) Herbert-Ingersoll ran into serious debt. To give some idea of how serious this debt was, I once overheard the chief accountant telephoning a supplier to ask if we might pay just £20 on account from an outstanding invoice with a total value of only £100 or so.

Then some light appeared at the end of the tunnel. A British government development grant was obtained (from memory, it was in the region of £2m) to be shared between Alfred Herbert and Herbert-Ingersoll). Astonishingly, I was asked to investigate the extension of our offices, plus the provision of a new manufacturing stores area. So I embarked upon a series of visits to construction companies to obtain outline proposals and quotations. A contract was placed for the new extensions.

The end of the company was now in sight. It was only a matter of time. I moved to a management job with another company (not in the machine tool industry) helped by much appreciated support from my former Herbert-Ingersoll managers. They were a great team. So it was with much sadness that I later learned of the closure not only of Herbert-Ingersoll, but also of Alfred Herbert in Coventry. The sad final straw was the collapse of Ingersoll's main operations in Rockford. This was all a great tragedy. An enormous pool of talent and dedicated staff was left to find other employment at a time when job vacancies were not plentiful.

From the project controls point of view, an important contributing factor at Daventry was the failure of Herbert-Ingersoll to use the original American designs for their first project. They redesigned everything, which resulted in enormous expense and delay. This was an extreme (but not isolated) example of project benefits realization failure caused by the 'not invented here' syndrome.

References and further reading

Lloyd-Jones, R. and Lewis, M.J. (2006), *Alfred Herbert Ltd and the British Machine Tool Industry, 1887–1983*, Abingdon: Routledge.

Case B
THE CHANNEL TUNNEL PROJECT

Dennis Lock

Until some 8,000 years ago, England was connected to the continent of Europe by a chalk mass in the sea known as Doggerland, but that connection was lost at the end of the last ice age when the ice melted and sea levels rose and Britain became an island. Various methods for reconnecting England with the continent had been discussed for many years, not always with total public approval because the English are accustomed to being an island race and consider that the sea somehow provides us with a degree of national security. However, a business plan in the 1980s for a tunnel connecting Folkestone (Kent) with Coquelles in the Pas de Calais finally won approval from both the French and British governments. This brief case summary highlights two features common to many large capital projects, namely that business plans can be flawed and overoptimistic and also that the perception of project success or failure depends upon which group of stakeholders is consulted.

Outline project description

Although always referred to as either Eurotunnel or The Channel Tunnel, there are in fact three parallel tunnels. One tunnel carries eastbound trains, a similar tunnel carries westbound trains and a smaller diameter cross-linked service tunnel runs between the two main tunnels. Traction power is supplied by overhead high voltage electric cables. Three different types of services use the tunnel:

1 High speed passenger trains.
2 Rail freight services.
3 Drive on, drive off shuttle services which carry passenger cars and heavy goods vehicles in huge, specially designed rail wagons.

On the continent of Europe, high speed rail services connect with cities such as Paris, Amsterdam and Brussels. In Britain a high speed rail link (HS1) was successfully constructed between Ashford (Kent) and St Pancras International station in London. The tunnel length is about 31.5 miles. Much of the tunnel boring was through chalk, for which 11 specially constructed tunnel boring machines (TBMs) were used. This huge joint venture project involved one group of five British construction companies and another group of five French construction companies. The project was approved by the French and British governments in 1987 and work began in 1988. Several British and French banks were involved in arranging finance. The tunnel opened to freight traffic in June 1994, and to passenger trains in November 1994.

There is a detailed Wikipedia account of this project, which also covers much of the history, politics and previous proposals for linking Britain with mainland Europe. This report records that the total project cost on completion was £9 billion, amounting to an 80 per cent budget overrun of the original £5.5 billion estimate. The financing agencies and banks were very hard-hit, and huge debts had eventually to be written off.

There was also a high human cost. Tunnelling, quarrying and mining have always been hazardous operations. I was told by one senior construction manager from this project that the predicted fatality rate in tunnelling operations of this kind is one person per mile. However, the actual number of French and British people killed, as reported by Construction News, was 12 (CNPLUS, 1990).

Flaws in the business plan

Business plans are typically optimistic, with projected costs understated and the returns on investment overstated and predicted too early. At the time of writing, for instance, Britain's HS2 project for a high speed rail service linking London with the Midlands is running well over its original estimated cost. Those hoping to win approval for new projects, whether in-house or contracted, will naturally feel compelled to estimate the costs as low as possible and list the expected benefits as advantageously as they can. This can mean that, in competitive tendering, the organization that wins the contract has done so on the basis of cost estimates that were set too low. The outcome of such unfounded optimism is known as 'the winner's curse'. This expression is more usually attributed to the successful bidder at an auction, who finds out that the winning bid was higher than the real value of the goods in the auction lot.

From the project controls point of view, attempting to control costs within budgets that are set too low is frustrating and soul-destroying. Underpriced project contracts are particularly unpleasant for project control practitioners, who know from the outset that the best they can do is to achieve optimum project performance and limit the extent of the inevitable financial loss and schedule overrun.

One criticism of the original Channel Tunnel project proposals is that the predicted market share of cross-channel traffic that would be captured by the tunnel was overstated. Competitors in cross-channel traffic were the sea ferries and airline operators. What the tunnel business plan predictions did not take into account was that these competitors would react aggressively to competition from the new

Channel Tunnel services by introducing low budget fares, thus retaining more of their business share than had been expected.

Success or failure?

The story of this ambitious project highlights an important aspect of projects controls, namely that whether or not a project can be considered as an unqualified success depends very much upon where you are positioned among the project stakeholders and end users.

For the investment banks that lost staggering sums of money when the financing debts were eventually written off, this project was hardly a great success. For the pre-existing cross-channel services (particularly the sea ferries) the Channel Tunnel has taken a slice of their potential trade, destroying their previously long-held monopoly. For those whose family members were seriously injured or killed during tunnel construction this project was clearly calamitous.

However, for business travellers and freight operators, the speed, comfort and connectivity of all the tunnel services is clearly an important and greatly appreciated advantage. For families and the general public, one has only to travel back to London on an evening train and see the tired but happy children clutching their Disneyland Paris trophies to know that they view the Channel Tunnel project as a great success.

References and further reading

Channel Tunnel, Wikipedia, accessed 27 December 2019.

CNPLUS (1990), 'Special Report on Channel Tunnel – Death Toll in Transmanche Link's Operation is Too High', *Construction News*.

Case C
LOSING AND REGAINING CONTROL

Dennis Lock

When a project manager is placed at the head of a team that is entirely responsible for carrying out all stages of a project, controlling the project is relatively easy. With a dedicated project team in place there should be no conflict of interest. Everyone can work to the common aim of delivering the project on time, within budget and to specification. However, not all project managers have the luxury of complete control. The two true tales that follow tell what can go wrong when project managers have to rely on the goodwill and competence of others over whom they have no line authority.

Background and organization

A company in Britain was involved in the design, manufacture and installation of hospital operating theatres. These were of modular construction and could be installed within existing buildings. That gave the hospital authorities the huge advantage of achieving sterile conditions without having to invest great sums in new buildings. Different sizes were available, but every operating theatre was octagonal in shape, held together by a frame consisting of eight supporting legs and an octagonal top frame. Each vertical leg was designed to be bolted to the top frame.

Every frame component was manufactured by welding together heavy gauge square section tubular steel. Everything was finished to a high specification, durable polyurethane enamel that was baked on. The frame components were shipped as separate pieces, to be bolted together by the specially trained assembly crew. The whole assembly was known as the spider for obvious reasons. Figure C.1 shows the idea.

These modular theatres were available in a range of different sizes so that the hospital authorities could order what they needed to suit their available space. All sizes

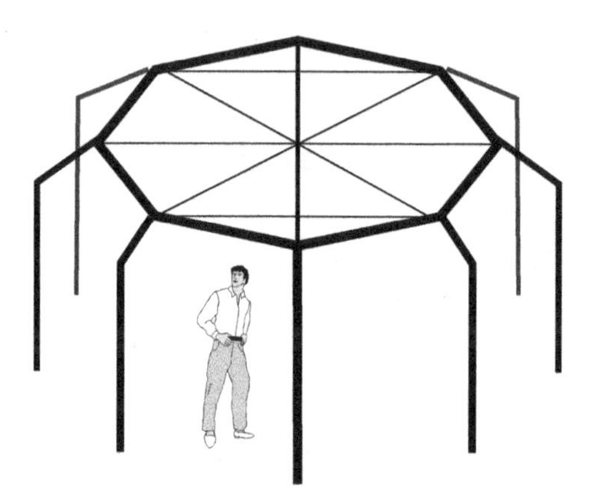

Figure C.1 'Spider' support frame for a hospital modular operating theatre.

used the same support legs, but required different top frames. Project management was entrusted to the company's medical division, where I was contract control manager, reporting to the divisional manager.

The design concept was sound, and the company had a good reputation for high quality and excellent customer service. It was also a happy workplace, with very good employer-employee relationships. However, there was a one serious weak link in the organization from the project controls point of view. This was the common services division that manufactured the main hospital theatre structural components and was also responsible for picking, packing and shipping everything that the assembly teams would need at the hospital sites. This common services division had its own manager and was a law entirely unto itself.

The day came when two of these operating theatres were to be erected inside a hospital in northern Europe. The local civic authority treated this first day as a big occasion. Local dignitaries joined the hospital management and the press to watch our design team set up the spider frame for the first theatre. The top frame for the first theatre was lowered by a crane on to the eight supporting legs, each of which was held in place by a member of the construction team. Under full public gaze, it soon became apparent that our common services division had picked, packed and shipped two top frames of the wrong size.

The project controller's revenge

Tension between the common services division and medical division was always high. One evening, outside normal working hours when the factory was deserted, my manager took me on a tour of the factory. He pointed out that some of our hospital theatre components (legs for the spiders) had not been moved from the manufacturing area to the paint shop. They had been there for weeks, leaning against a wall in their slightly rusting state.

My manager hinted at what I might do to stir things up and I did it. During the next day's lunch break I collected a strong member of my staff and we went together into the deserted factory. From there we carried one of the heavy spider's legs into the office of the common services division's manager and laid it across the desk, incidentally scratching the polished desktop in the process.

Then we retired back to medical division and awaited the inevitable explosion, which was not long in coming. Of course the common services manager had been humiliated. He had to have members of his own staff remove the offending object. There was an almighty row. Fingers were wagged at me 'Dennis, you really should not have done that!' But I survived unscathed and thenceforth we had no further problem and far better service from common services division. Time healed the wounds and we all became friends again.

The important relationship between organizational structure and ability to control

The case described above is a classic example of how the project controller's task is made more difficult when reliance on tasks being performed correctly depends on only goodwill. The organization in this case was a matrix, where the project management had no direct line authority over the common services division. Control is always easiest in a task force or team organization, when everyone is working towards common goals.

In this case example, control might have been easier if the components were being manufactured by an external supplier. In that case the supplier would have been motivated to perform well to some extent in the knowledge that failure would result in the loss of custom, with future orders going to competitors. When the project controller is dependent entirely on the goodwill of a common service department over which he/she has no line authority, there are only two courses of action that can be recommended, which are:

1. make a personal appeal to the common services manager or, if that fails;
2. escalate the difficulty by appealing to higher management.

INDEX